3rd International Conference on Design and Manufacturing Engineering (ICDME 2018)

MATEC Web of Conferences Volume 221 (2018)

Melbourne, Australia
16 - 18 July 2018

Editors:

L.H. Chen
Y. Kondo

ISBN: 978-1-5108-7406-0

Printed from e-media with permission by:

Curran Associates, Inc.
57 Morehouse Lane
Red Hook, NY 12571

Some format issues inherent in the e-media version may also appear in this print version.

Printed by Curran Associates, Inc. (2018)

For additional information, please contact EDP Sciences – Web of Conferences at the address below.

EDP Sciences – Web of Conferences
17, Avenue du Hoggar
Parc d'Activité de Courtabœuf
BP 112
F-91944 Les Ulis Cedex A
France

Phone: +33 (0) 1 69 18 75 75
Fax: +33 (0) 1 69 28 84 91

contact-edps@webofconferences.org

Additional copies of this publication are available from:

Curran Associates, Inc.
57 Morehouse Lane
Red Hook, NY 12571 USA
Phone: 845-758-0400
Fax: 845-758-2633
Email: curran@proceedings.com
Web: www.proceedings.com

TABLE OF CONTENTS

Preface

The 3rd International Conference on Design and Manufacturing Engineering (ICDME2018) was successfully held in Monash University, Melbourne, Australia during July 16-18, 2018. These proceedings consist of the selected papers that will be presented at ICDME 2018. The aim as well as objective of ICDME 2018 is to present the latest research and results of scientists related to in Design and Manufacturing Engineering topics. This conference provides opportunities for the delegates to exchange new ideas face-to-face, to establish business or research relationships as well as to find global partners for future collaborations. We hope that the conference results will lead to significant contributions to the knowledge in these up-to-date scientific fields.

The conference covered developments and recent trends in the fields of Design and Manufacturing Engineering. The ICDME2018 conference reflected the technology needs of the world. This proceeding contains the reviewed papers presented at the ICDME 2018 and covered the most important issues related to the design and manufacturing engineering with focusing on the Functional Material Design and Analysis, Product Design and Quality Control, Vehicle Engineering and Manufacturing, Mechanical Engineering and Control Technology and Aerospace Engineering.

With this selection the Conference Committee has assembled a comprehensive program offering 4 keynote speeches and 2 plenary speeches from renowned scientists from Australia and China. The topics of these speeches were *Soft Plasmene Nanosheets: From Design to Applications, Development of heatpipe plate technologies for high performance cooling applications, Future Fibres Research and Development, Progress in rolling technology and development of novel composite materials, Smart Materials for Chipless RFID Sensors: Revolution in Identification and Sensing in the New Millennium,* and *Design and Engineering of Sustainable Packaging Materials with Cellulose Nanofibres,* respectively. These speeches have shown the most up-to-date research results in the field of design and manufacturing engineering. The program also includes 6 parallel technical sessions with 42 oral presentations, focused on Materials Science and Engineering, Mechanical Design Manufacturing and Automation, Material Forming and Control Technology, Power Machinery Engineering, Engineering Education and Management, and Digital Manufacturing System and Quality Control.

We would like to thank all members of the conference committees, as well as the reviewers for their advice, which has certainly helped to improve the quality, accuracy, and relevance of each paper selected for the conference program and this volume. We extend our thanks to every author and participant who contributed to the success of the conference. Looking forward to meet you again in next year.

Prof. Wenlong Cheng, Monash University, Australia
Assoc. Prof. Warren Batchelor, Monash University, Australia
On behalf of Conference Committees
July 16-18, 2018
Melbourne, Australia

Statement of Peer review

In submitting conference proceedings to MATEC Web of Conferences, I certify to the Publisher that I adhere to the **Policy on Publishing Integrity** of the journal in order to safeguard good scientific practice in publishing.

1. All articles have been subjected to peer review administered by the proceedings editors.
2. Reviews have been conducted by expert referees, who have been requested to provide unbiased and constructive comments aimed, whenever possible, at improving the work.
3. Proceedings editors have taken all reasonable steps to ensure the quality of the materials they publish and their decision to accept or reject a paper for publication has been based only on the merits of the work and the relevance to the journal.

Title, date and place of the conference

2018 3rd International Conference on Design and Manufacturing Engineering (ICDME2018).
Monash University, Melbourne, Australia;
July 16-18, 2018.

Proceedings editor(s):

Li-Hao Chen

Date and editor's signature 8/Oct/2018

Statement of Peer review

In submitting conference proceedings to MATEC Web of Conferences, I certify to the Publisher that I adhere to the **Policy on Publishing Integrity** of the journal in order to safeguard good scientific practice in publishing.

1. All articles have been subjected to peer review administered by the proceedings editors.
2. Reviews have been conducted by expert referees, who have been requested to provide unbiased and constructive comments aimed, whenever possible, at improving the work.
3. Proceedings editors have taken all reasonable steps to ensure the quality of the materials they publish and their decision to accept or reject a paper for publication has been based only on the merits of the work and the relevance to the journal.

Title, date and place of the conference

2018 3rd International Conference on Design and Manufacturing Engineering (ICDME2018)
Monash University, Melbourne, Australia;
July 16-18, 2018.

Proceedings editor(s):

Yasuo KONDO

Date and editor's signature

Oct. 09. 2018

Optimization of Molding Parameters Using Taguchi Method to Increase the Electrical Conductivity and Tensile Strength of Conductive Polymer Composites

H Suherman[1], P Pulungan[1], Y Yovial[1], and Irmayani[2]

[1]*Department of Mechanical Engineering, Universitas Bung Hatta, Padang, Indonesia*
[2]*Department of Industrial Engineering, Universitas Ekasakti, Padang, Indonesia*

Abstract. The focus of this research is to increase the electrical conductivity and tensile strength of conductive polymer composites (CPCs) materials using Taguchi method. The efforts made is by optimizing the molding parameters, using two different size of conductive fillers, ie G25 (25 μm) and G13 (13 μm) in producing CPCs material. The molding parameters used are molding time and molding temperature. S/N ratio is use to obtain the optimum molding parameters, ie the larger is better. The results showed that Taguchi method L9 (2^3) succeeded in increasing the electrical conductivity and tensile strength of G25/G74/epoxy and G13/G74/epoxy composites. The highest electrical conductivity and tensile strength is on G13/G74/epoxy composites, ie 3.51 S/cm and 155.50 N/mm^2 respectively.

1. Introduction

The electrical conductivity and mechanical properties such as tensile strength are important properties to be possessed by a conductive polymer composite material. To obtain the optimum value for these both properties, it is necessary to optimize the parameters of the molding process. The parameters of the molding process used to produce the conductive polymer composites (CPCs) material are greatly affect the properties of the CPCs material produced [1-3]. Suherman et al. [1] conducted research on the effect of content, molding temperature, and molding pressure on the electrical conductivity of carbon nanotub (CNTs)/graphite/epoxy composite. The results showed that the incorporation of CNTs and graphite as a conductive materials produced a synergistic effect in improving electrical conductivity, as well as molding temperatures and molding pressure. The influence of the molding parameters of CNTs/graphite/epoxy nanocomposites to the in-plane and through-plane electrical conductivities has been observed by Suherman et al. [2]. They reported that the incorporation of CNTs and graphite materials had a positive effect in improving the electrical conductivity values of in-plane and through-plane directions, to a CNT content until 5 weight percentage (wt.%). The Taguchi method has widely used to obtain the optimum value of a molding parameter. Thanapat Sangkharat and Surangsee Dechjarern [3] conducted research on spinning process design. Ahmer et al. [4] examined the effect of time on the use of fly ash. Yizong et al. [5] analyzed the effect of molding parameters on the injection molding process using polyestyrene matrix material. Sapana et al. [6] studied the benefits of benzeneacetic acid removal using CaO2 nanoparticles. This research use the Taguchi method L9 (2^3) to optimize the molding parameters consisting of molding time and molding temperature using a second conductive filler with different sizes and shapes to increase the electrical conductivity and the tensile strength of the resulting conductive polymer composites.

2. Experimental

2.1 Materials

The conductive fillers used were graphite with three different sizes, ie 74 μm (G74), 25 μm (G25), and 13 μm (G13). G74 which has a larger particle size, is used as the main filler, whereas G25 and G13 which has a smaller particle size are used as the second filler. The conductive fillers used as main and second fillers have different shapes, namely flakes and sphere shapes. G74 is obtained from Asbury Carbon America, while G25 and G13 are obtained from FRIway Industry, China. Epoxy resin at 6 Poise viscosity is obtained from US Composites. The description of particle size, the main and the second conductive fillers, as well as the viscosity of the epoxy resin used are obtained from the manufacturer.

2.2 The manufacturing of G25/G74/epoxy and G13/G74/epoxy composites

The composition based on weight percentage (wt.%) of the main, second conductive fillers and epoxy resin as the matrix are shown in Table 1. The manufacturing process of G25/G74/epoxy and G13/G74/epoxy composites is by stirring resin and hardener using a mechanical stirring machine (IKA RW 20 Digital) for 10 minutes at 25 rpm rotation. Furthermore, the main and second conductive fillers (G25, G13 and G74) as the predefined compositions are combined with epoxy resin mixture. The G25/G74/epoxy and G13/G74/epoxy mixture are poured into an aluminum mold. The mold is then put into the oven for 90 minutes at 150°C.

Table 1. Composition of G25/G74/epoxy and G13/G74/epoxy composites

CPCs material	Composition (wt.%)		
	Second fillers	Main filler	Epoxy resin
G25/G74/epoxy composites	G25 : 5, 7.5, 10	G74 : 45, 42.5, 40	50
G13/G74/epoxy composites	G13 : 5, 7.5, 10	G74 : 45, 42.5, 40	50

2.3 Characterization

The electrical conductivity and tensile strength of G25/G74/epoxy and G13/G74/epoxy composites material produced by different molding process parameters were measured using ASTM C 61 [7] and ASTM D3039, respectively. Scanning electron microscopy (SEM) with type S-3400 N (Hitachi) use to see the spread of main and second conductive fillers in epoxy resin.

2.4. Control factors and levels of molding parameters

The control factors of molding parameters are molding time (A_{mt}), and molding temperature (B_{mt}). Table. 2 show two control factors with three levels of molding parameters. The molding time levels (A_{mt}) are 60, 75, and 90 min, while the levels of molding temperature (B_{mt}) are 110, 130 and 150°C.

Table 2. Factors and levels of molding parameters

Symbol	Factors	Unit	Level		
			1	2	3
A_{mt}	Molding time	Minute	60	75	90
B_{mt}	Molding temperature	^{0}C	110	130	150

3. Results and discussion

3.1 Characterization of main and second conductive fillers

Figure 1. shows the main and second conductive filler type used to produce G25/G74/epoxy and G13/G74/epoxy composites. SEM images shows that G74 as the main conductive filler has a flake shape. While G25 and G13 have flake and sphere shape.

3.2 The electrical conductivity of G25/G74/epoxy and G13/G74/epoxy composites

Table 3 shows the electrical conductivity generated using the molding parameters based on the Taguchi array L9 (2^3) in generating G25/G74/epoxy and G13/G74/epoxy composites. All parameters of the molding parameters used in this study show that G25/G74/epoxy and G13/G74/epoxy composites have different electrical conductivity. This indicates that all of the molding parameters used have an effect on the dispersion and the formation of conductive networks in the matrix polymer [8-10].

Table 3. shows that the size (25 μm and 13 μm), as well as different forms (sphere and flake) of conductive fillers produce different electrical conductivity. The 9 runs performed on two different sizes and shapes of conductive fillers showed that the smaller particle size (13 μm) of the same shape produce the highest electrical conductivity of 2.92 S/cm (run 1). This is due to the same shape of the smaller secondary conductive filler is more effective in filling the cavity formed by the main conductive filler.

Table 3. Taguchi array L9 (2^3) of molding parameters

Run	Factor and level		Electrical conductvity (S/cm)	
	A (molding time)	B (molding temperature)	13 μm	25 μm
1	60	110	2.92	1.41
2	60	130	2.26	2.72
3	60	150	2.4	2.37
4	75	110	2.57	1.49
5	75	130	1.28	1.13
6	75	150	2.78	2.07
7	90	110	2.44	1.61
8	90	130	2.74	1.58
9	90	150	2.86	1.30

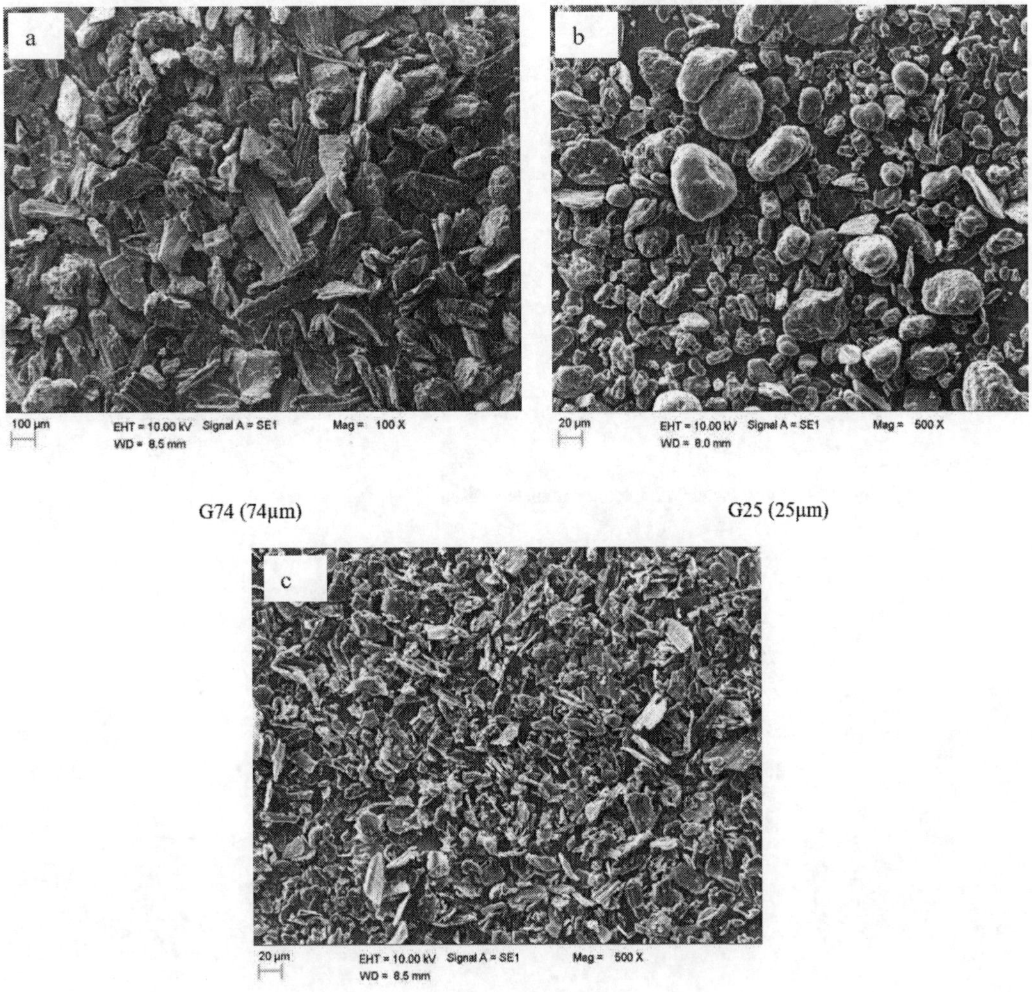

G13 (13µm)

Figure 1. SEM images: (a) G74, (b) G25 and (c) G13

3.3 Taguchi response on molding paremeters based on S/N ratio

Figures 2 and 3 show the S/N ratio of the molding parameters using particle size 13 and 25µm as the second conductive filler. The optimum parameters at A_3 (90 minutes heating time), B_3 (molding temperature 150 ^{0}C) were obtained on the molding parameters of 13 µm particle size (flake shape). While in particle size 25µm (sphere shape), the optimum molding parameter obtained are A_1 (moding time, 60 minutes), and B_3 (molding temperature, 150^{0}C). Thus, the optimum molding parameters recommended by Taguchi array L9 (2^3), (larger is better) then followed by specimens of G25/G74/epoxy and G13/G74/epoxy composites.

The Taguchi recommendations for two conductive fillers and tensile strength of different sizes and shapes are shown in Table 4. The highest electrical conductivity based on Taguchi recomendation was obtained on G13/G74/epoxy composites of 3.51 S/cm and 2.80 S/cm in G25/G74/Epoxy composites. So as tensile strength, the value obtained by a smaller conductive filler (13 µm) of 155.50 N/mm^2 is greater when compared with a larger second filler (25 µm) of 140.42 N/mm^2. The results obtained are in line with Antunes et al. [11] which states that the second conductive filler will be effective if the particle size is smaller if the conductive filler more than one.

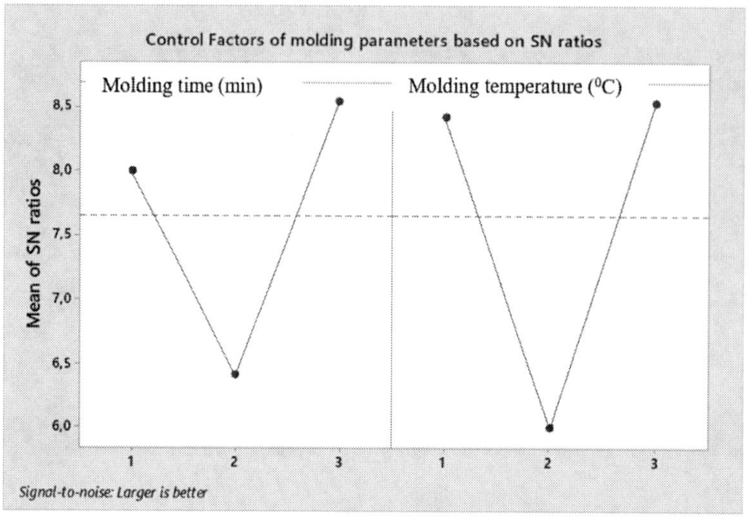

Figure 2. Control factors of molding parameters based on S/N ratio for particle size 13μm

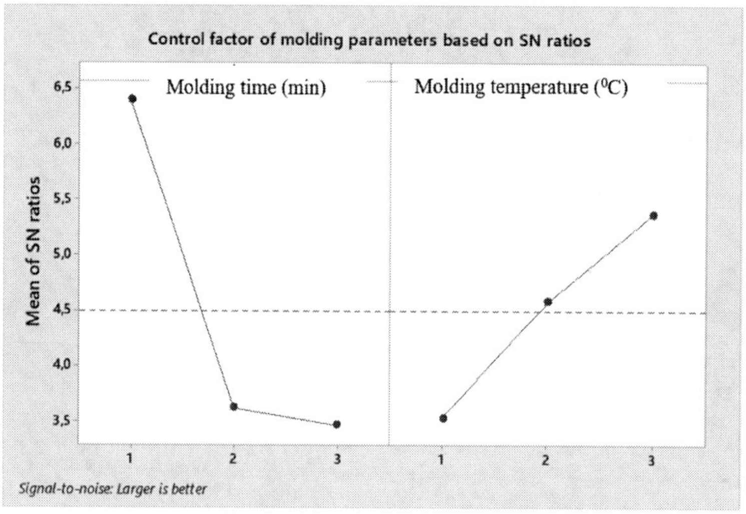

Figure 3. Control factor of molding parameters based on S/N ratio for particle size 25 μm

Table 4. Taguchi recommendation of molding parameters based on Taguchi array L9 (2^3)

No	G13/G74/epoxy and G25/G74/epoxy composites	Particle size of conductive filler	Electrical conductivity (S/m)	Tensile strength (N/mm^2)
1	Run 1	13 μm (flake shape)	2,92	145,50
2	Taguchi recomendation		3,51	155,50
3	Run 1	25μm (sphere shape)	1,41	127,72
4	Taguchi recomendation		2,80	140,42

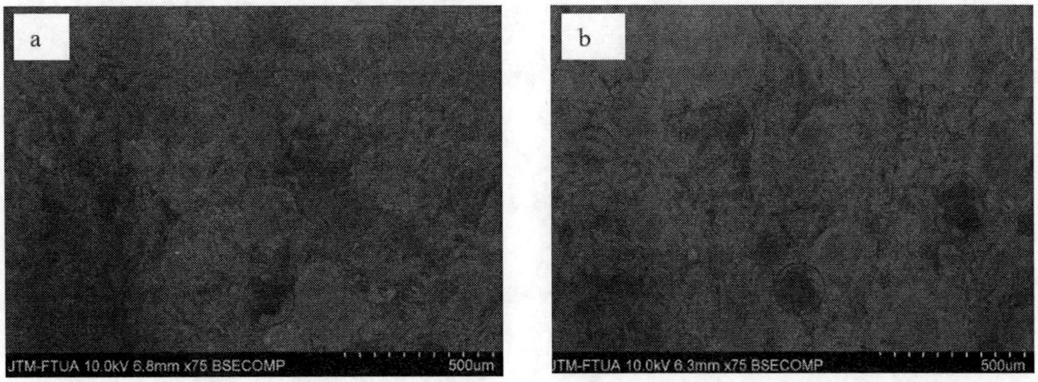

Run 1, particle size of 13µm Taguchi recomendation, particle size of 13µm

Figure 4. SEM images of fracture surface of G13/G74/epoxy composites (a) Run 1, (b) Taguchi recommendation

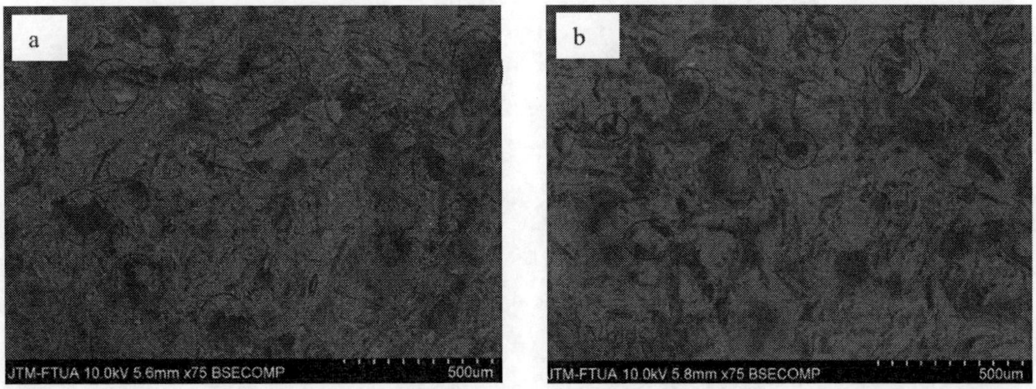

Run 1, particle size of 25µm Taguchi recommendation, particle size of 25 µm

Figure 5. SEM images of fracture surface of G25/G74/epoxy composites (a) Run 1, (b) Taguchi recommendation

Figure 4. shows that run 1 of G13/G74/epoxy composite (fig.4a) has more voids compare than the G13/G74/epoxy composite (fig.4b) generated from Taguchi recommendations. Similarly, Fig. 5 shows voids formed on G25 / G74 / epoxy composite materials (fig.5a) is more than the composite materials made based on Taguchi recommendations (fig 5b). This causes the electrical conductivity and tensile strength at run 1 lower than the Taguchi recommendation [12, 13].

4. Conclussion

The Taguchi method L9 (2^3) has been used to optimize the molding parameters of G25/G74/Epoxy and G13/G74/Epoxy composites. This method successfully increases the electrical conductivity and tensile strength the composite. Both the second conductive fillers used (G25 and G13) have succeeded in increasing the electrical conductivity and tensile strength for G25/G74/epoxy and G13/G74/epoxy composites. The G13/G74/epoxy composite increased the electrical conductivity and tensile strength from 2.92 S/cm to 3.51 S/cm (20.21%) and 145.5 N/mm2 to 155.50 N/mm2 (6.87%). While on G25/G74/epoxy composite, the electrical conductivity increased from 1.41 S/cm to 2.80 S/cm (98.58%) and tensile strength increased from 127.72 S/cm to 140.42 S/cm (9.94%). This condition indicates that the Taguchi method L9 (2^3) used effectively improves the electrical conductivity and tensile strength of the resulting G13/G74/Epoxy and G25/G74/Epoxy composites.

Acknowledgment

The authors gratefully acknowledge the financial support given for this work by the Directorate General of Higher Education through the applied research of excellent university under contarct number: 023-34/LPPM-Penelitian/Hatta/II-2018.

References

1 H. Suherman, A.B. Sulong, J. Sahari. Int. J. Mech. Mater. Eng. **5**, 74 (2010)
2 H. Suherman, A.B. Sulong, J. Sahari. Ceram. Int. **39**, 1277 (2013)
3 T. Sangkharat, S. Dechjarern. Procedia Eng. **207**, 1713 (2017)
4 A.A. Siyal, K.A. Azizli, Z. Mana, H. Ullah. Procedia Eng. **148,** 302 (2016)
5 T. Yizong, Z.M. Ariff, A.M. Khalil. Procedia Eng. **184,** 350 (2017)
6 S.M. Sapana,L. Kailas, Wasewar. J. Appl. Res. Technology **15,** 332 (2017)
7 B.K. Kakati, D. Sathiyamoothy, A. Verma. Int. J. Hydrogen Energy **35,** 4185 (2010)
8 P.C. Ma, M.Y. Liu, H. Zhang, S.Q. Wang, R. Wang, K.Wang , Y.K. Wong, B.Z. Tang, S.H. Hong, K.W. Paik, J.K. Kim. ACS Appl.Mater.Interfaces **1,** 1090 (2009)
9 N. Hu, Z. Masuda, G. Yamamoto, H. Fukunaga, T. Hashida, J. Qiu. Composites Part A **39,** 893 (2008)
10 H. Suherman, A.B. Sulong, M.Y. Zakaria, N.R. Rajendra Royan, J. Sahari. Songklanakarin J. Sci. Technol. **40**, 105 (2018)
11 R.A. Antunes, C.L. Mara, de Oliveira, Ett. Gerhard, Ett. Volkmar. J. Power Sources **196**, 2945–2961 (2011)
12 H. Suherman, A.B. Sulong, J. Sahari. Adv. Mater. Res. **264-265,** 559 (2011)
13 N.R. Rajendran Royan, A.B. Sulong, J. Sahari, H. Suherman. Journal of Nanomaterials **2013** Article ID 717459 (2013)

Empirical Modeling of Average Cutting Speed during WEDM of Gas Turbine Alloy

Bhupinder Singh[1] and Joy Prakash Misra[1]

[1]*Dept. of Mechanical Engineering, National Institute of Technology, Kurukshetra, India*

Abstract. Cutting speed (CS) is a key performance measure to achieve optimal utilization of the WEDM process. However, input process parameters of WEDM and combination of wire and workpiece material greatly hamper CS and hence productivity and machining efficiency. Therefore, it is essential to pick the right combination of parameters and wire and workpiece material to obtain better CS. In this paper, four process parameters: T_{on}, T_{off}, S_v, and I_p were chosen to develop an empirical model for CS during WEDM of nimonic 263 to provide a guideline to the potential users of the technique. This paper describes the response surface methodology (RSM) based mathematical modeling for average cutting speed. Furthermore, analysis of variance (ANOVA) was applied to find out significant process parameters and it was depicted that pulse on time and peak current were the major parameters affecting CS. In addition, WEDMed surfaces were analysed through FE-SEM at various discharge energy levels. The WEDMed surfaces appeared in the form of micro-cracks, craters, spherical droplets and the lump of debris. It is obvious from the current investigation that input parameters have the significant influence on cutting speed. The key features of experimental procedure are also highlighted in this paper.

1 Introduction

WEDM is one of the extensively accepted non-conventional machining processes employed to manufacture components with complex shapes, profile and sharp edges that are difficult-to-machine by other traditional and non-traditional machining processes [1] and hence, this process has wide application areas including aerospace industry, medical implants, electronic industry, automobile industries, etc. It is an electro-thermal production type advanced machining process in which removal of material occurs due to melting and vaporization owing to a series of sparks between workpiece and wire electrode (thin copper, brass or tungsten of diameter 0.05-0.3 mm) submerged in a dielectric fluid (deionized water). Both electrodes are connected to a pulsed DC power supply. There is a gap between the wire and the workpiece usually ranges from 0.025 mm to 0.05 mm and is continuously maintained by a computer-controlled positioning system. The wire is kept under tension by tensioning device to overcome the inaccuracies in the machined parts. The material is removed by the series of electrical discharges between wire and workpiece in the presence of deionized water. At the point, ionization occurs and electricity flows through the ionized column of dielectric fluid between the positive and negative electrode. Once ionization takes place, the dielectric becomes heated from the flow of electricity and then changes into a gas known as plasma. Under this condition electrons rapidly pass through the ionized plasma in the form of a spark. The positive ions from the workpiece attracted to the negatively charged electrode and the negatively charged electron collide with the workpiece and produce a spark, and this spark melts and vaporizes the small amount of material from the workpiece [2–4]. The details of the mechanism of material removal, process parameters, the influence of process parameters on measures of process performance are available in the literature [5–8]. Figure 1 illustrates the schematic of WEDM process.

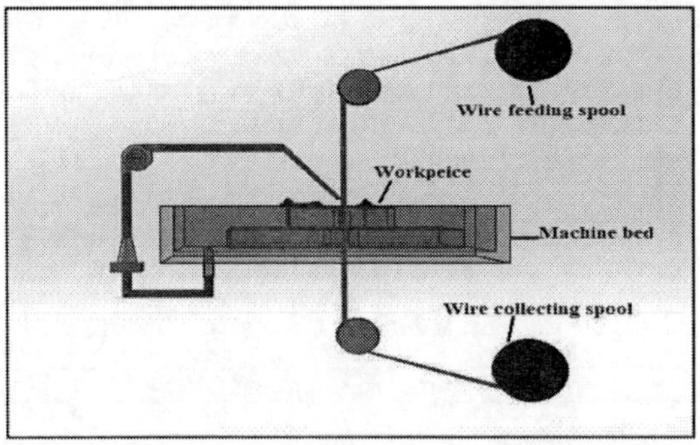

Figure 1. Schematic of WEDM process (Guitrau, 1997).

2 Experimental procedure

2.1 Experimental setup and details

The experimental investigation has been conducted on 734 sprintcut WEDM (*Make: Electronica Machine Tool*). This machine setup comprises of four major sub-components: power supply system, dielectric system, positioning system, and wire drive mechanism. The detailed functions of each sub-element are available in reference [1], [2]. From literature survey, it has been observed that most of the research work carried out on steel and steel alloys, aluminium alloys and titanium alloys [6], [9], [10]. Less research work has been reported on nickel-base super alloys and therefore it is very much essential to explore WEDM of nickel-base super alloys owing to its industrial importance in gas turbine and aerospace industry. Traditionally, combustor components have been made-out of nimonic 263 sheets. So, in this research work nimonic 263 has been selected as workpiece material. Rectangular nimonic 263 sheet of 100 mm (length), 100 mm (width), 10 mm (thickness) was used for the experimentation work and 5mm x 5mm square cuts are taken on the workpiece. Half hard-brass wire (ϕ =0.25 mm) with 900 N/mm^2 is used as cutting tool, while de-ionized water was used as a dielectric. The process parameters of WEDM process are broadly classified into four groups: power supply related parameters, electrode related parameters, workpiece related parameters, and dielectric related parameters.

2.2 Design of experiments

In this study an effort has been made to find out the relationship between input and response parameters by developing empirical model for CS. This experimental study consists of four factors: T_{on}, T_{off}, S_v, I_p and each factor is having three levels. The input process parameters values and their levels for the main experiments were decided on the basis of pilot experiments. Which are conducted by using OFTA. It is obvious from the results of pilot experiments that, variation of wire feed rate, wire tension and servo feed have some insignificant effect on the CS. Therefore, these parameters were kept fixed during experimentation. The dielectric fluid temperature, conductivity, dielectric pressure and dielectric chilling medium (-2°C) were also kept constant. In this study, experimental runs are designed and planned according to the Box-Behnken design (BBD) of response surface methodology (RSM). Hence, total twenty-nine runs including five replications of centre run are required [11]. The parametric combinations for different trial runs are presented in Table 3. Table 1 and Table 2 shows the levels and values for input and fixed process parameters.

Table 1. Ranges and levels of input process parameters.

Input process parameters				
Parameters	**Ranges**	**Levels**		
		L1	**L2**	**L3**
Pulse-on time	105-131 µs	105	118	131
Pulse-off time	30-60 µs	30	45	60
Peak current	40-220 A	40	130	220
Servo voltage	20-80 V	20	50	80

Table 2. Values of fixed process parameters.

Fixed process parameters		
Parameters	**Value**	**Fixing criteria**
Dielectric fluid	De-ionized Water	Literature review and Pilot experiments
Peak Voltage	2 m/c unit	
Water Pressure	1 m/c unit	
Wire Feed	8 m/c unit	
Wire Tension	12 m/c unit	
Servo Feed	2200 m/c unit	
Workpiece material	Nimonic 263	Industrial application

3 Results and discussions

Table 3 shows the order, parametric combination, and design of experiments based on the coded surfaces and corresponding results for cutting speed. It is evident from Table 3 that maximum value of CS of 3.74 mm/min is obtained for experiment no. 25 (T_{on}: 118 µs; T_{off}: 45 µs; I_p: 130 A; and S_v: 50 V) while experiment no. 2 (T_{on}: 105 µs; T_{off}: 60 µs; I_p: 130 A; and S_v: 50 V) provides the lowest value of CS of 0.49 mm/min and % of change is 86.89 %. It is also evident that wire breakage has encountered in four experiments (03, 12, 19 and 22) due to wrong parametric combinations.

Table 3. Experimental design and corresponding results.

Ex. No.	Values of input process parameters				
	T_{on} (µs)	**T_{off} (µs)**	**I_p (A)**	**S_v (V)**	**CS (mm/min)**
01	105(-1)	30(-1)	130(0)	50(0)	1.65
02	105(-1)	60(+1)	130(0)	50(0)	**0.49**
03	131(+1)	30(-1)	130(0)	50(0)	WB
04	131(+1)	60(+1)	130(0)	50(0)	2.32
05	118(0)	45(0)	40(-1)	20(-1)	1.75
06	118(0)	45(0)	40(-1)	80(+1)	0.99
07	118(0)	45(0)	220(+1)	20(-1)	2.44
08	118(0)	45(0)	220(+1)	80(+1)	2.69
09	105(-1)	45(0)	40(-1)	50(0)	0.55
10	105(-1)	45(0)	220(+1)	50(0)	1.6
11	131(+1)	45(0)	40(-1)	50(0)	0.99
12	131(+1)	45(0)	220(+1)	50(0)	WB
13	118(0)	30(-1)	130(0)	20(-1)	2.15
14	118(0)	30(-1)	130(0)	80(+1)	2.12
15	118(0)	60(+1)	130(0)	20(-1)	3.03
16	118(0)	60(+1)	130(0)	80(+1)	1.04
17	105(-1)	45(0)	130(0)	20(-1)	0.91
18	105(-1)	45(0)	130(0)	80(+1)	0.74
19	131(+1)	45(0)	130(0)	20(-1)	WB
20	131(+1)	45(0)	130(0)	80(+1)	3.11
21	118(0)	30(-1)	40(-1)	50(0)	1.92
22	118(0)	30(-1)	220(+1)	50(0)	WB
23	118(0)	60(+1)	40(-1)	50(0)	0.68
24	118(0)	60(+1)	220(+1)	50(0)	2.12
25	118(0)	45(0)	130(0)	50(0)	**3.74**
26	118(0)	45(0)	130(0)	50(0)	3.66
27	118(0)	45(0)	130(0)	50(0)	3.69
28	118(0)	45(0)	130(0)	50(0)	3.54
29	118(0)	45(0)	130(0)	50(0)	3.64

Quadratic model is providing the lowest p value. Hence quadratic model is selected for analysis of variance for the present study. Table 4 presents mean square, the sum of square, the degree of freedom, F value and p value for different terms of the quadratic model. The model F-value of 148.95 with corresponding p-value of <0.0001 implies that the model is highly significant as shown in Table 4. There is only a 0.01% chance that a "Model F-Value" this large could occur due to noise. Values of "Prob > F" less than 0.05 indicate model terms are significant. In this case T_{on}, T_{off}, I_p, $T_{on} \times I_p$, $T_{off} \times S_v$, $S_v \times I_p$, T_{on}^2, T_{off}^2, S_v^2, I_p^2 are significant model terms. Values greater than 0.1 indicate the model terms are not significant.

Table 4. ANOVA for average cutting speed.

Source	SS	DF	MS	F-Value	P> F
Model	28.87	14	2.06	148.95	< 0.0001
A-T_{on}	4.28	1	4.28	308.91	< 0.0001
B-T_{off}	1.94	1	1.94	140.28	< 0.0001
C-S_v	0.026	1	0.026	1.88	0.2006
D-I_p	2.68	1	2.68	193.25	< 0.0001
AD	0.42	1	0.42	30.45	0.0003
BC	0.96	1	0.96	69.37	< 0.0001
CD	0.26	1	0.26	18.42	0.0016
A2	3.16	1	3.16	228.20	< 0.0001
B2	2.80	1	2.80	202.44	< 0.0001
C2	3.29	1	3.29	237.71	< 0.0001
D2	4.27	1	4.27	308.22	< 0.0001
Residual	0.14	10	0.014		
Lack of Fit	0.14	6	0.023		
Pure Error	0.000	4	0.000		
Cor Total	29.01	24			

In WEDM, cutting speed changes due to the combination of wire tool and workpiece material, the polarity of the voltage to apply, the duration of the spark, discharge energy during the spark and so on. The average cutting speed value is plotted against the process inputs to investigate the effects of input process parameters on a measure of process performance. Figure 2 illustrates the effects of T_{on}, T_{off}, I_p and S_v on CS. It is evident that T_{on} is most significant effect on CS. It is clear from the figure 2 that CS continuously increases by increasing T_{on} and decreases by decreasing T_{on}. It has been also observed that I_p is having some similar effects on CS. On the other hand, it has been observed that T_{off} and S_v is having some adverse effects on CS. Where CS values decreases by increasing the T_{off} and S_v, whereas by decreasing T_{off} and S_v some increment was noticed in CS. T_{off} plays a vital role in CS, at the lower level of T_{on} with the increase in T_{off} level CS will be lower compared with the higher level of T_{on}. Alternatively, a Smaller value of servo voltage narrows down the spark gap, which leads to more number of sparks per unit time. It speeds up the machining rate and thus the amount of material removal from the workpiece and tool [7].

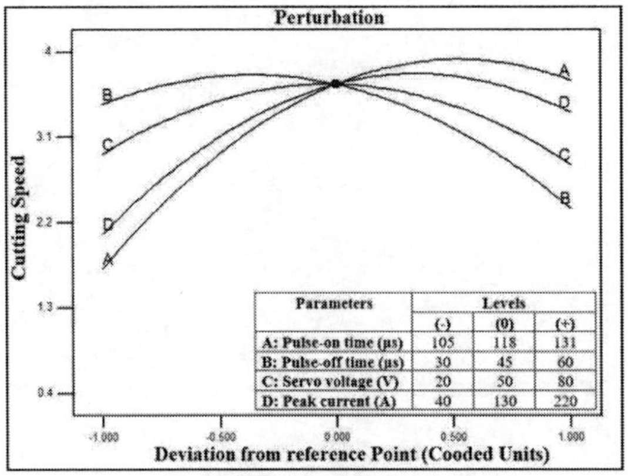

Figure 2. Effects of T_{on}, T_{off}, S_v and I_p on cutting speed [X-axis in coded value]

It is evident from analysis of variance table that the input parameters also have interaction effects on process performance characteristic. Figure 3 illustrates the interaction effects of: (a) I_p-T_{on} vs CS; (b) S_v-T_{off} vs CS; and (c) I_p-S_v vs CS. After that, a regression model of CS has been developed (in actual value) as illustrated using equation (1). The coefficient of determination (R^2) value of the developed model is found as 0.9952; is established the creditability of this model.

$$\begin{aligned} CS = {} & +3.66 + 0.99 \times A - 0.55 \times B - 0.052 \times C \\ & + 0.64 \times D - 4.375 \times A \times B \\ & + 0.11 \times A \times C + 0.46 \times A \times D \\ & - 0.49 \times B \times C - 0.069 \times B \times D \\ & + 0.25 \times C \times D - 0.95 \times A^2 \\ & - 0.76 \times B^2 - 0.79 \times C^2 \\ & - 0.94 \times D^2 \qquad (1) \end{aligned}$$

(A= Pulse-on time; B= Pulse-off time; C= Peak current; D= Servo Voltage)

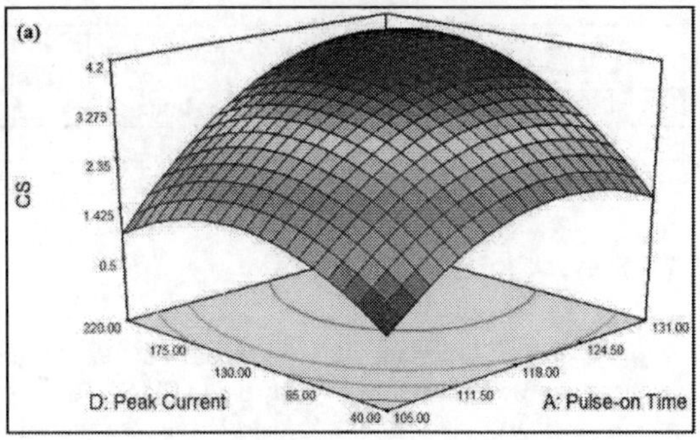

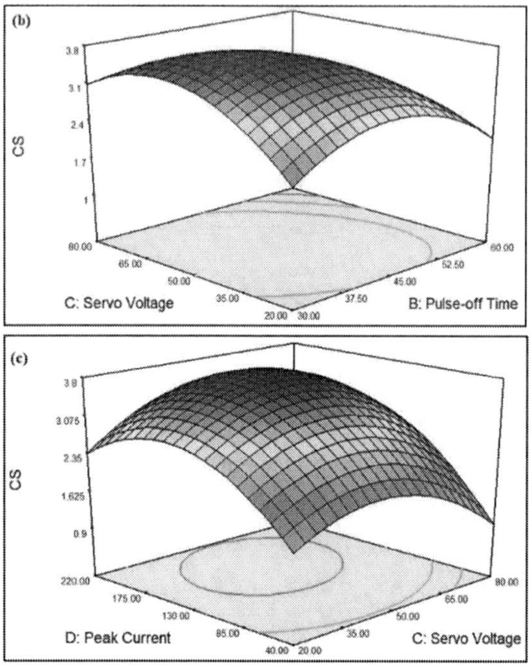

Figure 3. Interaction effects of input parameters on Cutting Speed: (a) I_p-T_{on} vs Cutting Speed; (b) S_v-T_{off} vs Cutting Speed; and (c) I_p-S_v vs Cutting Speed.

Finally, optimization is carried out using to find the optimum combination of process parameters to conduct confirmation experiments. Results of confirmation experiments are presented in table 5. Moreover, the values of CS obtained in confirmation experiments are compared with the corresponding predicted values to quantify the absolute percentage of error in prediction and thus, the prediction capability of developed model is evaluated.

Table 5. Experimental design and corresponding results.

Modified Parameters				Cutting Speed		% of Error
A	B	C	D	Predicted	Experimental	
118	37	58	138	3.75	4.22	0.11
130	32	53	144	3.90	4.52	0.13

3.1 Surface integrity analysis

Figure 4 shows the surface characteristics of the WEDMed surface of nimonic 263, obtained by using FE-SEM. Three specimens were selected for microstructure observation at low CS, average CS and higher CS. From the investigation of WEDMed surface, it is obvious that, molten metal is deposited in the form of lump of debris during WEDM. Figure 4 (a) shows the WEDMed surface at experimental condition corresponding to low discharge energy (LDE), where the parameters are set at their lowest values. Figure 4 (a) appears to be smoother as discharge craters are extremely small with the small amount of debris, fewer micro-voids are present on the surface due to LDE. Figure 4 (b) shows the surface topography at medium discharge energy (MDE), which produces the more and deep craters on the WEDMed surfaces as seen in obtained micrographs. On the other hand, at higher T_{on} and I_p, the discharge energy per duty cycle increases, which produces the wider and deeper craters, pockmarks and lump of debris on the WEDMed surfaces as seen in FE-SEM micrograph 4 (c). It is obvious from figure 4 (c) that the size of bombard is deeper and wider due to high T_{on}, which subsequently results in a high CS (3.74 mm/min) [18]. Figure 4 (d) shows the CS values for all levels: LDE, MDE and at HDE.

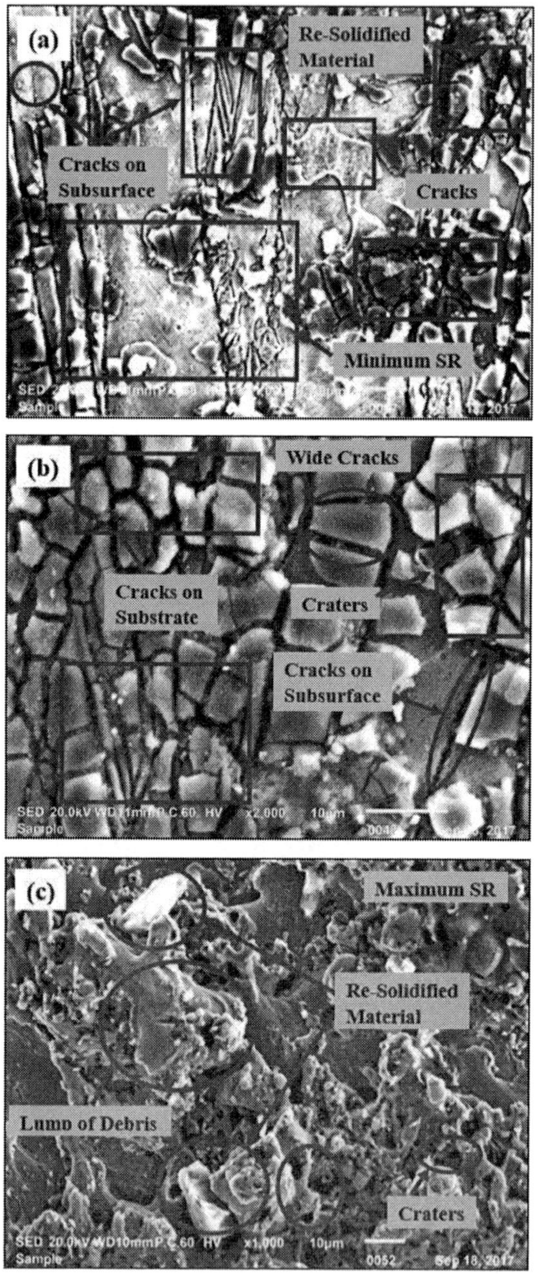

Figure 4. Microstructure of the WEDMed surfaces: (a) at low CS (Experiment no. 2); (b) at medium CS (Experiment no. 24); and (c) at high CS (Experiment no. 25).

4 Conclusions and future scope

In this study a parametric study, modeling and optimization of WEDM process of nimonic 263 has been fulfilled based on BBD were conducted to develop empirical model of process. Following conclusion has been drawn from the analysis of the results:

1. It has been observed that factor T_{on} is most significant input process parameter for CS, followed by I_p. It was also noticed that T_{off} and S_v have less significant effects on responses.
2. It is also observed that input parameters are having an interaction effect on cutting speed. It was also depicted CS increases with increase in I_p and T_{on} and with decrease in T_{off}.
3. Higher values of T_{on} and I_p leads to wire breakage. In addition, if the S_v is too low, an amount of discharge energy was absorbed by wire tool and debris in the gap cannot be flushed properly by the dielectric, it results in the arcing and wire breakages occur. Four experiments were not performed due to wrong parametric setting which results in wire breakage.
4. Surface topography reveals that WEDMed surfaces at low CS appears to be smoother as discharge craters are extremely small with the small amount of debris, fewer micro-voids are present on the surface due to low discharge energy during WEDM operation. While, Large amount of debris, craters and significant number of micro-cracks were observed on eroded surface of wire used at high CS.
5. The EDS analysis reveals, the highest percentage of migrated element is Cu: 18.22% and Zn: 6.78% is for experiment no. 25 (T_{on}: 118 μs; T_{off}: 45 μs; I_p: 130 A; and S_v: 50 V).
6. The developed empirical model is found best suited as it displays the very low value of prediction error (~1%) and hence, can be utilized by future users for prospective application of the technique. T_{on}: 130 μs, T_{off}: 32 μs, I_p: 144 A and S_v: 53 V are found optimum for the maximized value of CS.
7. Future study could consider the tribological and microstructure study of WEDM machined miniature parts: gear, spline etc. for obtaining better insight of the process.

References

1. E. Jameson, "Electrical discharge machining," (2001).
2. B. Fleming, "The EDM how-to book." p. 141, (2005).
3. K. H. Ho, S. T. Newman, S. Rahimifard, and R. D. Allen, "State of the art in wire electrical discharge machining (WEDM)," Int. J. Mach. Tools Manuf., **44**, no. 12–13, pp. 1247–1259, 2004.
4. B. Singh and J. P. Misra, "A Critical Review of Wire Electric Discharge Machining," Ann. DAAAM Proc. Int. DAAAM Symp., pp. 249–266, (2016).
5. N. Tosun, C. Cogun, and a. Inan, "The Effect of Cutting Parameters on Workpiece Surface Roughness in Wire EDM," Mach. Sci. Technol., **7**, no. 2, pp. 209–219, (2003).
6. T. Singh, J. P. Misra, and B. Singh, "Experimental Investigation of Influence of Process Parameters on MRR during WEDM of Al6063 alloy," Mater. Today Proc., **4**, no. 2, pp. 2242–2247, (2017).
7. A. Kumar, V. Kumar, and J. Kumar, "Parametric Effect on Wire Breakage Frequency and Surface Topography in WEDM of Pure Titanium," J. Mech. Eng. Technol., **1**, no. 2, pp. 51–56, (2013).
8. P. Shandilya, P. K. Jain, and J. P. Misra, "Experimental investigation during wire electric discharge cutting of SiCp/6061 aluminum metal matrix composite," Ann. DAAAM Proc. Int. DAAAM Symp., **21**, no. 1, pp. 1091–1092, (2010).
9. N. Tosun, C. Cogun, and G. Tosun, "A study on kerf and material removal rate in wire electrical discharge machining based on Taguchi method," J. Mater. Process. Technol., **152**, no. 3, pp. 316–322, (2004).
10. R. Chalisgaonkar and J. Kumar, "Multi-response optimization and modeling of trim cut WEDM operation of commercially pure titanium (CPTi) considering multiple user's preferences," Eng. Sci. Technol. an Int. J., **18**, no. 2, pp. 125–134, (2014).
11. R. H. Myers, D. C. Montgomery, and C. M. Anderson-cook, "Response Surface Methodology," Wiley, **3**, pp. 1–1247, (2009).

Effect of Intermetallic Compound Deposits on Wear Resistance and Microhardness on AISI 1020 Steel Substrate Using GTAW Process

J.S.Gill[1], Sikandar Singh Dhiman[2]

[1] *Associate Professor, Department of Mechanical Engineering, S.L.I.E.T, LONGOWAL (Punjab, India).*
[2] *Welding Engineer, M/s Multitex Filtration Engineers Limited, Greater Noida (U.P, India)*

Abstract. Enhancing the oxidation resistance along with sufficiently improved mechanical properties of low carbon steel has always been a challenging task for surfacing industry. Modern automobile industry is looking for joining the dissimilar ferrous and non-ferrous metals to reduce the weight of the overall structure and not compromising the strength of the fabrications. This paper deals with the deposition of iron-aluminium intermetallic over a low carbon substrate using gas tungsten arc welding (GTAW) process. Oxidation resistance of the iron and aluminium metal powders deposits in varying ratios and few mechanical and metallurgical properties such as microhardness, microstructure and wear resistance were investigated and are reported in the present paper.

1. Introduction

Surface of a given material can be improved by depositing the filler metal of the desirable properties. During hardfacing process surface is made relevantly harder than the substrate material for surface improvement. This may lead to extra expenditure on the production costs of components, but it leads to an improved surface both in terms of strength achieved and improved life of the components. It may be applied to a new part during production to increase material's surface properties as well as it may be used to restore a worn-down surface [1]. Surfacing is one of the most useful and economical way to improve the surface properties and wear behaviour of a component [2]. Surface properties and quality achieved of the surfaced component depends upon the selected surfacing alloys and the welding process used for the surfacing. In the present work low carbon steels were subjected to hardfacing by depositing a layer of iron-aluminium (Fe-Al) intermetallic using GTAW process in autogenous mode. Fe-Al intermetallic compounds are regarded as promising materials for industrial applications because of their low cost, low density (5.56 g/cm^3), high specific strength, high temperature strength, lighter weight than steels as well as excellent oxidation and corrosion resistance [2-4].

2. Experimentation

Fine Iron (300 mesh size) and Aluminium powders (120 mesh size) in the varying ratios such as 2:3, 1:1, and 3:2 were thoroughly mixed. Paraffin wax 3% and acetone to adequate quantity were added in the mixture to form a paste applied to the grooved specimens of AISI 1020 low carbon steel which was selected as the substrate material. The base metal composition is shown in Table 1 given below:

After removing rust, oxide film and other impurities the specimens of 150 × 35 × 25 were machined a groove or a slot of 3 mm deep and 10 mm wide on the shaper machine as shown in the Figure 1 given below:

Table 1. Base Metal Composition

C (%)	Mn (%)	Cr (%)	Mo (%)	Ni (%)
0.114	0.6	0.034	0.013	0.018
Si (%)	S (%)	Al (%)	Cu (%)	Fe (%)
0.043	0.015	0.019	0.042	Balance

The slots were fabricated to guide the application of pastes. Pastes of the iron and aluminium metal powders mixed in the required ratios as mentioned above were categorized as C1, C2 and C3whereas letter C designate about the composition of the filler material. The specimens were heated to 50°C in muffle furnace [6] and held for one hour after the application of paste and hence dried. Figure 1 given below shows the actual photograph of the specimen before welding.

Figure 1 Actual Photograph of The Specimen before Welding

Some trials were conducted to conduct the feasibility studies of the hardfacing deposits by varying different welding conditions. The welding parameters for the experimentation were selected on the basis of results obtained from these trial experiments. GTAW process was used to Industrially pure argon gas was used for experimentation with gas flow rate of 10 litres/min, average arc travel speed was maintained around 3 mm/s. Thoriated tungsten non consumable electrode with diameter of 2.4 mm was used in GTAW process in autogenous mode Electrode polarity used was selected to be DCEN (Direct Current Electrode Negative) the arc voltage used during welding was 16 V whereas the welding current was around 120 A. The following tests were performed to investigate the wear behaviour of the weld deposits.

2.1 Oxidation test

Oxidation test was performed to check the resistance of surfaced material against oxidation at high temperature by heating the specimens to the sufficient temperature and holding them for sufficiently long-time intervals.

2.2 Micro-hardness test

Micro-hardness measurements were carried out on VHN (Vickers Hardness Number) Scale. Micro-hardness readings along x-axis and y-axis were taken from surfaced area of all specimens. Specimens of the different compositions for micro-hardness test were prepared by polishing them with emery papers of grade 100 to grade 1500, and with alumina paste on velvet cloth on disc polishing machine. The average values of micro-hardness have been reported in this paper.

2.3 Pin on disc wear test

Wear test was carried out on pin-on-disc wear testing machine as per ASTM G99 standards. Cylindrical pins of 8 mm diameter with spherical tip of radius 4mm and length of 35 mm was selected.

2.4 Micro-Structural analysis

Micro-structral changes have been studied to know the morphology of structural growth with addition of variable composition of intermetallic compound [4].

3. Results and discussions

The oxidation resistance of the surfaced area was evaluated by heating the specimens for 800°C and holding for 4 and 8 hours in furnace and then cooled in air for 12 hours. All the samples were thoroughly washed with acetone before and after performing the experiment and then weighed. Oxidation resistance has been correlated to the weight loss [5,6 & 7].

The results after conducting the test has been collected and presented in the Table 2 and graphically presented in Figure 2 given below. It was found that the effect of higher temperature as well as longer exposure time at that temperature has significant effect on the weight loss and thus oxidation resistance of the surface. It is also worth reporting that the composition C1 found to be most resistant to oxidation as compared to the other two compositions.

Table 2 Weight Loss of Specimens after Oxidation Test

Specimen designation (according to composition)	Weight loss (mg) for (800°C/4 hours)	Weight loss (mg) for (800°C/8 hours)
C1	9.32	9.98
C2	11.14	14.81
C3	19.21	22.37

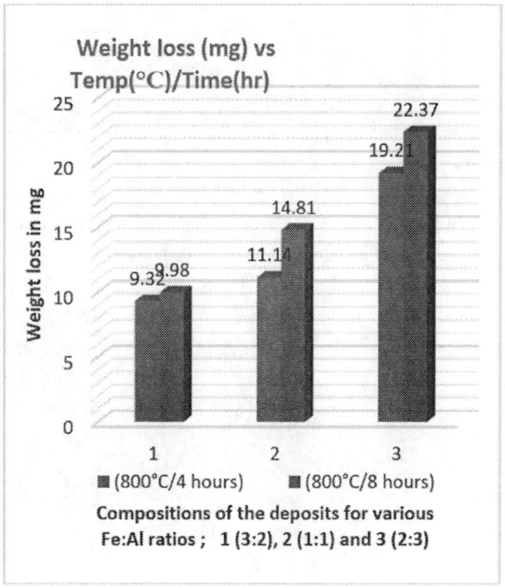

Figure 2 Graphical Representation of Weight Loss after Oxidation

Microhardness testing was conducted on the specimens on Vickers hardness tester under the constant load of 500 gm. The hardness values have been taken along the x-x axis with difference of 1 mm distance and y-y axis with difference of 0.5 mm distance. Then, average value of each composition has been calculated. The results are shown in Table 3 and the graphical representation is shown in Figure 3. Wear test was also done to access the resistance offered by surface rubbing against some hard metal due to which pin on disc tests wear tests were conducted. Wear tests were performed under normal atmospheric conditions. All the specimens prior and after the test were cleaned with acetone and weighed with digital electronic balance having accuracy of ±0.001 g. Difference between initial weight and final weight gave the total weight loss after sliding wear test. Wear rate is considered by total weight loss during experiment. Table 4 shows the results performed in sets

$$\text{wear rate (mg/hr)} = \frac{\text{(Final wt.} - \text{Initial wt.)}}{\text{Time (minute)}} \times \mathbf{60} \times \mathbf{1000}$$

Table 3 Microhardness Results of The Specimens

Composition	Weld bead profile for Micro hardness	Micro hardness values (VHN)		Micro hardness avg. value (VHN)
C1		y-y axis	x-x axis	422
		424.1	418.7	
		426.2	421.5	
		415.7	425.8	
C2	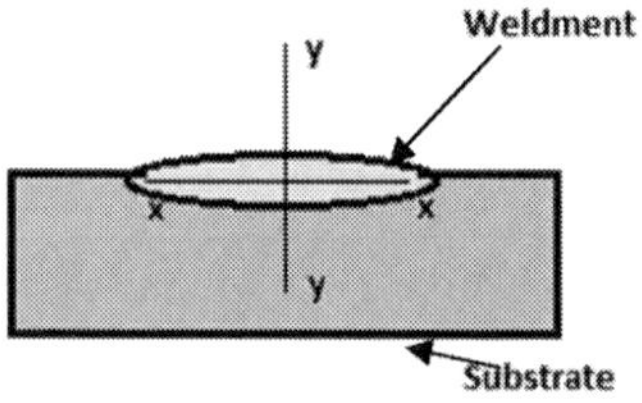 (a)	y-y axis	x-x axis	385
		382.5	390.4	
		392.4	383.2	
		378.6	387.3	
C3		y-y axis	x-x axis	396
		405.3	392.2	
		394.1	402	
		387.4	398.6	
Base metal				242

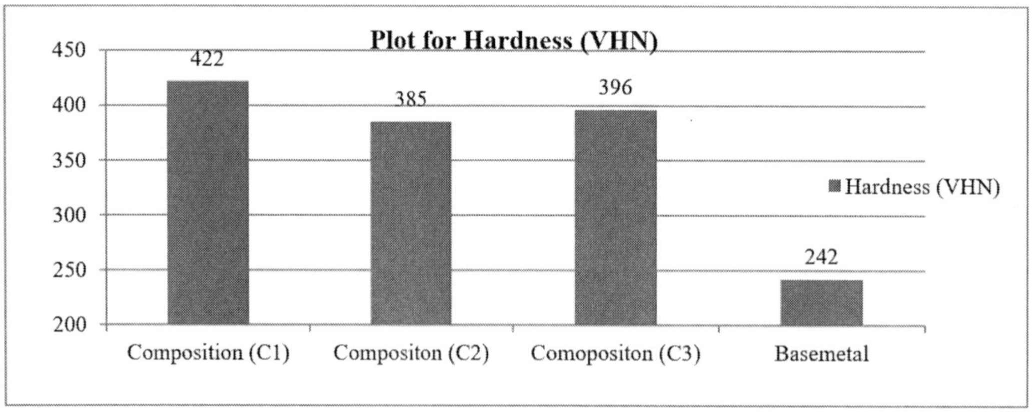

Figure 3 Graphical Representations of Microhardness Results

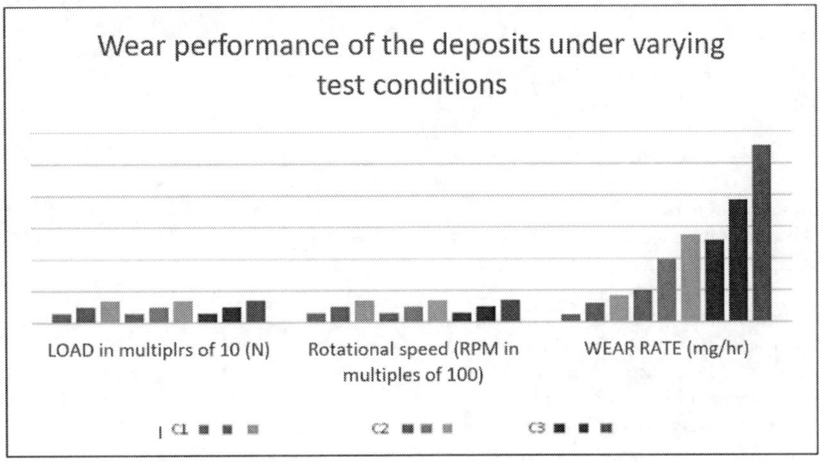

Figure 4 Wear Resistance of the Fe-Al Intermetallic Deposits

Table 4 Wear Rate Results at Different Loads and Rotational Speeds

Specimens	LOAD (N)	Rotational speed (RPM)	WEAR RATE (mg/hr)
C1	30	300	2.4
	50	500	6.0
	70	700	8.4
C2	30	300	10.1
	50	500	19.9
	70	700	27.5
C3	30	300	25.71
	50	500	38.56
	70	700	55.67

The samples were polished as per the ASTM standards for further metallurgical investigations. Few of the optical micrographs are presented below in Figure 5.

Composition (C1) (Fe:Al/60:40)	Composition (C2) (Fe:Al/50:50)	Composition (C3) (Fe:Al/40:60)

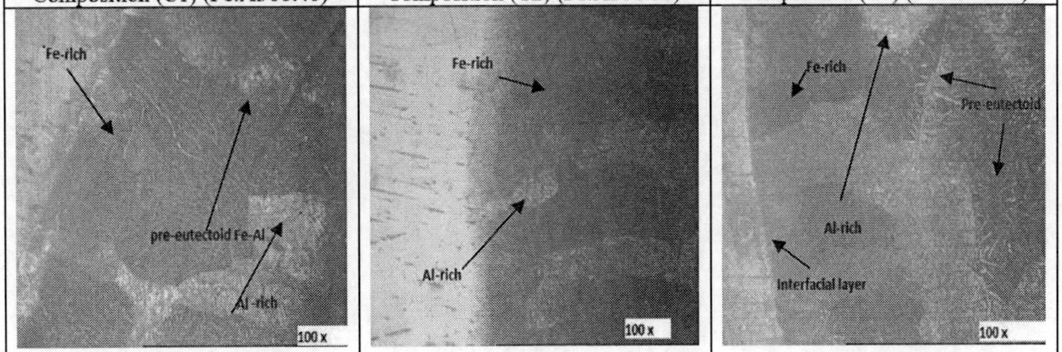

(a) Microstructure of weldment of composition (C1)	(b) Microstructure of weldment of composition (C2)	(c) Microstructure of weldment of composition (C3)

Figure 5 Optical Micrographs of Fe-Al Intermetallic Deposits of Varying Compositions

4. Conclusions

From above mentioned results and discussions it can be concluded that surfacing by GTA welding process shows 100% feasibility of filler material used for experimentation and few important conclusions can be made:

Weight loss of the specimens due to oxidation results show that both the temperature and the time duration of exposure play significant role by affecting the oxidation resistance.

Further it is found that composition C1 in which Fe was mixed in excess as compared to C2 and C3 showed remarkable results in wear resistance and microhardness,

Composition C2 in which iron and aluminium metal powders {Fe:Al= (1:1)} mixed in equal ratio showed reasonable microhardness, and reasonably high wear resistance and can be recommended for industrial use. The ease of the deposition and the surface texture obtained by this composition mixture was found to be best amongst all other composition mixtures used in the present study.

References

[1] Tatsuya SAKIYAMA et.al, Dissimilar Metal Joining Technologies for Steel Sheet and Aluminum Alloy Sheet in Auto Body, *NIPPON STEEL TECHNICAL,* REPORT No. **103**, 91-98 (2013).

[2] A Kostka et.al, On the formation and growth of intermetallic phases during interdiffusion between low-carbon steel and aluminum alloys, *Acta Materialia* **59**, 1586-1600 (2011).

[3] J. Luo et al., Study on the new joining technology of aluminum alloy/Q235 steel dissimilar metal by longitudinal electromagnetic hybrid TIG welding-brazing method, *IEEE*, **Vol. 11**, pp. 1698-1701 (2011).

[4] Mitsuhiro Watanabe et.al., Growth Manner of Intermetallic Compound Layer Produced at Welding Interface of Friction Stir Spot Welded Aluminum/Steel Lap Joint, *Materials Transactions*, **Vol. 52, 5**, 953-959 (2011).

[5] John A Taylor et.al., Iron-Containing Intermetallic Phases in Al-Si Based Casting Alloys, *Procedia Material Science* **1**, 19-31 (2012).

[6] Erdemir et al., Formation of Fe-Al intermetallic coating on low carbon steel by novel mechanical alloying technique, *Powder technology*, **Vol. 247**, pp. 24-29 (2013).

[7] Gang Zhang et.al, Analysis and modelling of the growth of intermetallic compounds in aluminium-steel joints, *RSC Adv.* **7**, 37797-37805 (2017).

Process Parameter Optimization for Abrasive Water Jet Machining of Titanium Alloy Using Meta-Heuristic Algorithms

Vishal S Sharma[1], Amit Kumar[2], Munish Kumar Gupta[3], Neeraj Bhanot[4]

[1&2]I & P Deptt., Dr. B.R. Ambedkar NIT Jalandhar, Punjab, India
[3]MED, LCET, Ludhiana, INDIA
[4]Department of Quantitative methods and Operation Management, Indian Institute of Management, Amritsar, Punjab, India
Address correspondence to Vishal S Sharma, sharmavs@nitj.ac.in

Abstract. Recently, the trend of optimization algorithms for improvements of surface quality and productivity characteristics in abrasive water jet machining of titanium alloy (Ti-6Al-4V alloy) has become increasingly more widespread in various industrial sectors i.e., aircraft and automobile Industries. Here, the present research attempts to select the ideal or best AWJM process parameters by implementing the well known meta-heuristic algorithm i.e., Teacher learning based optimization method (TLBO). The AWJM experiments as per the Taguchi L9 orthogonal array were performed on Ti 6Al-4V titanium alloy by considering jet transverse speed, stand-off distance and abrasive flow as the input parameters. Then, the influence of process parameters on surface roughness and material removal rate has been performed by means plot and ANOVA analysis. After that, the results are optimized with the TLBO method. The overall results indicate that the TLBO method is an efficient method used to find the optimal results with very short interval of time i.e., within 3 sec.

1. INTRODUCTION

Abrasive water jet machining (AWJM) is a modern trend of cutting solution that is largely implemented in the machining of difficult-to-cut materials such as nickel and titanium based alloys [1]. The cutting action in AWJM takes place due to the erosion action of high velocity abrasive particles along with the jet of water, respectively. The AWJM process expresses the major advantages such as no contact tool that might break during machining, less heat-affected zone and low cutting forces in various industrial applications [2]. However, the involvement of large process parameters such as jet traverse speed, stand-off distance and abrasive flow rate makes the AWJM performance difficult at large scale application [3]. Therefore, the optimization of AWJM process parameters is an emerging solution for drastically improvement performance in terms of surface quality and productivity[2]. The optimization of process parameters have been performed in two manners i.e., single response optimization and multi-response optimization. In the context of single response optimization, the Taguchi based S/N ratio analysis has been successfully implemented around the world in various applications such as machining, rapid prototyping, material processing etc [4–6]. However, the Taguchi method is not preferred where more than one response or process parameter has been involved. Therefore, to tackle this situation, the implementation of multi-response optimization is a fruitful option and provides a very useful result for practical applications of process.

In the recent years, several research efforts have been made on multi-response optimization of process parameters by implementing different methods such as response surface methodology, desirability approach, multi-criteria decision methods and meta-heuristic algorithms such as particle swarm optimization, genetic algorithm, teacher learning based optimization etc [7–11]. In the first work, Lan performed the optimization of turning parameters by using TOPSIS and Taguchi methods [12]. Secondly, Gauri et al. successfully applied the TOPSIS method for determine the optimal parameters setting in ultrasonic process [13]. Similarly, the application of PSO in micro-machining of medical devices is reported by Vazquez et a.l [14]. On same context, Ciurana et al. used the ANN model integrated with PSO for obtaining the ideal parameters in pulsed laser micromachining for hardened AISI H13 steel [15]. Likewise, Ulutan and Ozel implemented the physics based simulation model of PSO for prediction of responses during turning of nickel alloys [16]. In another work, the optimization of grinding process parameters using PSO has been carried by Pawar et al. [17].

From the current literature, it has been interestingly noticed that the trend of multi-response optimization of process parameters by using discussed method is currently a hot topic of research. However, it has been observed that the implementation of these optimization methods in AWJM of process parameters using meta-heuristic algorithms is still not reported in literature. Therefore, this study expresses the first attempt to investigate the multi-response optimization of AWJM process parameters in mostly used Ti-6Al-4V titanium alloy using teacher learning based meta-heuristic algorithm, respectively. In this regard, the range of AWJM process parameters i.e., jet traverse speed, stand-off distance and abrasive flow have been carefully selected and the optimization of surface roughness and material removal rate have been performed. Thereafter, the effect of process parameters on quality and productivity characteristics has been investigated by S/N ratio plots. In the end, the optimization is performed by subjected algorithm.

2 MATERIALS AND METHODS

2.1 Workpiece material

The AWJM experiments were performed on mostly used grade of titanium alloy i.e., Ti-6Al-4V alloy (size: 575mm × 55m ×12mm; hardness: 35HRC), shown in Figure 1. It is also known as the "workhorse" of the titanium industry because it is by far the most common Ti alloy, accounting for more than 50% of total titanium usage. Besides, this is referred as a poor machinability material due to its various inherent characteristics such as low thermal conductivity, high work hardening capability and chemical affinity etc. The chemical composition of used Ti-6Al-4V alloy is presented in Table 1.

Fig. 1. Workpiece material i.e., Ti-6Al-4V titanium alloy

Table 1. The compositional range for Ti 6Al-V4 alloy

Carbon (Maximum) 0.10 %	Titanium Balance
Aluminum 5.50 to 6.75 %	Vanadium 3.50 to 4.50 %
Nitrogen 0.05 %	Iron (Maximum) 0.40 %
Oxygen (Maximum) 0.020 %	Hydrogen (Maximum) 0.015 %
Other, Total (Maximum) 0.40 %	

2.2 Abrasive water jet machining centre

The OMAX Jet Machining centre (model no 2652, shown in Figure 2), a precision abrasive jet machining has been used for performing the experimental runs. The AWJM tool is operated under specific software control and able to cut complex parts out of most materials including metal, plastic, glass, ceramics, stone and composites using standard CAD drawing files. The machine has X-Y motion system, high-pressure plumbing and swivels, a high-pressure nozzle, and an abrasive feed system (hopper).

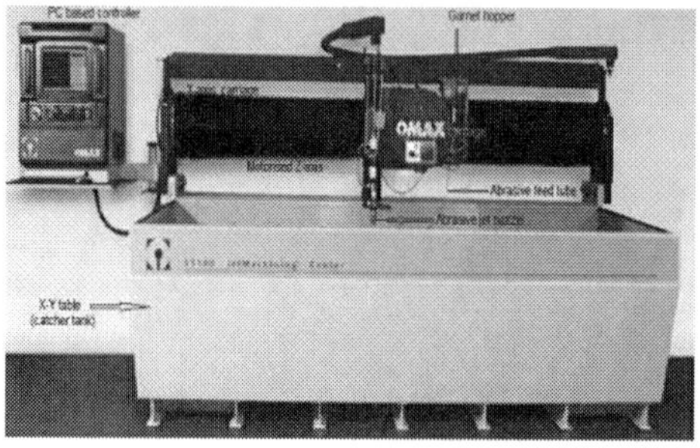

Fig. 2. Abrasive Water Jet Machining Centre

2.3 Measurement of responses

The present work consists of two output responses i.e., surface roughness and material removal rate. For measurement of surface roughness values, the Surtronic S-100 Series portable roughness tester has been used. The average surface roughness parameter (Ra) is selected for experimental readings. Similarly, for material removal rate, the mass of work pieces were measured before and after the trials. Then, the time of experiment was recorded from the machine and thus MRR (gram/min) was calculated after dividing the difference of mass (gram) by time (minutes).

2.4 Design of Experiments

In order to efficiently perform the experiments, the Taguchi L9 orthogonal array design has been selected. The three different process parameters at three levels i.e., pressure (levels: 1500, 2000, 2500 Bar), stand of distance (SOD) (level: 1.5, 2, 2.5 mm), Jet Transverse Speed (JTS) (level: 40, 50, 60 mm/min) are selected as input values and combinations are varied in conformance with the L9 orthogonal array, as furnished in Table 2 and 3. The other important parameters that were kept constant during tests are Abrasive Flow Rate (0.356 Kg/min), jet impact angle at neutral nozzle position (90°), orifice diameter (0.3556 mm), nozzle length (76.2 mm), nozzle diameter (0.762 mm), abrasive material (80 mesh garnet particles with the density of 4100 kg/m3) and average diameter of abrasive particles (0.18 mm). The garnet consists of chemically 36% FeO, 33% SiO2, 20% Al2O3, 4%MgO, 3% TiO2, 2% CaO and 2% MnO2.

Table 2. List of parameters and their levels

Factors	Level 1	Level 2	Level 3
Pressure (bar)	1500	2000	2500
Jet Transverse Speed (mm/min)	40	50	60
Stand-off distance (mm)	1.5	2.0	2.5

Table 3. Experimental results

Input Parameters			Output Parameters	
Pressure (bar)	**Jet Transverse Speed (mm/min)**	**Stand-off distance (mm)**	**Material Removal Rate (cm^3 /min)**	**Surface Roughness (µm)**
1500	40	1.5	1.352182	2.100
1500	50	2	1.589574	2.075
1500	60	2.5	1.817500	2.300
2000	40	2	1.433636	2.250
2000	50	2.5	1.651333	2.067
2000	60	1.5	1.980418	2.000
2500	40	2.5	1.442000	2.100
2500	50	1.5	1.658889	2.125
2500	60	2	1.908355	1.800

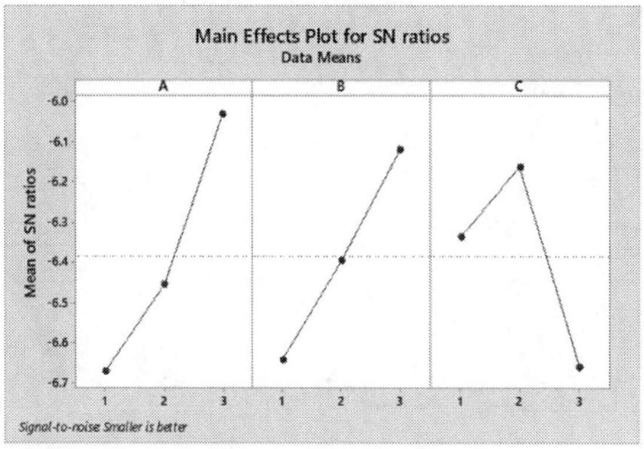

Fig. 3. Main effect plots for surface roughness values
* A- Pressure, B- Jet Traverse Speed (JTS), C- Stand Off Distance (SOD)

Table 4. ANOVA for SR (R_a)

Source	DF	Adj SS	Adj MS	F-Value	%
Pressure	2	0.03474	0.01737	0.39	21.17
JTS	2	0.08835	0.01022	1.54	54.53
SOD	2	0.02057	0.01029	0.23	13.05
Error	2	0.02034	0.04417		
Total	8	0.16409			

3 RESULTS AND DISCUSSIONS

In this section, firstly the influence of process parameters on surface roughness and material removal rate has been analyzed. After that, the multi-response optimization has been performed by the TLBO method. Lastly, the confirmation runs were executed to confirm the validity of meta-heuristic algorithm.

3.1 Influence on surface roughness

In order to check the statistical significance of each individual process parameter, initially the analysis of variance (ANOVA) test has been applied on the experimental data. The analysis of variance (ANOVA) test for surface roughness (Ra) is shown in Table 3. It is evident that, Jet transverse speed alone influences Ra significantly. Since the F – Value of Jet transverse speed is 1.54 and having percentage contribution of 54.53 %, therefore this parameter is considered to be most significant for Ra. Further, the main effect for means are successfully plotted to study the influence of each factor on surface roughness values, as shown in Figure 3. It has been remarkably seen that the values of surface roughness increases with the change in pressure (1500 to 2000) and transverse speed (40 to 60). Besides, the stand-off distance shows the opposite results as compare with the pressure and jet transverse speed. The surface roughness values initially increases by changing the values of stand-off distance 1.5 to 2.0 and decreases with change in stand-off distance from 2.0 to 2.5.

3.2 Influence on material removal rate

The ANOVA for MRR is shown in Table 5. It is evident that Jet transverse speed alone has very dominating effect on MRR. Since the F – Value of Jet transverse speed is 160.97 and having percentage contributions as 84.32%, therefore this parameter is considered to be most significant for MRR. Furthermore, the main effect plot for MRR has been shown in Figure 4. It has been observed from the main effect plots that the material removal rate is increases with the pressure from 1500 to 2500 and jet transverse speed from 40 to 60, respectively. On the other hand, the material removal rate decreases with change in stand-off distance from 1.5 to 2.5. This is attributed to the scientific reason that the increase in pressure and jet transverse speed increases the velocity of carrier gas that conveying the abrasive particles, hence more material is removed from the workpiece. On the contrary, increase in stand-off distance at certain point reduces MRR because of the reduced taper in the machined groove and kerf width, respectively.

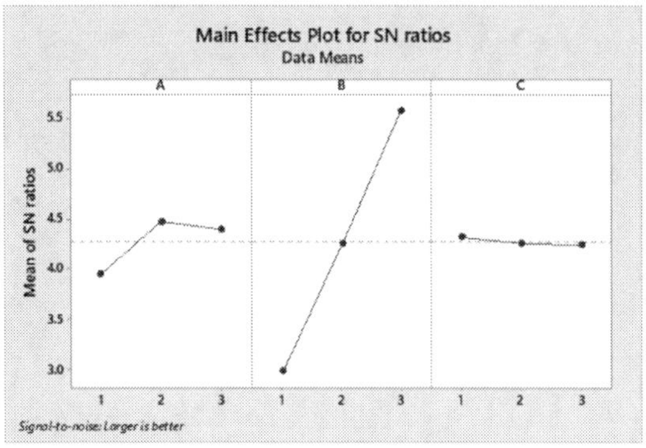

Fig. 4. Main effect plots for material removal rate values
* A- Pressure, B- Jet Traverse Speed (JTS), C- Stand Off Distance (SOD)

Table 5. ANOVA for MRR

Source	DF	Adj SS	Adj MS	F-Value	%
Pressure	2	0.017707	0.008853	7.80	4.58
JTS	2	0.365310	0.182655	160.97	84.32
SOD	2	0.001170	0.000585	0.52	3.01%
Error	2	0.002269	0.001135		
Total	8	0.386456			

3.3 Multi-response optimization of process parameters using meta-heuristic algorithm

In the current work, the well known meta-heuristic algorithm i.e., teacher learning based optimization (TLBO) method has been selected to perform the multi-response optimization [18]. The TLBO algorithm, originally developed by R.V. Rao, is successfully implemented to solve the constrained and unconstrained optimization problem in different areas [19]. They also compared the efficiency of TLBO algorithm with other methods of optimization. It was reported that, the TLBO algorithm provides better results over several evolutionary techniques in less number of iterations. The algorithm does not require any specific algorithm control parameters. The boundary conditions for the parameters are collected from the Table 2. The detailed information and the implementation procedure are deeply discussed elsewhere.

Working of TLBO: To achieve the desired goal, the AWJM process parameters with respect to the surface roughness and material removal rate has been optimized in this part. For this purpose, initially a combined objective is generated (as given by Equation (1)) and the multi-response optimization is performed for the combined objective.

$$Max\ C.O. = W_1 * (Ra/Ra_{min}) - W_2 * (MRR/MRR_{max}) \quad (1)$$

Where, W1 and W2 are the weights (i.e., 0.5 each) given to the surface roughness and MRR values. Then, the regression equation of combined objective function is established for TLBO fitness function, as given in Equation 2.

C.O = 0.776 − 0.000026 Pressure + 0.00464 Jet Transverse Speed + 0.0156 Stand-off Distance (2)

Further, to find out the optimum values, the algorithm are executed a number of times. Table 6 presents the optimal solution along with the best, worst and average readings using TLBO method. Similarly, Figure 5 shows the convergence characteristics graph of obtained optimal results. It has been observed that the success rate of TLBO method is 95 % and convergence time is only 3 seconds for each iteration. This is because there is no specific parameters are used for proper function the algorithm.

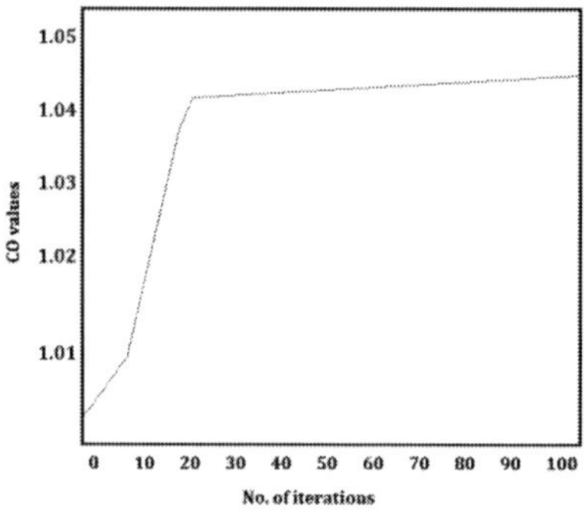

Fig. 5. Convergence characteristics plots for CO values

3.4 Confirmatory experiments

This is the final step usually performed to check the validity or adequacy of the developed optimization models. The obtained AWJM optimal parameters are given as: 1500 bar pressure, 60 mm/min jet transverse speed and 1.5 mm stand-off distance, respectively. It has been obtained that the experimental values obtained at these optimal setting show very close relation with each other. The error below 5% confirmed that the meta-heuristic algorithms i.e., TLBO shows very promising results in terms of AWJM process parameters. Hence, the TLBO is an effective and time saving optimization methods used for determining the ideal conditions.

Table 6. Optimized parameters for maximization of CO value

Parameters	TLBO
Pressure (bar)	1500
Jet Transverse Speed (mm/min)	60
Stand-off distance	1.5
Best Solution	1.04425
Average	1.03426
Worst	1.02436
Average time (sec)	4.32
Confirmatory experiments	1.04536
Error	4.32%

4 CONCLUSIONS

This research presents the experimental investigation complimented with the surface quality and productivity optimization in AWJM of Ti-6Al-4V titanium alloy. The following conclusions can be drawn for effective machining of Ti 6Al 4V by AWJM process as follows:

- From the optimization results, the success rate of TLBO method is 95 % and convergence time is only 3 seconds for each iteration.
- The ANOVA analysis showed a high degree of fit with respect to the high correlation value.
- From the ANOVA analysis, jet transverse speed (JTS) plays a major role on influencing material removable rate (MRR) by 84.32% as observed in ANOVA F test.
- In case of Surface roughness, jet transverse speed plays major significance of about 54.53%. Stand-off distance is having least significance influence on SR.
- The obtained AWJM optimal parameters are given as: 1500 bar pressure, 60 mm/min jet transverse speed and 1.5 mm stand-off distance, respectively.

References

1. Sasikumar KSK, Arulshri KP, Ponappa K, Uthayakumar M (2016) A study on kerf characteristics of hybrid aluminium 7075 metal matrix composites machined using abrasive water jet machining technology. Proc Inst Mech Eng Part B J Eng Manuf 232:690–704. doi: 10.1177/0954405416654085
2. Jagadish, Bhowmik S, Ray A (2015) Prediction of surface roughness quality of green abrasive water jet machining: a soft computing approach. J Intell Manuf 1–15. doi: 10.1007/s10845-015-1169-7
3. Chaturvedi V, Singh D (2015) Multi Response Optimization of Process Parameters of Abrasive Water Jet Machining for Stainless Steel AISI 304 Using VIKOR Approach Coupled with Signal to Noise Ratio Methodology. J Adv Manuf Syst 14:107–121. doi: 10.1142/S0219686715500080
4. Gupta MK (2015) Optimization of machining parameters for turning AISI 4340 steel using Taguchi based grey relational analysis. Indian J Eng Mater Sci 22:679–685.
5. Gupta MK, Singh G, Sood PK (2015) Modelling and Optimization of Tool Wear in Machining of EN24 Steel Using Taguchi Approach. J Inst Eng Ser C 96:269–277. doi: 10.1007/s40032-015-0175-z
6. Biswas SA, Datta S, Bhaumik S, Majumdar G (2009) MULTI-RESPONSE OPTIMIZATION : A CASE STUDY IN. 2009:26–28.
7. Thakur DG, Ramamoorthy B, Vijayaraghavan L (2009) Optimization of Minimum Quantity Lubrication Parameters in High Speed Turning of Superalloy Inconel 718 for Sustainable Development. World Acad Sci Eng Technol 54:224–226.
8. Neşeli S, Yaldız S, Türkeş E (2011) Optimization of tool geometry parameters for turning operations based on the response surface methodology. Measurement 44:580–587. doi: http://dx.doi.org/10.1016/j.measurement.2010.11.018
9. Guodong L, Yong L, Quancun K, Hao T (2016) Selection and Optimization of Electrolyte for Micro Electrochemical Machining on Stainless Steel 304. Procedia CIRP 42:412–417. doi: 10.1016/j.procir.2016.02.223
10. Yildiz AR, Öztürk F (2010) Hybrid Taguchi-Harmony Search Approach for Shape Optimization. In: Geem ZW (ed) Recent Adv. Harmon. Search Algorithm. Springer Berlin Heidelberg, Berlin, Heidelberg, pp 89–98
11. Gupta MK, Sood PK, Sharma VS (2016) Optimization of machining parameters and cutting fluids during nano-fluid based minimum quantity lubrication turning of titanium alloy by using evolutionary techniques. J Clean Prod 135:1276–1288. doi: 10.1016/j.jclepro.2016.06.184
12. Lan T-S (2009) Taguchi optimization of multi-objective CNC machining using TOPSIS. Inf Technol J 8:917–922.
13. Gauri SK, Chakravorty R, Chakraborty S (2011) Optimization of correlated multiple responses of ultrasonic machining (USM) process. Int J Adv Manuf Technol 53:1115–1127. doi: 10.1007/s00170-010-2905-y
14. Vazquez E, Ciurana J, Rodríguez CA, et al. (2011) Swarm intelligent selection and optimization of machining system parameters for microchannel fabrication in medical devices. Mater Manuf Process 26:403–414.
15. Ciurana J, Arias G, Ozel T (2009) Neural network modeling and particle swarm optimization (PSO) of process parameters in pulsed laser micromachining of hardened AISI H13 steel. Mater Manuf Process 24:358–368.
16. Ulutan D, Özel T (2013) Multiobjective optimization of experimental and simulated residual stresses in turning of nickel-alloy IN100. Mater Manuf Process 28:835–841.
17. Pawar PJ, Rao R V, Davim JP (2010) Multiobjective optimization of grinding process parameters using particle swarm optimization algorithm. Mater Manuf Process 25:424–431.
18. Venkata Rao R, Kalyankar VD (2013) Parameter optimization of modern machining processes using teaching-learning-based optimization algorithm. Eng Appl Artif Intell 26:524–531. doi: 10.1016/j.engappai.2012.06.007
19. Venkata Rao R, Kalyankar VD (2011) Parameter Optimization of Machining Processes Using a New Optimization Algorithm. Mater Manuf Process 27:978–985. doi: 10.1080/10426914.2011.602792.

Investigation on the Particle Growth of Rutile TiO_2 Suppressed by Manganese

Aripin[1], Edvin Priatna[1], Nundang Busaeri[1], Rudi Priyadi[2], I Nyoman Sudiana[3], I Made Joni[4], Svilen Sabchevski[5]

[1]*Department of Electrical Engineering, Faculty of Engineering, Siliwangi University, Jl. Siliwangi 24, Tasikmalaya 46115, Indonesia*
[2]*Department of Agrotechnology, Faculty of Argiculture, Siliwangi University, Jl. Siliwangi 24, Tasikmalaya 46115, Indonesia*
[3]*Department of Physics, Faculty of Mathematics and Natural Sciences, University of Haluoleo, Kendari, Southeast Sulawesi 93132, Indonesia*
[4]*Nano Technology and Graphene Research Center (NTGRC), Padjadjaran University, Bandung, West Java 45363, Indonesia*
[5]*Lab. Plasma Physics and Engineering, Institute of Electronics of the Bulgarian Academy of Sciences, 72 Tzarigradsko Shose Blvd., Sofia 1784, Bulgaria*

Abstract. In this paper, we present the results of an investigation on the crystallite growth of rutile TiO_2 and the formation of Mn–O–Ti bonds in a TiO_2/manganese composition. The effects of manganese loading from 20 wt% to 80 wt%, into TiO_2 at a temperature of 1200°C on the structural properties, have been studied. The material's properties have been characterized on the basis of the experimental data obtained using X-ray diffraction (XRD), Fourier transform infrared (FTIR) spectroscopy and Scanning Electron Microscopy (SEM). It has been found that an increase of the loading by manganese up to 80 wt% leads to a decrease of the rutile crystallites size and an increase of the amount of Mn–O–Ti bonds. The analysis based on the interpretation of both XRD patterns and FTIR spectrum allows one to explain the reduction of the crystallite size of rutile TiO_2 by an increase of the manganese loading.

1 Introduction

The titanium dioxide (TiO_2) is an important multifunctional material due to its numerous applications, for example in solar cells, photo-catalyst devices, gas sensors, in various semiconductor devices (as a wide-bandgap n-type semiconductor), etc. It has three polymorphs, namely anatase, brookite, and rutile. Rutile is the most common and abundant stable phase of TiO_2 to which other phases transit when heated to a temperature above 600°C [1]. Rutile TiO_2 has many advantageous characteristics that are appropriate for various applications. For example, the high dielectric constant of rutile (close to 100) and high electric breakdown strength (> 100 kV/cm) make it a suitable dielectric material for high energy density capacitors [2]. The property of high refractive index has been used in both optical and anti-reflection coatings as well as in beam splitters [3]. The high thermal expansion of rutile is favorable for giving rise to a change in the semiconductor bandgap of dye-sensitized solar cells [4]. The surface properties of rutile such as the specific surface area, surface structure, active reaction sites are appropriate for uses in various photocatalytic processes [5,6] except the particle size of rutile, which is too large. This limits the utilization of such attractive and cost-effective material as a photocatalyst. All these facts and considerations motivated us to look for an alternative method to reduce the particle size of rutile TiO_2. One such way is to dope TiO_2 with metal ions.

Manganese ores can be used for doping TiO_2 due to their unique structures, uniform pores, and channels. In a previous study, Reddam et al., [7] have investigated the particles size and BET surface area of rutile of TiO_2/manganese composite at 600°C. It has been found that the sizes of the rutile crystallites decrease from 41.35 to 31.01 nm for undoped TiO_2 and TiO_2/manganese composite, respectively. Paul et al., [8] have prepared TiO_2/manganese composite with 5 wt% manganese. The sizes of the rutile crystallites have been found to be 15 nm and 6 nm for undoped and manganese doped TiO_2, respectively. In another study, Ettireddy et al., [9] have studied a TiO_2 supported manganese oxide catalysts and its specific surface area decreases from 54 to 49 m^2/g when the content of SiO_2 increases from 2.5 wt% to 10 wt%. These studies have demonstrated that the manganese in TiO_2/manganese could effectively limit the growth of crystallites. In the present work, we have studied the effects of the increased content of manganese and the temperature on the quantity of Mn-O-Ti bonds and their relationship to the size of the rutile crystallites.

2 Experimental Procedure

2.1 TiO_2 and Manganese composite preparation

 https://doi.org/10.1051/matecconf/201822101005

The manganese ores were obtained from the artisanal mining area in Tasikmalaya, Indonesia. The materials for the samples were selected manually, cleaned by washing with water, and then dried in the sun. The dried manganese ores were crushed into small pieces with an area of approximately 1 cm^2. Then, they were grinded using a jar mill to pass through a 150 mesh sieve. Table 1 shows the chemical composition of the prepared powder samples using crystalline TiO_2 powder and manganese ore. The sample of TiO_2/manganese with a chemical composition of 20 wt% manganese (M20T) was prepared by dissolving 16 g of TiO_2 in a solution containing 50 mL ethanol and 50 mL deionized water, and then was stirred at 80 °C for one hour. Next, 4 g of the manganese ore powder was added slowly into the solution containing Ti^{2+} ions under continuous stirring at 50 °C for 6 hours. The mixed solution was heated at 80 °C until the solvent completely evaporated. The resultant TiO_2-manganese powder was dried in an oven at a temperature of 120 °C for 2 hours and then stored in a desiccator for a further treatment. The samples M40T, M60T and M80T were prepared following the same procedure.

Table 1. Chemical composition of samples of the synthesized TiO_2-manganese composites

Sample	Manganese [g]	TiO_2 [g]
20 wt% Mn (M20T)	4	16
40 wt% Mn (M40T)	8	12
60 wt% Mn (M60T)	12	8
80 wt% Mn (M80T)	16	4

2.2 Sintering the synthesized TiO_2-manganese composites

The powder samples were placed into a porcelain crucible and then inserted into a horizontal furnace, where they were sintered with a controlled heating rate of 10°C/min up to the temperature of 1200°C in air. Then the temperature was kept constant at 1200 °C for 2 hours. The cooling was performed by natural convection after turning off the electric furnace and leaving the samples inside.

2.3 Method for characterization of the synthesized TiO_2-manganese composites

The experiments on the X-ray diffraction (XRD) of the samples have been carried out using a Smartlab X-ray diffractometer with filtered Cu Kα radiation at a wavelength of 0.15418 nm. The accelerating voltage and the applied current were 40 kV and 30 mA, respectively. The diffraction patterns were registered over the range from 10° to 90° at a scan rate of 0.01 °/s. Fourier transform infrared (FTIR) spectra were obtained using a Varian 800 FTIR spectrometer (Scimitar Series model) in the wavenumber range of 400–4000 cm^{-1} with a spectral resolution of 4 cm^{-1}. All FTIR measurements were carried out at room temperature in a specular reflectance mode using the KBr pellet technique. The SEM images of fracture surface of samples were taken by using a scanning electron microscope (SEM) at a magnification of 1000-10,000 times and at accelerating voltage of 10-15 kV.

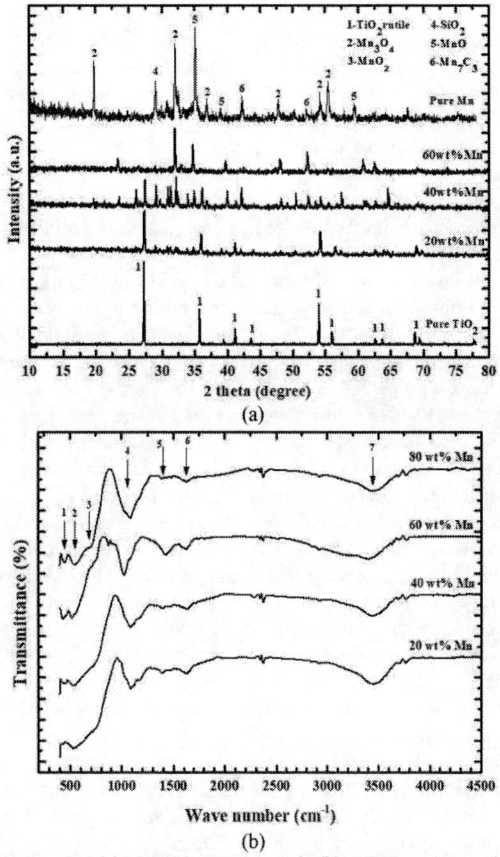

Figure 1. (a) XRD patterns and (b) FTIR spectra for TiO_2/manganese composite with varying manganese content and sintered at 1200°C.

3 Results and Discussion

Figure 1a shows the XRD pattern of TiO2/manganese composite with a varying loading of manganese and sintered at 1200°C. In the sample of pure TiO_2, some peaks appear at 2θ = 27.3°, 35.8°, 40.9°, 44.5°, 54.5°, 56.3°, 62.5°, 63.6°, and 68.7°. These peaks correspond to the (110), (101), (111), (112), (211), (220), (002), (311), and (301) planes of the rutile TiO_2 (R) (JCPDS 21-1276) [10]. This result is in agreement with the findings of the study of phase formation of TiO_2 xerogel composites carried out by Aripin et al. [11, 12]. Then, for the sample of manganese ore (pure manganese), it consists of manganese oxide mainly in the form of pyrolusite (MnO2), quartz (SiO_2), hausmannite (Mn_3O_4), manganosite (MnO), and manganese carbide (Mn7C3). Hausmannite (peak 2) appears at 2θ = 18.31°, 31.59°, 36.29°, 48.37°, 54.78°, and 55.92° (JCPDS 24-0734), manganosite (peak 5) at 2θ = 34.39°, 36.69°, and 59.12° (JCPDS 00-001-1206), and manganese carbide (peak 6) at 2θ = 41.47° and 51.18° (PDF 01-075-1498). In the sample with 20 wt% Mn, the rutile TiO_2 (peak 1) is formed as a major crystalline phase while quartz (SiO_2) (peak 4), hausmannite (Mn_3O_4) (peak 2) and manganosite (MnO) (peak 5) as minor phases. As the loading increases, the peak intensity of rutile TiO2 decreases while the other crystalline phases tend to increase. With the increase of the loading up to 60 wt% Mn, crystalline rutile TiO_2 gradually disappear. The crystallite size of the rutile phase has been evaluated using the Scherrer's equation [13] for the peak at 2θ = 27.3°. The calculated size of rutile crystallites decreases from 70.81 nm to 27.18 nm, when the content of TiO_2 increases from pure TiO_2 to 60 wt% manganese. This means that the crystallite size of rutile can be correlated to the manganese content in the composite.

Figure 1b shows the FTIR spectra for TiO_2/manganese composite with varying manganese and sintered at 1200°C. For all samples, the band (peak 1) located at 484.1 cm^{-1} indicates the presence of α-Mn_2O_3 phase. The peak 2 that appears at 533 cm-1 corresponds to the Mn–O–Ti vibrational mode [7]. The peak 3 at 696 cm^{-1} can be assigned to Ti–O vibrational modes of rutile. The band (peak 4) identified at 1020 cm^{-1} can be attributed to pyrolusite (MnO_2). These results are consistent with the XRD analysis reported above and are in quantitative agreement with the results reported in the literature for other MnO_2 nanoparticles [14, 15]. The absorption bands (peak 5 and 6) observed at about 1645 and 1386 cm^{-1} were attributed to the bending H–O–H bond groups of adsorbed water molecules. The broad band (peak 6), at 3452 cm^{-1}, is attributed to the O–H stretching of hydroxyl groups that are present on the surface of the material [16]. It has been found that the integrated intensity of Mn–O–Ti bonds increases with increasing the amount of manganese ore. The high quantity of Mn–O–Ti bonds with smaller amounts of TiO2 indicates that more manganese ore particles facilitates the aggregation of TiO_2, and the manganese ores and TiO_2 phases are not separated from each other in the composite. In this case, the presence of manganese ore particles impedes direct contact of TiO_2 particles through Mn–O–Ti bonds, resulting in greater suppression of the phase transformation from anatase to rutile. Eventually, this leads to the formation of smaller particles of rutile TiO_2 with smaller amounts of TiO_2.

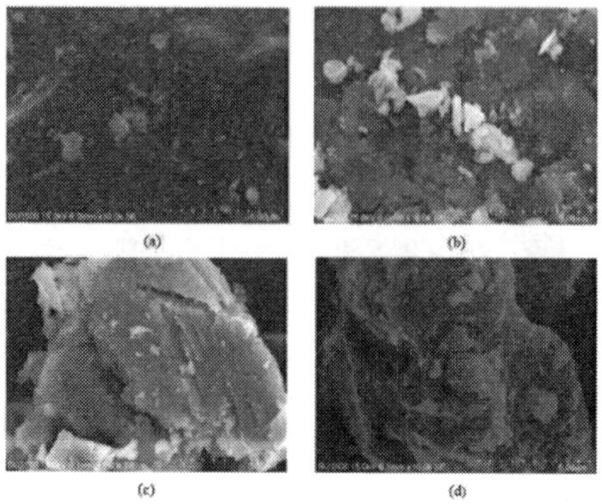

Figure 2. SEM images of TiO_2/manganese composite with (a) 20wt%Mn, (b) 40wt%Mn, (c) 60wt%Mn, and (d) 80 wt%Mn and sintered at 1200°C.

Figure 2 shows SEM images of TiO_2/manganese composite with a varying manganese content and sintered at 1200°C. The SEM image for samples of 20 wt% and 40 wt% illustrates that the micrographs are characterized by a dense grains with different texture and morphology and unevenly particle size distribution. The irregular-shaped grains with dark contrast are associated with the presence of TiO_2/manganese. The small grains of SiO_2 with bright texture is dispersed on the surface of manganese. For samples with 60 wt% and 80 wt% of Mn, the microstructure of the samples includes large particles with an irregular shape. The larger particles, which can be seen in this figure are probably aggregates of the smaller particles, because based on the FTIR results for the case with a higher concentration of manganese, the presence of a big number of manganese particles facilitates the aggregation of TiO_2. It should be mentioned also that the manganese ores and the TiO_2 phases are not separated from each other in the composite.

4 Conclusion

A novel TiO_2/manganese composite has been produced successfully from TiO_2 and manganese ore by controlling the loading with manganese. The results presented in this paper show that an incorporation of manganese into TiO_2 gives an appreciable effect on the crystallite size of TiO_2 and on the formation of Mn-O-Ti bonds at a temperature of 1200°C. An interesting phenomenon has been observed analyzing the XRD pattern of the samples with a large loading with manganese. It has been found that the large amounts of manganese results in a significant decrease in the size of rutile crystallites. Such interpretation is supported by the observed large quantities of Mn–O–Ti bonds in TiO_2/manganese composite.

Acknowledgments

This research was supported by a fund of the Siliwangi University through the Project of Research for Inovasi Perguruan Tinggi in 2017 with contract number: 1140/D3/PL/2017.

References

1. A. Nilchi A, Janitabar-Darzi S, and Rasouli-Garmarodi S 2001 *Mater. Sci. Appl.* **2** 476.
2. Marinel S, Choi D.H, Heuguet R, Agrawal D, and Lanagan M 2013 *Ceram. Inter.* **39** 299.
3. Wang X, Wu G, Zhou B, and Shen J. Mater. **6** 2819.
4. Alivov Y and Fan Z.Y 2009 *Appl. Phys. Lett.* **95** 16.
5. Mahltig B, Haufe H, Kim C,W, Kang Y. S, Gutmann E, Leisegang T, and Meyer D.C 2013 *Croat. Chem.Acta* **86** 143.
6. Jin Q, Arimoto H, Fujishima M and Tada T *Catal.* **3** 444.
7. Abdelouahab-Reddam H, Abdelouahab-Reddam Z, El Mail R, and Aaliti A 2014 *Intern. J. Innov. Appl. Stud.* **8** 660.
8. Paul S, and Choudhury A 2012 *Intern. J. Inov. Res. Develop.* **1** 24.
9. Ettireddy P, Ettireddy N, Mamedov S, Boolchand P, Smirniotis P 2007 *Appl. Catal. B Environt.*, **76** 123.
10. JCPDS-International Centre for Diffraction Data, 1997 PCPDFWIN **1.30**.
11. Aripin H, Joni I.M, Mitsudo S, Sudiana I.N, Priatna E, Busaeri N, and Sabchevski S 2017 *Intern. J. Techn.* **8** 1507.
12. Aripin A, Mitsudo S, Sudiana I.N, Busaeri N, Sunendar B, Sabchevski S 2016 *Mater. Sci. Forum* **872** 81.
13. Duy P.P, Kim K.K, Cao V.T, Van Q.V, and Thi T.V 2014 *Intern. J Mater. Sci. Appl.*, **3** 147.
14. Yang X.J, Liu Z, Makita M, and Ooi K 2004 *Chem. Mater.* **16** 5581.
15. Yang R, Wang Z, Dai L, and Chen L 2005 *Mater. Chem. Phys.* **93** 149.
16. Yang S and Gao L 2006 *J. Amer. Ceram. Soc.* **89** 1742.

Optimizing prefabricated construction techniques in UAE as a solution to shortage of middle-income housing

Mohamed Elkaftangui[1], Basem Mohamed[1]

[1]*Abu Dhabi University, Architectural Department, P.o.Box 59911, Abu Dhabi University, United Arab Emirates*

Abstract. Housing is a major subject in the Middle East and North Africa (MENA) region, JLL, the world's leading real estate investment and advisory firm, highlights the social and economic impact of a marked shortfall in the required supply of middle-income housing, suggesting that significant efforts are needed to address the current imbalance. For an equal standard and life style in UAE and KSA, the gap of sale and rent prices raises many questions. The middle-income sector of the market represent almost 40 percent of all households in the UAE. Several factors that have contributed to the current shortage of affordable housing, high land values, High capital costs for associated infrastructure development, low adoption of prefabricated construction techniques have contributed to higher construction costs, lower financial returns, making such developments less attractive for developers, limited access to suitable finance for low income families, due to generally immature mortgage markets. Resolving the shortage of middle-income housing requires a concentrated effort involving government agencies, private developers and other stakeholders. This Study will investigate the UAE Housing market to propose solutions for a better progress of the precast sector and focusing on the promotion of sustainable and industrial approaches to construction.

1 Introduction

The United Arab Emirates (UAE) faces a current estimated demand from 390,000 low income Emirati households [1]. The importance of the middle-income sector of the market should not be under estimated, with over 820,000 middle-income households, representing almost 40 percent of all households in the UAE. Only 22 percent of residential units launched in 2016 to date in Dubai are "affordable" to "middle-income" households while we have not seen any residential units being launched this year that meet the meaning of 'affordable'.

The UAE Vision 2021 articulates the importance of family and community. The vision articulates the importance of protecting the environment, preserving culture, providing access to infrastructure and facilities to create well rounded lives for Nationals. Efforts were intensified in 2010 through the Emirati Family Housing Program to create 'complete communities' of over 13,000 villas with regard to the above requirements within four years [2].

Many programs and initiatives have been proposed to meet the demand, in 2014, the leader developer of Abu Dhabi, Aldar has delivered to the Government of Abu Dhabi, 4,857 villas at Al Falah, bringing the total number of units developed or delivered to 6,744 homes for UAE nationals.

In the same year, other 1,887 units have been delivered to the Abu Dhabi Government authorities, these developments consist of 839 villas in Watani and Al Raha in Abu Dhabi, 448 villas at Sila'a in the Western Region, and 600 villas at Ghuraibah in Al Ain.

Another development by Abu Dhabi-government-owned builder Aldar, launched on 2017, consists of 1,020 villas on Yas Island and 996 villas in Al Falah and 275 Villas at Al Ghuraibah in Al Ain.

In August 2015, and according to Dubai Municipality's director, two communities have been announced, consisting of 1,400 and 1,000 villas, for nationals in Wadi Safa and Al Aweer 1, and the distribution of land to eligible citizens in Wadi Safa. In November 2017, the Abu Dhabi government has announced a new housing project, called Al Riyadh City, as part of the capital's drive to expand public housing policies, The project, said to be Abu Dhabi's largest ever housing development, is expected to lodge 200,000 UAE citizens and occupy 45% of the city's total residential land area.

These projects have a total developmental value of Dhs 5.7bn and are being fully funded by the Abu Dhabi government, the developer said in a statement. The UAE, like other Gulf Arab countries, spends a large amount on social welfare for its citizens. Providing free or low cost housing to Emiratis is one of the welfare measures widely adopted by the state.

The Emirates Development Bank (EDB), which was launched with Dh10 billion authorised capital, is set to spearhead the development of affordable homes with a commitment to deliver 30,000 homes for the UAE citizens from 2015 to 2021.

The huge existence of a precast concrete industry and the numerous successful building projects achieved using precast concrete, for the whole or just a part of the structure, is proof that the technique is practical and economical, and certainly present the best solution to meet the expectations of UAE vision 2021 for fast housing deliverable in term of quantity and quality.

In this research, we highlight the reality of the supply and demand of the UAE residential sector and analyse what actions has been undertaken by the UAE government to meet the growing demand of housing units, we also analyst the practice of the precast concrete activity in UAE to enhance the mechanism of development of the precast concrete

2 Precast construction in UAE

Precast concrete comprises roughly 30 per cent of the UAE construction market and its share is increasing, the competition for precast is from ready-mix, steel structures and building blocks.

According to the statement of the Abu Dhabi Chamber of Commerce and Industry, the UAE now has a huge, advanced infrastructure in precast, sophisticated factories using the latest technologies in production lines. The total value of the precast concrete market has been valued at around $1.6bn, there are over 25,000 labourers and technicians working in the precast industry in the UAE in at least 29 precast manufacturing facilities, with the construction cost of each of these factories ranging from $50m to $350m.

Prefabrication has been proven to provide up to 52% savings in the global construction waste. The global movement for the reduction of carbon emission in the construction industries of the world is geared towards the creation of environmentally responsible building technologies. Several studies propagate prefabrication techniques and usage of precast concrete elements in construction, as a significant contributor to positive environmental and economic impacts especially in highly commercialised regions like the UAE.

Using precast concrete offers many general advantages as high loading capacity, longer span for slab and beams, less formwork but we emphases the following specific advantages in relation with UAE context:

- Faster construction time: 80-90% of the recent repetitive massive housing projects in the UAE are made of precast concrete as an efficient alternative to the traditional building techniques, according to Gulf Precast leader of the precast concrete manufacturers in UAE, a regular medium size development of 650 villas 500 m2 each (example mentioned is: Akoya and Akoya oxygen villas in Dubai), required a total duration of 6 months for engineering, assembly and the final delivery. On the Al Falah villa development in Abu Dhabi, 4,126 villas were completed in 24 months using precast, while only 730 villas were completed in 20 month using traditional cast-on-site.
- Cost reduction: comparison is based on the scale of the development; for small scale projects it would reach up to 2x the traditional construction technique. While, in repetitive, big scale developments the price can reach the same price of the traditional construction techniques with the time benefit in both cases. The use of modular components with precast superstructures and the incorporation of a pre-fabricated bathroom pod installed at the precast erection stage illustrates a clear time saving of approximately 8 programme days per villa, a significant advantage of early completion of bathroom units is a reduced dependence of bathroom MEP final-fix installations and associated testing & commissioning inspections on ceramic tiling completion [3].
- Better quality and more durable products: the precast concrete industry offers an extraordinarily diverse range of unique and creative building technologies, which enables the manufacturers globally develop initiative building and infrastructure products to architects [4]. The UAE manufactures produces mainly holowcore slabs, beams and columns, wall panels with a maximum standard of 4.7m. / 8.5m. Due to crane capacities, table size and transport, GRC and GRG. The precast concrete offers the possibility to have less weather dependency and to maintain the high precast concrete quality, the UAE hot and humid climate does not always work in favour of site-cast concrete.
- Sustainable: made of natural raw materials (stones, gravels, sand, cement), locally available almost everywhere in UAE, precast concrete minimizes the whole life cycle impact on the environment when compared with other construction materials. Precast concrete meet the requirement of Estidama rating system [5] which is the UAE version of the LEED, with a minimum mandatory U Value of 0.4 in Dubai and 0.32 in Abu Dhabi.

3 Defining failures in the practice of the precast concrete activity from design to construction

The practice of the precast concrete activity in UAE meets some difficulties during the execution of super projects. These problems are mainly generated during the phases of the design and the construction which affect the quality of final product or delivery time.

There are two methods of housing provision for Nationals in the Emirate of Abu Dhabi, who may choose between the following options:

- Land and loan: Eligible Nationals will be provided with a plot of land and financing to build a house. 'Bayti' (My Home) initiative for Abu Dhabi citizens, which was developed to ease the burden on Abu Dhabi Emirati families utilising Abu Dhabi Housing Authority housing loans, provides 58 structurally and architecturally pre-approved housing designs. The number of rooms ranges between 4 and 8 bedrooms, with the possibility of future expansion. It streamlines the process of delivering housing to the required quality, on time and on budget. The initiative has also shortened build time, with full construction to be completed within a maximum period of two years. The expected cost, and interest-free loan amount, for each house is AED 2 million. This initiative and mean of delivery are not part of the precast concrete activity but generally cast in site and considered as an individual construction.

- Allocation: The government provides eligible Nationals a house that is fully constructed and serviced by all relevant infrastructure and utilities. Houses are likely built by a master planner/master developer, this mass construction where thousands of houses to be delivered through different governmental programs as mentioned in the introduction.

In this research, we select the most important project "Al Falah" delivered recently in Abu Dhabi and highlight and analyse some facts relevant to the execution of this type of projects and the precast activity 'Figure 1', 'Figure 2'.

Al Falah Community development in Abu Dhabi consists of five 'villages' presenting 4,857 villas and a range of community amenities, providing sustainable neighbourhoods serving modern needs and lifestyles whist preserving and promoting the traditional Emirati identity and preferred style of living.

Figure 1. Site Aerial view of Al-Falah housing project.

Figure 2. One of the precast concrete family houses in Al-Falah development

3.1. Design and Users

Although the government's commitment large effort to improve Emirati housing, "Abu Dhabi Municipality-permit Department" registered in 2016 a total number of 8,769 permits of modifications and addition permits in Emirati housing in Abu Dhabi, presenting 2181 addition and 5737 modification permits. In Al Falah new community of 4869 houses the number of modifications and addition permits in (2014-2015-2016) is (3272) permits presenting 67.2% of the houses.

According to the owners and the permits contents [6], the majorities of modifications could be summarized as follow:

- Modification of double height entrance Lobby, as it seem to be generally not preferable.
- The location of ground water tank and pump room is contradicting with landscape layout.
- The location of fire hydrants are not coordinated with villas Main gate/entry.
- The main gate is not providing enough the privacy required in Emirati house.
- The French window is not a suitable design for Emirati.
- The colour and arrangement of toilets and ceramic cladding are not suitable.
- Tight space between the driver's room and the fence is really tight and need to be relocated.
- Living rooms is too small.
- Water, electricity and HVAC need monitoring neglected according to water and electricity bills.

3.2. Construction and Management

In Al-Falah project, the critical challenge has been to constantly build five villas per day by using precast fabrications to reduce labor costs by up to 25%, compared to conventional concrete. The following critical areas for the delivery solutions applied by El-Seif Engineering Contracting for the Construction of 2,069 Villas across two villages of the Al Falah development are highlighted below:

- Co-ordination conflicts: particularly where milestone completion dates are combined and shared access required at multiple phases; from concurrent villa-erection vs. deep drainage, to wearing course asphalting vs. villa finishes, snagging and handover, amongst others.
- Many conflicts appeared as a result of variable terrain levels and required interconnections between villa and infrastructure-contractor constructed drainage components.

4 Reflexions and Conclusion

A range of factors influence the success of the precast concrete activity and the construction of big scale fast track projects. Attention for design and construction must escape failures listed above and maximize the advantages that the precast concrete construction may offer. The design must be based on an understanding and response to cultural factors relevant to national housing, this include privacy, family size average between 7-10 people, family expansion.

Environmental factors influence the National housing by their physical environment including the landscape variety, different climates and regions. These considerations and the early stage involvement of the future owners of tenants must definitely reduce the permits modifications.

Use of universally accessible Building Information Modelling (BIM) technology can resolve co-ordination conflicts presented above, using software like Revit or Tekla allow all architectural, structural and electro-mechanical design processes occur at the same time since 'modules' are parts of an entire building. Buildings are virtually tested, enabling asset managers to evaluate building performance prior to manufacturing and construction. The built BIM model can be combined with Facility Management (FM) software enabling all project teams manage and control the assets.

A 'cloud-based', 3D modelling service would likely facilitate more timely review of design modifications for such conflicts and perhaps eliminate many on-site if combined into an Early-Contractor-Involvement (ECI) design environment. A real time model visualization would allow for more immediate design and construction-phase planning and decision making by all stakeholders.

Modular bathroom units were not exploited on the Al Falah project, although they offer considerable advantages and savings of approximately 8 programme days per villa in time delivery, quality constancy and minimal materials wastage [3].

Given the climatic conditions in the UAE, the need for a shading devises can be considered inevitable. An approach for the design of a shading systems stems from the plasticity of concrete as a material, and supported by a computational strategy with specific focus on opening size in relation to allocated function must be undertaken to develop the actual precast wall panels used in UAE whish neglect the opening treatments and the opening protection [7], this process must be initiated with a series of assumptions derived from examining residential design trends in the UAE, with specific focus on opening size in relation to allocated function. Providing the possibility to

Modularity and Flexibility of the precast concrete construction provides adaptability to keep costs down, the development of this construction method permits for basic shapes to be combined in multiple ways that allow precast systems to be pre-built.

An exploration work with the ambitions of rethinking the design and production of precast concrete façade panels focusing on the UAE Industry is required to design and optimize an integrated facade system generating a self-shading element in response to specific contextual conditions comprising the optimization of window extrusion and rotation to find the adequate solution [8].

Figure 3. Wall Panels Taxonomy

This optimization can follow cardinal orientations; East, South East, South, South West, and West, Resulted in a catalogue of 30 variations of optimized panels, corresponding to all possible locations, functions and orientations.

The role of digital fabrication in the form of 3D printing could be anticipated, recently, advancements in Additive Manufacturing techniques, such 3-Dimentional printing have been a vital area of exploration since its development in the mid-1990s. We believe that 3D printing of concrete allows for producing highly customized building components and could create a new era in the construction industry when adapted to specific contextual and cultural environments.

All stakeholders need to be educated on the importance and benefits of precast concrete construction. Builders most likely will be interested in initial costs, time savings, construction methods, reduced job site impact and planning. Owners most likely will be more interested in long-term value, design flexibility, life cycle costs, safety and initial costs.

Modular solutions, new precast technologies are well appropriate to low and midrise residential developments from townhouses and villas to apartment buildings and communities. The efficient use of precast along with the suitable building methodologies and technologies allows for the efficacious delivery, in affordable stages, in a fraction of the conventional construction cost and time.

References

1. J. Gavin, J. Lasalle, jll's Year in Review report for UAE – JLL MENA, (2017). http://www.jll-mena.com/mena/en-gb/research/397/jll-2017-year-in-review-report-for-uae
2. https://government.ae/en/about-the-uae/the-uae.../the-federal-national-council-
3. J. Eyre, R. Hart, G. Connolly. Optimising delivery solutions for social & affordable housing, Abu Dhabi, UAE. Al Falah Community Project, a social housing case study (2004). Available on (www.conference.net.au/cibwbc13/papers/cibwbc2013_submission_35.pdf)
4. S. Kumar, "Precast Take the Lead". (2011). Available from <http://www.constructionweekonline.com/article-17403-precast-takes-the-lead/> (accessed 13 Feb. 2018).
5. https://www.upc.gov.ae/en/publications/...guidelines/pearl-rating-manual-estidama
6. Abu Dhabi Municipality, Permit Department. https://www.abudhabi.ae/portal/public/en/departments/adm
7. B. Bell, "Topocast, Parametric Precast". Available from <http://topocastlab.com/portfolio/parametric-precast/> (accessed 7 Feb. 2018).
8. B. Mohamed, M. Elkaftangui, R. Zureikat. "Towards Rethinking the Precast Concrete Industry in the UAE", *Learning prototyping and Adapting*, CADRIA 2018, pp287-296.

Laser cleaning for rust removal on mild steel: An experimental study on surface characteristics

Vishnu Narayanan, Ramesh K Singh and Deepak Marla

Department of Mechanical Engineering, IIT Bombay, Maharashtra, India

Abstract. Laser cleaning is a relatively new and promising industrial cleaning method. Despite its potential as an alternative to conventional de-rusting methods, very few studies have been done on rust removal. The current study is aimed at investigating experimentally the effect of various important laser parameters on properties of surface after laser processing. Rusted mild steel samples were irradiated with moving pulsed laser beam. Parameters like scan speed, number of passes, etc. were varied and the change in properties such as surface profile, roughness, hardness, etc. of the processed surface were observed along with depth of removal. The depth of removal was found to increase with an increase in laser power, and number of scans but decreased with an increase in scan speed. The surface got discoloured and deformed at lower speeds and higher powers. The surface roughness after processing hardly showed any variation with the initial roughness values before cleaning but depended on the scanning path of the laser. The microhardness values after laser processing varied with path traced by the laser beam which points to microstructure changes in the samples due to laser processing.

1 Introduction

Laser beams can be used for removal of paint [1], rust [2], grease [3], etc. from surfaces and also for cleaning of dies used in molding [4]. This method is non-contact as well as non-destructive. Laser cleaning doesn't need any abrasive or chemical media. The suitable laser type and optimum parameters are different for each application [5]. The actual physical process depend on the operating conditions and various parameters which determine the outcome of the cleaning process. So these parameters have to be chosen carefully to get the desired outcome.

Several previous studies had been done to identify the optimum working conditions for laser cleaning of rust. A lot of them focussed on cleaning and conservation of archeological artefacts [6]. Other works focussed on finding the threshold fluence for a particular laser equipment, optimum working conditions for certain application [7] or to find the mechanisms behind the cleaning process [7]. Koh and Sarady [6] conducted experiments to study the ideal wavelength and fluence for laser cleaning of corroded artefacts. It was concluded that TEA CO_2 laser was better compared to Nd:YAG laser since risk of damage was higher with Nd:YAG laser. Meja et al. [7] conducted experiments to study effect of wavelength, fluence, pulse repetition rate and environment of laser cleaning of rust from iron. Daurelio et al. [2] studied the rust removal rate for different fluence and pulse repetition rate. Wang et al. [8] conducted experimental study on de-rusting of steel blocks using Nd:YAG laser. The parameters for complete removal of rust was studied and the change in roughness, hardness and corrosion resistance with change in fluence were studied. Chen et al. [9] studied the effectiveness of fibre laser in removal of rust, grease, etc. and concluded that laser cleaning is a good alternative for conventional methods of surface preparation in ship building. Ali et al. [3] studied the change in depth of removal, roughness hardness, etc. with change in peak pulse power and pulse repetition rate.

The literature survey pointed out that the surface damage and type of morphological changes that happen during laser cleaning of rust has not been adequately studied. In addition, effect of each laser parameter on the surface characteristics after laser processing has not been reported properly. In order to obtain desired surface properties and avoid damage, a knowledge on the relative importance of each parameter on surface characteristics post laser processing will be helpful to choose the optimum parameters.

The objective of this study was to conduct experiments in order to observe the change in surface roughness and the surface damage with the change in laser parameters during laser cleaning of rust. This will help in predicting the optimum conditions for laser cleaning of rust. In this work, rusted mild steel samples are de-rusted with nanosecond pulsed laser beam of wavelength 1064nm. The samples were scanned with the laser and the effects like depth of material removal, change in surface properties, and morphology after laser irradiation were studied.

2 Experiments

https://doi.org/10.1051/matecconf/201822101007

In this study, a nanosecond pulsed laser with an average power of 30 W and a wavelength of 1064 nm, with a fibre delivery system was used. It has a pulse width of 80 ns and pulse repetition rate was kept at 50 kHz for all the experiments. It was equipped with a galvo-scanner coupled with an F-theta lens, using which the laser was scanned on the surface of samples.

Experiments were conducted to study the effect of scan speed and number of passes on the surface of rusted and pure mild steel surfaces. When rust is removed by laser cleaning, the pure steel layer beneath the rust layer would also get subjected to laser irradiation as process nears completion. Therefore, it is important to know the adverse effects of process parameters that are optimal for effective rust layer removal. In order to understand the heating effects occurring underneath the rust layer and quantify the damage, experiments were also conducted on pure steel samples. Lines were scanned on the mild steel samples at different speeds keeping all other parameters same and the samples were analysed using Zeta optical profile analyser and the effects like depth of removal, and change in surface profile were studied. Galvo-scanner can be used to scan a particular area by following a predefined hatch pattern. This option was used to scan samples with multiple scan patterns as shown in figure 1 and investigate whether the path followed by laser during the cleaning process affected the material properties.

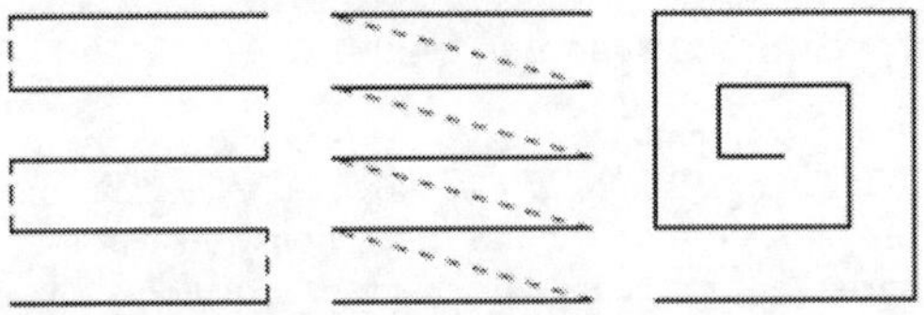

Figure 1. Paths followed by the laser beam

The change in surface roughness of the scanned area was measured with a Zeta optical profile analyser. Vickers hardness for the cleaned sample was measured and change in hardness after laser processing with different scan paths were found.

3 Results and discussion

The experiments were conducted for different set of speeds and different number of scans. The cleaned surface was analysed with a Zeta optical profile analyser to measure the material removal and change in surface morphology was observed.

3.1 Effect of scan speed

Figure 2 shows the change in depth of ablation with a change in scan speed. The experiments were conducted to study the change in surface profile when the sample is scanned with the laser at different speeds ranging from 2 mm/s to 1000 mm/s. The lines traced were studied with a Zeta profilometer. It was observed that at lower speeds (less than 100 mm/s), the surface was deformed and discoloured to dark brown by the laser beam. Whereas, at speeds above 200 m/s the surface did not show any discolouration and relatively flat surface was transformed into a wavy profile with a depression at the centre and elevations on either side after laser irradiation. The wavyness and discolouration decreases with an increase in scan speed but so does the depth of removal. The depth of removal follows a steady decrease with an increase in scan speed initially till it reaches 500 mm/s, and is observed to remain almost constant thereafter.

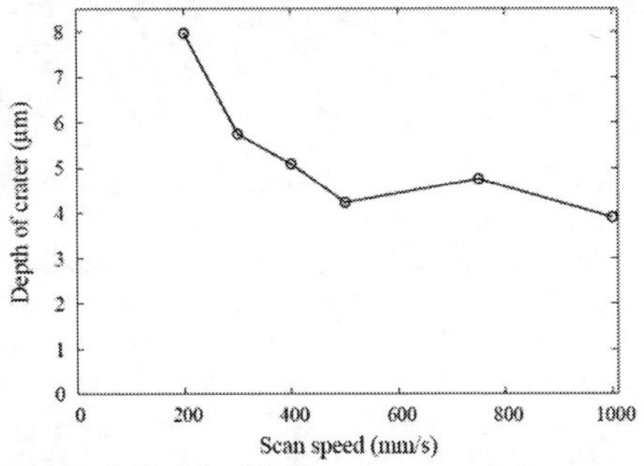

Figure 2. Depth of crater vs. scan speed

3.2 Effect of number of scans

Figures 3 and 4 show the surface profiles after 15 scans at 100 mm/s and 500 mm/s, respectively. The study showed that for lower speeds, multiple scans deform the surface in a disordered way. At lower scan velocity with multiple scans, it was observed that the processed surface is irregular with peaks and valleys forming at random positions on the irradiated region (see figure 3).

But for higher scan speeds, the surface gets deformed into a wavy pattern with a crater in the irradiated zone and humps formed on either ends of the crater (see figure 4).

Figure 3. Surface after 15 scans for 100 mm/s

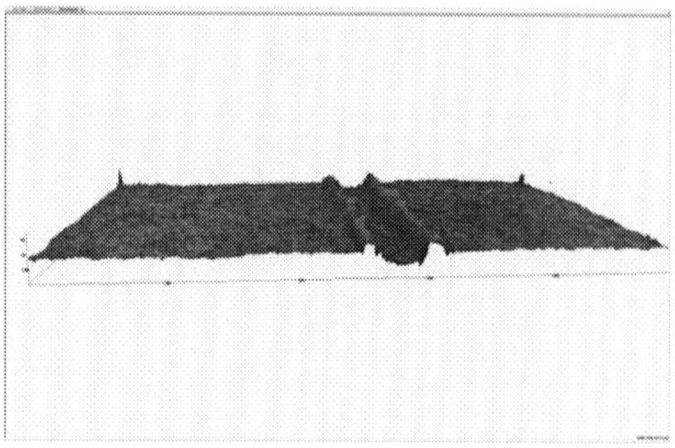

Figure 4. Surface after 15 scans for 500 mm/s

As seen in figure 4, the processed surface for high speeds can be observed to be of wavy cross section. The general wavyness of the surface remains the same for any number of passes. Figures 5 and 6 show the relationship between number of passes with the average depth of depression and the average height of peak on the surface, respectively.

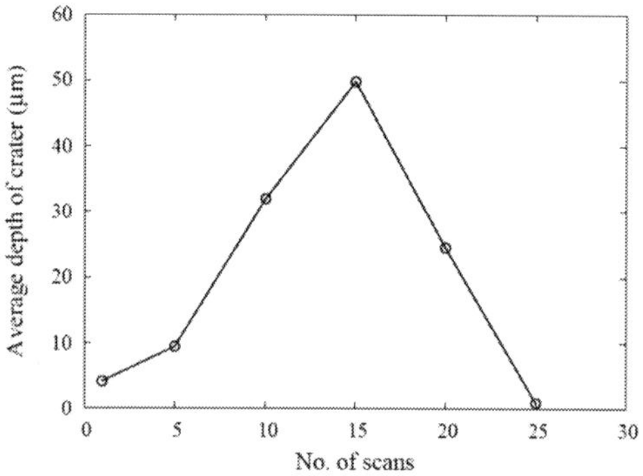

Figure 5. Average depth crater vs. number of scans

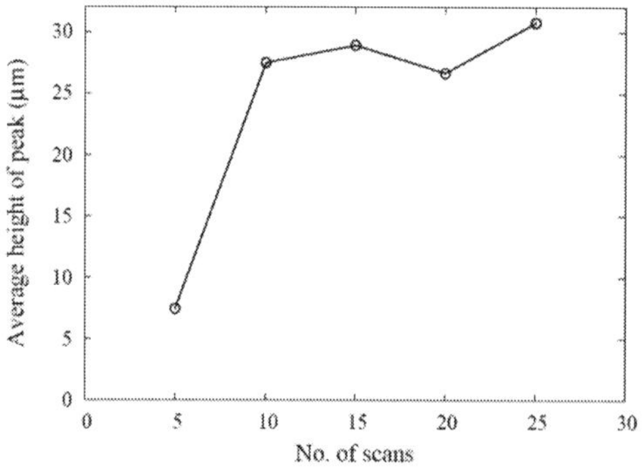

Figure 6. Average height of peak vs. number of scans

The surface shows a clear wavy profile with the peaks becoming taller with each scan for first 10 scans and remains almost same with a further increase in number of scans as seen in figure 6. But the depth of valleys (observed at central region of laser irradiated zone), increases for the first 15 scans and then decreases as seen in figure 5.

3.3 Effect of laser power

Unlike the scan speed, depth of removal increased with an increase in laser power or fluence (figure 7).

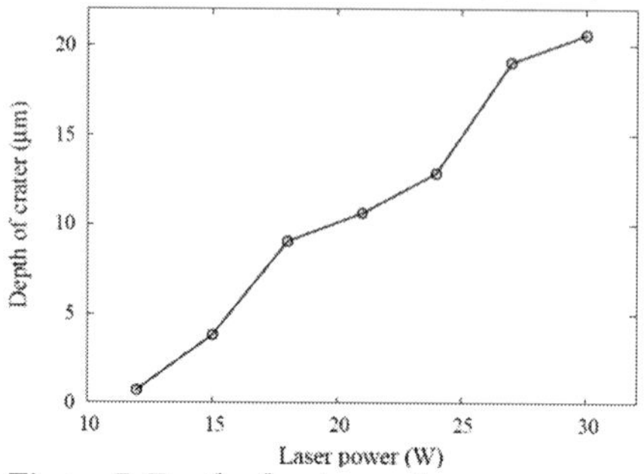

Figure 7. Depth of crater vs. Laser power.

The height of peak and depth of valley increased steadily with an increase in laser power. For lower powers the processed surface looked clean, but as power was increased the surface appeared darker after irradiation. The discolouration and change in surface topology is possibly due to a combined effect of corrosion and deformation initiated by the laser beam. Further studies can shed light on a deeper understanding of the phenomena observed during the experiments.

3.4 Effect of scanning path

Three scan patterns (figure 1) were selected and used to study the effect of path followed by laser in material properties of surface after inteaction with the laser. The surface was scanned once with each pattern. The study was repeated for three scans for every pattern. Each pass was timed 10 seconds apart. The roughness and hardness of this area was measured and compared with initial roughness before laser processing for rusted and pure samples.

The study showed that the surface roughness values were more or less the same irrespective of the initial condition of the sample. A pure MS surface (S_a = 0.919 , S_z = 6.56) and a rusted MS surface (S_a = 26.77 , S_z = 346.82) had similar surface roughness values after cleaning. Roughness values after cleaning were S_a = 1.347 to 1.463 and S_z =.12.43 to 12.51 for pure samples. In comparison the surface roughness of the cleaned rusted samples were S_a = 1.413 to 1.646 and S_z = 8.955 to 12.15. The final surface profile and surface roughness was observed to be unrelated to the initial surface profile but was dependent on scan patterns. So the surface roughness of the processed parts is found to be independent of the initial surface roughness before laser cleaning.

Surface hardness was also studied for laser cleaned surfaces. The hardness was found to have increased significantly after laser treatment. It was observed that the hardness varied according to scan path. The hardness was also diffferent in different parts of the sample scanned by a pattern. The significant change in hardness after laser processing points that the microstructure changes depend on the scan path amongst other factors.

4. Conclusion

In this work, experiments were done on rusted and pure mild steel samples to study the effect of laser parameters in de-rusting process. For laser cleaning of rust there exists a range of parameters for which cleaning is satisfactory. For lower speeds and higher number of scans, the surface gets damaged by the laser. The damage seems to be a combined effect of corrosion and deformation. Further study could reveal the explicit phenomena involved. Single pass at higher speeds give a well cleaned surface. Multiple passes caused changes in surface topology which was random for low speeds and followed a particular pattern for high speeds. Path followed by laser was found to have affected the surface hardness and surface roughness of the sample. The hardness of the cleaned surface was enhanced but roughness values seemed to be independent of the initial roughness and varied only with change in laser parameters. Further studies have to be done to understand and know more about these phenomena. Precisely, the effect of scan path on the microstructure changes, and the ability to control it in order to get desired hardness values for the processed part has to be studied futher.

Acknowledgments

The authors wish to gratefully acknowledge that this research was funded by the DST-Swarnajayanti Fellowship award [DST / SJF / ETA – 02/2014–2015] and Technology Systems Development Program [DST / TSG /AMT / 2015 / 226 / G], Government of India. The authors would also wish to thank Mrs. Bhargavi Ankamreddy for her help in conducting experiments.

References

1. X. Li, T. Huang, A. W. Chong, Opto-electronic engineering **44** 340–44 (2017)
2. G. Daurelio, G. Chita, M. Cinquepalmi, Appl. Phys. A Mater. Sci. Process **69**, S543–46 (1999)
3. S. N. Ali, Z. A. Taha, T. S. Mansour, *Advances in Condensed Matter Physics* **2014** (2014)
4. Y. F. Lu, W. D. Song, M. Hong, M. Zhong, Q. Chen, T. Chong, Jpn. J. Appl.Phys. **39** 4811–13 (2000)
5. J. F. Asmus, Technology and Conservation **3** 14-18 (1978)
6. Y. Koh, I. Sárady, Journal of Cultural Heritage **4** 129–33 (2003)
7. P. Meja, M. Autric, P. Alloncle, P. Pasquet, R. Oltra, J. P. Boquillon Appl. Phys. A Mater. Sci. Process. **69** 687–90 (1999)
8. Z. Wang, X. Zeng, W. Huang, Surf. Coatings Technol. **166** no. 1 10–16 (2003)
9. G. X. Chen, T. J. Kwee, K.P.Tan, Y.S. Choo, M. H. Hong, J. Laser Micro Nano Eng. **7** 249-53 (2012)

Interaction between plastic deformation and hydrogen damage behavior of 30CrMnSiNi2A steel

Lei Fu, Hongyuan Fang

State Key Laboratory of Advanced Welding and Joining, Harbin Institute of Technology, Harbin,150001, China

Abstract. The interaction between plastic deformation and hydrogen damage behavior of 30CrMnSiNi2A steel was investigated by pre-strain tensile tests and hydrogen charging by electrochemical method. This paper mainly contains two parts. The plastic deformation was restrained by hydrogen-charged, and the effect of hydrogen brittleness damage behavior was accelerated by pre-plastic deformation measure. Tensile pre-strian tests with hydrogen charging at current density from 0 to 50 mA/cm^2 for 120 min were performed at room temperature. Both rate of reduction in area and elongation were decreased due to the transition from ductile to brittle fracture by hydrogen charging, which meant the ability of plastic deformation was reduced by hydrogen. With hydrogen concentration increasing, yield strength also increased indicating that the plastic deformation forming conditions of steel were improved by hydrogen. Hydrogen content increased with pre-strain measured by glycerol gas collection method. Due to the pre-strain measure before hydrogen charging, the reduction of area and elongation were further reduced, while the strength was unexpectedly low. It was because pre-strain promoted the formation of hydrogen-induced crackings (HIC). This proved that the plastic deformation promoted the generation of hydrogen damage.

1 Introduction

The problem of hydrogen embrittlement in industrial applications has attracted enormous attention for years, which is particularly severe in the high-strength steels. Hydrogen embrittlement attributes to the degradation of mechanical properties of mild steel after corrosion [1]. Blades of the Maraging steel is sensitive to hydrogen embrittlement phenomenon proved by experiments [2]. The chemical composition of high-Mn steels and hydrogen embrittlement sensitivity are closely related as the material's stacking fault energy and phase stability were also affected simultaneously [3]. When the martensitic advanced high-strength steel is in low pH in 3.5% NaCl, hydrogen embrittlement susceptibility increased with increasingly negative potentials [4]. For achieving high resistance to hydrogen embrittlement, the high carbon manganese steel can be introduced nano-twins microstructures [5]. The addition of Ti content in hot-stamped boron steels is another method that can be used to enhance the resistance to hydrogen embrittlement [6]. Refining microstructure of quenched-partitioned-tempered low carbon steel also improves the hydrogen embrittlement resistance [7].

Susceptibility to hydrogen embrittlement changes with materials that have been mechanically treated. Hydrogen embrittlement sensitivity was decreased after laser peening due to the basis of compressive residual stress and dislocation changes [8]. Some documents also reported that pre-strain can also affect the hydrogen embrittlement sensitivity [9-11]. Pre-strain increased the resistance to hydrogen embrittlement in STS 310S by suppressing the transition of a ductile to brittle fracture [9]. As the pre-strain increases, the hydrogen embrittlement sensitivity of screw-thread steel bars increases first and then decreases [10]. However, pre-strain does not show sufficient effect on hydrogen embrittlement susceptibility of high-entropy alloy [11].

30CrMnSiNi2A is a kind of low-alloyed martensite ultra-high strength steel with remarkable application in many industries which is sensitive to stress corrosion cracking [12] and hydrogen embrittlement [13]. With respect to the 30CriMnSiNi2A steel, the effect of pre-strain on hydrogen embrittlement has rarely been investigated. Therefore, it is necessary to study the effect of pre-strain on hydrogen embrittlement of 30CriMnSiNi2A steel. This paper also helps explain the mechanism of hydrogen embrittlement for 30CriMnSiNi2A steel from the perspective of macroscopic experiments and microscopic fracture.

2 Experimental

2.1 Materials and samples

The tested steel used in this investigation, 30CrMnSiNi2, was a kind of low alloy ultra-high strength steel with the chemical composition listed in table 1.

https://doi.org/10.1051/matecconf/201822101008

Hot-rolled rods of 30 mm in diameter and 70 mm in length were austenitised at 890 °C for 25 min and then cooling slowly in the air. The test samples were taken along the hot-rolled rods longitudinal direction; their shape and dimension being illustrated in figure 1.

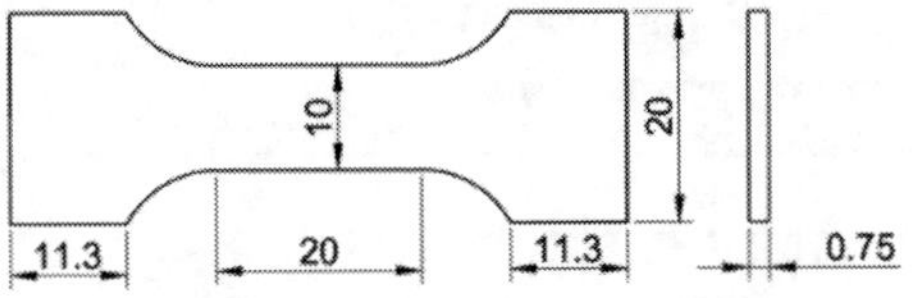

Figure 1. Dimension of tensile sample (mm).

2.2 Pre-strain test

Sample A was stretched to rupture directly in order to observe the origin mechanical properties and fracture morphology. The tensile elongation (failure strain) of the uncharged sample was 13.7% (as shown in figure 2). Samples (C2 and D2) were stretched to the engineering strain of 6.6% and 4.8% respectively (i.e. pre-strain) and uniform deformation was observed without necking obviously. Pre-strain tests were performed at rates of 2.8×10^{-3} s^{-1} by the tensile testing machine at room temperature.

Table 1. Chemical composition of high-strength steel (wt %)

C	Mn	Si	Cr	Ni	S	P	Fe
0.30	1.1	1.05	1.05	1.6	0.010	0.015	Bal

2.3 Electrochemical hydrogen charging and measuring test

In order to fill the samples with hydrogen atoms, electrolytic hydrogen charging method was utilized, which set the sample as the cathode, a platinum electrode as the anode, a constant current source as the power supply and the acid solution as the electrolyte. The acid solution of 0.5 mol/L H_2SO_4 was added 0.5 g/L $NaAsO_2$ as the toxic agent to promote a greater proportion of hydrogen atoms produced from cathode into the sample [14].

Sample B, C1 and D1 were charged with hydrogen at the current density of 5, 30 and 50 mA/cm^2 respectively for 10, 2 and 2 hours to observe the effect of hydrogen embrittlement. The sample B was charged hydrogen with current density of 5 mA/cm^2 for 10 hours to prove that small hydrogen charging current density had no significant effect even for long duration. The hydrogen charging parameters of time and current density were presented in table 2.

Hydrogen concentration in samples was measured by glycerol gas collection method in a water bath at 40 °C for 72 hours. Samples were measured under the same strain and hydrogen charging conditions as said before. The volume of diffusible hydrogen was measured and then converted to the concentration unit needed, as seen in figure 3.

It can be seen from figure 3 that as the hydrogen charging current density increases, the diffusion hydrogen content also increases. However, the hydrogen concentration of the sample after pre-strain treatment is significantly higher than the hydrogen concentration without pres-train. Several documents also reported that the content of hydrogen increased with pre-strain [10,15,16]. For screw-thread steel bars, the amount of hydrogen increases linearly with the exponent of pre-strain [10]. When a medium-carbon steel is preloaded, the quantity of hydrogen increases proved by thermal desorption test around 50 °C to 200 °C [15]. The reason for the increase in hydrogen quantity is that the pre-strain increases the number of hydrogen trap sites verified by heat-treated transformation induced plasticity steel [16].

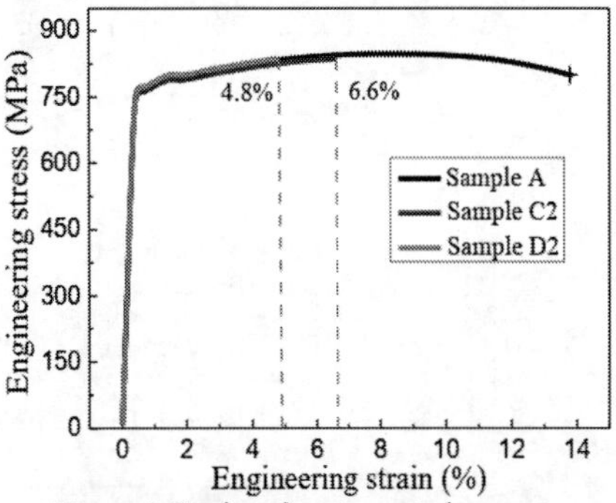

Figure 2. Engineering stress-strain curves.

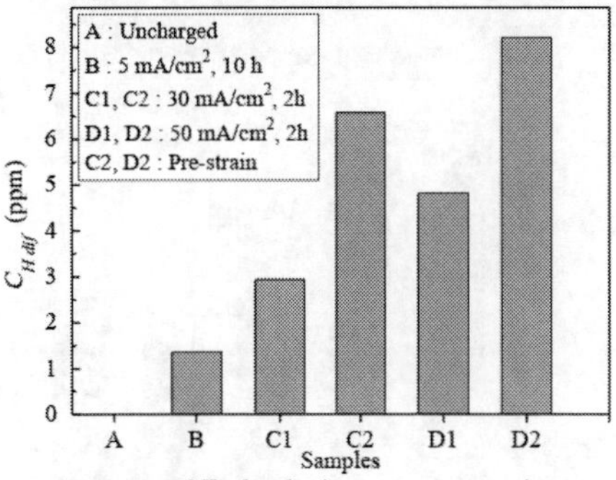

Figure 3. Diffusion hydrogen concentration.

2.4 Tensile test

Engineering stress-engineering strain curves are obtained after hydrogen charging and tensile tests, as seen in figure 4(a). Data of the rates of reduction in area (*Z*) is also displayed in figure 4(b). Results of the tensile tests performed on the samples with and without pre-strain before and after hydrogen (H) charging are displayed in table 2, from which the following three observations can be made.

Firstly, H charging does affect the ductility of 30CrMnSiNi2A steel in significant manner. Hydrogen charging causes the samples to become brittle, reduce the

https://doi.org/10.1051/matecconf/201822101008

reduction in area and decrease the elongation, that is, hydrogen reduces the plastic deformability. With the increase in the quantity of hydrogen, this change becomes more pronounced. Secondly, pre-strain accelerates hydrogen embrittlement, leading to a sharp drop in fracture strength. This may be due to the increase in hydrogen trap sites caused by plastic strain [16]. Finally, in samples without pre-strian, H charging leads to an increase in the yield strength (σ_s), that is, the conditions for the onset of plastic deformation increase. Several documents also reported that the quantity of hydrogen increased the yield strength. The yield stress of austenitic stainless steel increases substantially with the increase in hydrogen charging time [17, 18].

Table 2. Conditions of hydrogen charging and pre-stain data

	H-charged		Pre-strain		Total		C_H	σ_s
	I/S (mA/cm^2)	t (h)	Z (%)	ε (%)	Z (%)	ε (%)	(ppm)	(MPa)
A	0	0	0	0	10.0	13.7	0	767
B	5	10	0	0	9.2	10.8	1.36	802
C1	30	2	0	0	7.2	7.1	2.93	831
C2	30	2	4.6	6.6	4.8	6.9	6.58	--
D1	50	2	0	0	5.1	5.2	4.82	887
D2	50	2	3.4	4.8	3.5	5.2	8.21	--

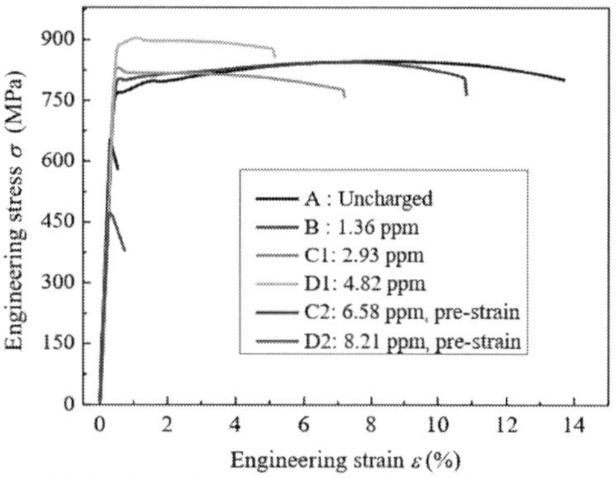

(a) Engineering stress-engineering strain curves

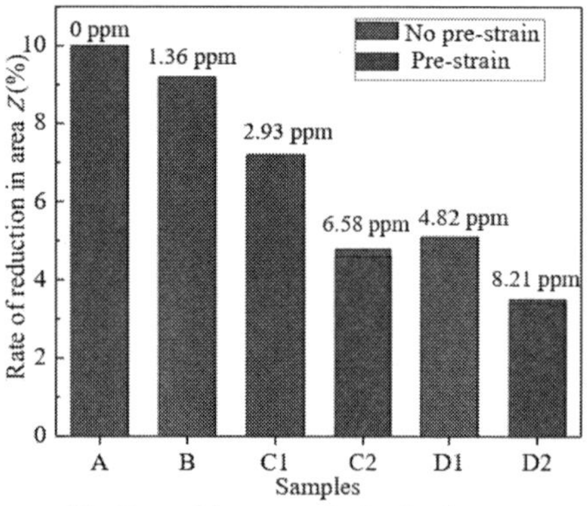

(b) Data of the rates of reduction in area

Figure 4. Change in mechanical properties after hydrogen charging.

3 Result analysis

3.1 Fracture morphology

The fracture surfaces of hydrogen free and hydrogen charged 30CrMnSiNi2A steel samples without and with pre-strain after tensile tests are displayed in figure 5 and figure 6, respectively. The low magnification image of fracture surface of the hydrogen free sample exhibits ductile fracture with apparent necking feature, as demonstrated in figure 5(a). The fracture surface consists of two regions, i.e., central crack initiation which is normal to tensile stress and propagation region (shear lip region) typically at 45° slope to the tensile stress direction which is the maximum shear stress direction. The high magnification image at the shear lip region show dimple fracture with a range of dimple sizes, as shown in figure 5(b).

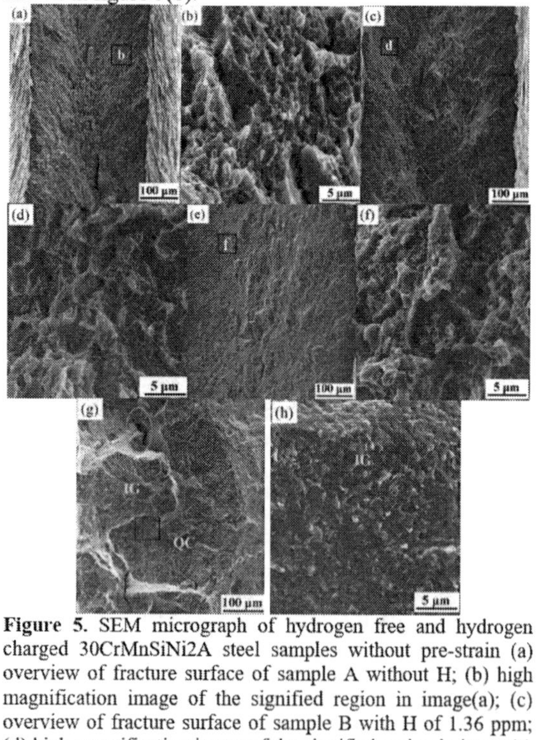

Figure 5. SEM micrograph of hydrogen free and hydrogen charged 30CrMnSiNi2A steel samples without pre-strain (a) overview of fracture surface of sample A without H; (b) high magnification image of the signified region in image(a); (c) overview of fracture surface of sample B with H of 1.36 ppm; (d) high magnification image of the signified region in image(c); (e) overview of fracture surface of sample C1 with H of 2.93 ppm; (f) high magnification image of the signified region in image(e); (g) overview of fracture surface of sample D1 with H

of 4.82 ppm; (h) high magnification image of the signified region in image(g). QC: quasi-cleavage; IG: intergranular fracture.

In the presence of hydrogen (1.36 ppm and 2.93 ppm), the fracture surfaces of sample B and C1 remain ductile with dimple fracture mode, as indicated by figures 5(d) and 5(f) respectively. However, the proportion of the shear lip region in the entire cross section decreases as the hydrogen concentration increases, as seen in figures 5(c) and 5(e), which corresponds to a reduction in the rate of reduction of the area due to hydrogen charged, as shown in figure 4(b). For sample D1 with H of 4.82 ppm, brittle fracture surface with a mixed fracture mode of intergranular fracture (IG) and quasi-cleavage fracture (QC) is observed, as illustrated in figures 5(g) and 5(h). This result shows that when the hydrogen quantity reaches a certain value, the 30CrMnSiNi2A steel undergoes a ductile-brittle transition, and the plastic deformability is further reduced.

For the samples with pre-strain, the fracture surfaces are divided into two diametrically opposed regions, as illustrated in figures 6(a) and 6(d). A part of the fracture surface has covered with IG and QC fracture (figures 6(b) and 6(e)), and fine microvoid coalescence fracture (MVC) occupies other area (figures 6(c) and 6(f)). Since the fracture strength of the pre-strained samples is significantly lower than the yield strength, there may be already hydrogen damage (hydrogen bubbling or hydrogen induced cracking) in the samples before the tensile test. The IG and QC fracture zone is featured by several semi-ellipsoidal pits, as shown in figures 6(a) and 6(d), which may be the hydrogen damage due to the sufficient hydrogen charged.

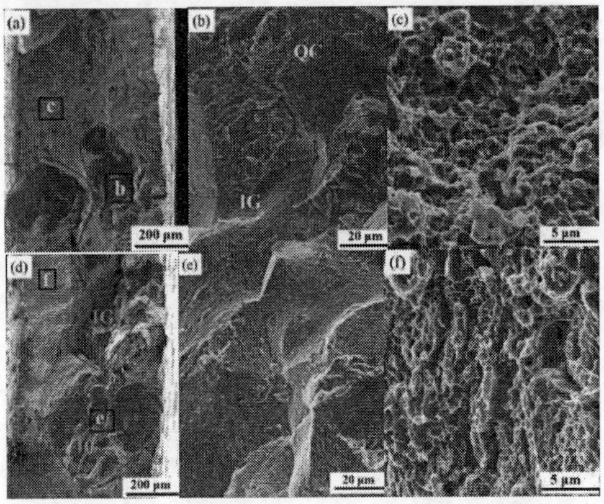

Figure 6. Microstructure of 30CrMnSiNi2A steel samples with pre-strain (a) overview of fracture surface of sample C2 with H of 6.58 ppm; (b) high magnification image of the signified region in image(a); (c) high magnification image of the signified region in image(a); (d) overview of fracture surface of sample D2 with H of 8.21 ppm; (e) high magnification image of the signified region in image(d); (f) high magnification image of the signified region in image(d). QC: quasi-cleavage; IG: intergranular fracture; MVC: micro-void coalescence.

3.2 Metallographic photos

Figure 7 shows the hydrogen-induced cracking (HIC) of the 30CrMnSiNi2A steel samples with pre-stain before tensile tests including transverse and longitudinal cracks. As hydrogen concentration increases, the number of cracks increases. The presence of cracks in the specimen leads to crack growth at low stress. This explains why the fracture strengths of pre-strained samples are lower than yield strengths. This result shows that the reason that the plastic deformation promotes hydrogen damage is to accelerate the generation of hydrogen-induced crackings.

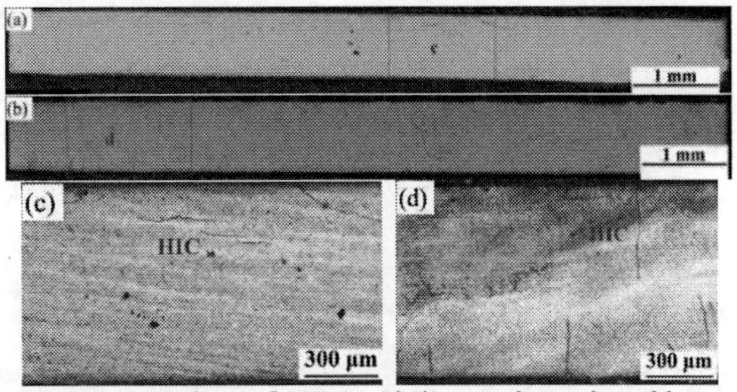

Figure 7. OM views of 30CrMnSiNi2A steel samples with pre-strain after hydrogen charging (a) overview of sample C2 with H of 6.58 ppm; (b) overview of fracture surface of sample D2 with H of 8.21 ppm; (c) high magnification image of the signified region in image(a); (d) high magnification image of the signified region in image(b). HIC: hydrogen-induced cracking.

4 Conclusion

To summarize, the interaction of plastic deformation and the hydrogen damage of 30CrMnSiNi2A steel was investigated, results of which show that the plastic deformation of 30CrMnSiNi2A steel is suppressed and hydrogen embrittlement is accelerated by pre-strains. This is because pre-straining does enhance the H solubility in the steel and then promote hydrogen induced cracking.

References

1. Li L, Mahmoodian M, Li CQ and Robert D 2018 *Constr. Build. Mater.* **170** 78–90
2. Barsanti M, Beghini M, Frasconi F, Ishak R, Monelli B D and Valentini R 2018 *Proc. Struct. Integ.* **8** 501–8
3. Koyama M, Akiyama E, Lee YK, Raabe D and Tsuzaki K 2017 *Int. J. Hydrog. Energy.* **42** 12706–23
4. Venezuela J, Blanch J, Zulkiply A, Liu Q, Zhou Q, Zhang M and Atrens A 2018 *Corros. Sci.* **135** 120–135
5. Khedr M, Li W, Zhu X, Zhou P, Gao S and Jin X 2018 *Mater. Sci. Eng.* A **712** 133–139
6. Kim HJ, Jeon SH, Yang WS, Yoo BG, Chung YD and Ha HY 2018 *J. Alloy. Compd.* **735** 2067–80

7. Yang J, Song Y, Lu Y, Gu J and Guo Z 2018 *Mater. Sci. Eng.* A **712** 630–636
8. Agyenim-Boateng E, Huang S, Sheng J, Yuan G, Wang Z and Zhou J 2017 *Surf. Coat. Tech.* **328** 44–53
9. Ji H, Park IJ, Lee SM and Lee YK 2014 *J. Alloy. Compd.* **598** 205–212
10. Li X, Wang Y, Zhang P, Li B, Song X and Chen J 2014 *Mater. Sci. Eng.* A **616** 116–122
11. Zhao Y, Lee DH, Kim WJ, Seok MY, Kim JY, Han HN and Suh JY 2018 *Mater. Sci. Eng.* A **718** 43–47
12. Liu J, Guo Q, Yu M and Li S 2014 *Chinese. J. Aeronaut.* **27** 1327–33
13. Wang G, Yan Y, Li J, Huang J, Su Y and Qiao L 2013 *Corros. Sci.* **77** 273–280
14. Azizi S, Salah M, Nefzi H, Khaldi C, Sediri F, Dhahri E and Lamloumi J 2015 *J. Alloy. Compd.* **648** 244–252
15. Nagumo M, Nakamura M and Takai K 2001 *Metall. Mater. Trans.* A **32** 339–347
16. Ronevich JA, De Cooman BC, Speer JG, De Moor E and Matlock DK 2012 *Metall. Mater. Trans.* A **43** 2293–301
17. Takakuwa O, Mano Y and Soyama H 2014 *Int. J. Hydrog. Energy.* **39** 6095–103
18. Daniel P A and Carl J A 1995 *Metall. Mater. Trans.* A **26** 2849–58

Nominal stress based equal fatigue bearing capacity design of HSLA steel V-shaped butt-welded joints

Xue Wen, Ping Wang, Zhibo Dong, Hongyuan Fang

State Key Laboratory of Advanced Welding and Joining, Harbin Institute of Technology, Harbin 150001, China

Abstract. Based on nominal stress, this paper proposed a fatigue life estimation method of V-shaped butt welded joints. Then, the fatigue bearing capacity of the base metal under the same fatigue load was used as design goal of the V-shaped butt welded joints and an equal fatigue bearing capacity (EFBC) design method for V-shaped butt welds was thus formed. High strength low alloy steel Q690 was chosen as the experimental material to carry out the equal fatigue bearing capacity design.

1 Introduction

With the continuous development of lightweight structures, the proportion of high-strength steel in mechanical structures is increasing [1-7]. High-strength low-alloy (HSLA) steel are widely applied in the fabrication of various structures, such as ship building, pressure vessel, offshore construction as well as submarine, owing to high strength, excellent toughness, and high strength-to-weight ratio. In welding HSLA steel, it imposes various problems, such as hydrogen-induced cracking, softening, and hardening of heat-affected zone (HAZ) because the steel is sensitive to thermal cycle [8, 9].

Under-matched welds are sometimes used in joining high strength steels with the aim of minimizing the hydrogen induced cold cracking tendency [10], reducing or even preventing costly preheating operations [11]. However, the low load carrying capacity of under-matched welds has been an obstacle to its wide engineering application [12, 13].

Currently, the fatigue design methods for welded structures given in various international standards include nominal stress method, structural stress method, effective notch stress method and fracture mechanics method [14-17]. The global stress method (also known as the nominal stress method) is the most simple and widely used approach to design weldments against fatigue. When either nominal stresses cannot be calculated unambiguously or a reference fatigue curve for the specific geometry of the welded detail being assessed is not available, then either hot-spot or local stress based approaches are recommended to be used. The structural hot-spot stress method is applied by determining, on the component surface, the linear-elastic stress states at either two or three reference points. Subsequently, by using these reference stress states, structural stresses are extrapolated to the weld toes at the hot spots. Structural stresses can be determined experimentally by using strain gauges attached to the component's surface at different distances from the weld toe. Obviously, this experimental procedure is not applicable when the area of interest in the vicinity of the weld is not accessible. This problem can be overcome by estimating the stress states at the extrapolation points via linear-elastic Finite Element (FE) models. The hot-spot method was originally developed to assess the fatigue behaviour of offshore structures, with its use being subsequently extended to other structural applications. The effective notch stress approach makes use of linear-elastic stresses determined at either the weld toe or the weld root by introducing a fictitious fillet having radius equal to 1 mm, with this strategy being applicable to welded joints having thickness larger than (or equal to) 5 mm. On the contrary, when the relevant thickness is lower than 5 mm, the effective notch stress approach is recommended as being applied by using a fictitious radius of 0.05 mm. This approach can be used to assess welded joint in which the fatigue crack initiation process takes place not only at weld toes, but also at weld roots [18]. Fracture mechanics may be used to assess the fatigue properties of welded joints in which no imperfections have been detected. In such cases it is necessary to assume the presence of an initial crack, for example based on prior metallurgical evidence, the detection limit of the used inspection method or fitting from fatigue data, and then to calculate the stress intensity factor [19]. However, whether the joint can have a fatigue bearing capacity not lower than that of the base metal through geometric design is yet to be studied.

From the stress-life curves of materials and structures, it can be seen that the fatigue life of materials and structures depends on the fatigue stress on their load-carrying sections. For a certain external fatigue load, the fatigue stress value is determined by the size of the load bearing cross-section. Therefore, by changing the size of

 https://doi.org/10.1051/matecconf/201822101009

the load-carrying section of the structure, it is possible to develop the fatigue capacity of the structure. If we find the relationship between the size of the bearing section and the fatigue life of the structure, and then let the fatigue life of the structure equal to the fatigue life of the base metal under the same fatigue load, a weld structure with equal fatigue bearing capacity with the base metal can thus be obtained.

This paper starts from the perspective of structural safety and reliability, estimating the fatigue life of the V-shaped butt-welded joint based on nominal stress. And then use the fatigue bearing capacity of the base metal under the same fatigue load as the design goal of the V-shaped butt welded joint. An equal fatigue bearing capacity design method for V-shaped butt welds was thus formed. High strength low alloy steel Q690 was chosen as the experimental material to carry out the equal fatigue bearing capacity design.

2 Nominal stress based fatigue life estimation of V-shaped butt-welded joints

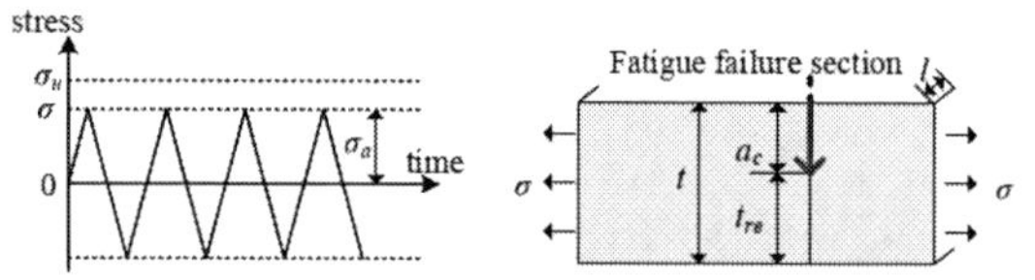

Figure 1. Fatigue failure process of homogeneous materials.

The fatigue failure process of homogeneous materials shown in Figure 1 can be described as: After the fatigue crack expand a certain length in the material, the stress value on the remaining bearing section of the specimen reaches the tensile strength σ_u of the homogeneous material [20,21], which leads to the instant rupture. The instant rupture section size t_{re} of homogeneous material can be approximated as:

$$t_{re} = \sigma \cdot t/\sigma_u \quad (1)$$

For the case of fatigue stress ratio -1 shown in Figure 1,

$$\sigma = \sigma_a = P/(t \cdot l) \quad (2)$$

From the fatigue failure process of homogeneous materials, it can be seen that the fatigue life of homogeneous materials N corresponds to the number of stress cycles before the fatigue crack propagating to the instant rupture section t_{re}. The dimension of the fatigue crack before extending to the instant rupture section can be referred to as the critical crack propagation size a_c.

$$a_c = \mathrm{t} - t_{re} \quad (3)$$

To estimate the number of stress cycles experienced by fatigue cracks extending to a certain size in a homogeneous material, a traditional method is to first divide the fatigue failure process into three phases, namely the crack initiation phase, the crack propagation phase, and the instant rupture phase. Then use the material strain-life ($\Delta\varepsilon$-N_i) curve to obtain fatigue crack initiation life N_i [22,23], and take the integral of Paris formula [24,25] to calculate the number of stress cycles N_p experienced by the fatigue crack when propagating to a certain size. Finally, add N_p to N_i to obtain the number of stress cycles experienced by fatigue crack when propagating to a certain size. The main problem of this method is that the definition of fatigue crack initiation is not uniform, which will bring certain calculation error, and it is inconvenient to use.

Since Wohler discovered the nominal stress-fatigue life (S-N) curve [26], it has undoubtedly become the most intuitive and effective way to characterize the fatigue properties of materials. Many researchers, including Wohler, summed up the functional relationship between nominal stress and fatigue life of homogeneous materials on the basis of a large amount of material fatigue data, and continuously improved it [27]. Most of the research results show that there is an exponential function relationship between the nominal stress on the load-bearing cross-section of the homogeneous material and the fatigue life in the high-cycle fatigue stage, that is:

$$(\sigma)^m \cdot N = C \quad (4)$$

Where m and C are material related parameters.

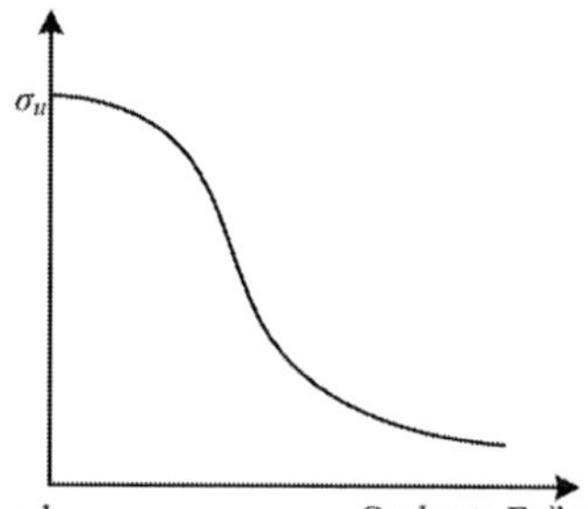

Figure 2. S-N curve of homogeneous materials

Based on nominal stress-fatigue life function of homogeneous materials, researchers can easily estimate the fatigue life of homogeneous materials under certain fatigue loads.

However, welded joints have heterogeneous mechanical properties and also exhibit highly heterogeneous microstructural variations in local regions[28-35]. Besides, due to the presence of the groove, the ratio of the base metal and the deposited metal on the fatigue bearing section of the welded joint is not a constant value. Therefore, the influence of the geometric size of the base metal and the deposited metal at the same fatigue bearing section on the fatigue life of the welded joint needs to be determined to achieve fatigue life estimation. This study found that the relationship between the geometric parameters of V-shaped butt weld joints (as shown in Figure 3) and its fatigue life is as follows:

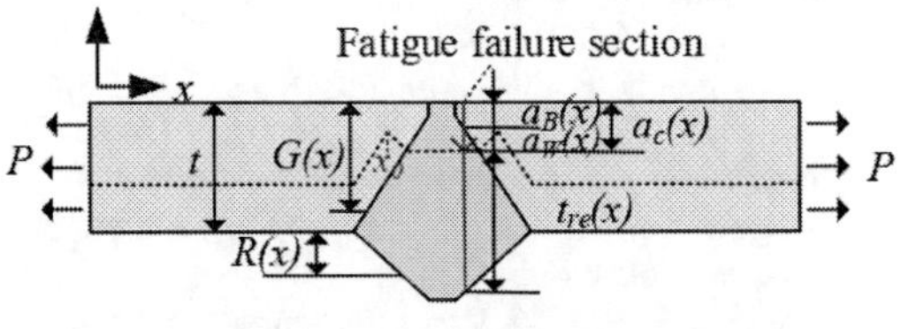

Figure 3. V-shaped butt-welded joint

$$N = \frac{C_W \cdot [t+R(x)-G(x)] \cdot [t+R(x)]^{m_W}}{\left[t+R(x)-\frac{P}{\sigma_b^W \cdot l}\right] \cdot (\frac{P}{l})^{m_W}} + \frac{C_B \cdot [t+R(x)+G(x)-t_{re}(x)] \cdot [t+R(x)]^{m_B}}{\left[t+R(x)-\frac{P}{\sigma_b^B \cdot l}\right] \cdot (\frac{P}{l})^{m_B}} \quad (5)$$

The instant rupture section of the V-shaped butt-welded joint t_{re} in equation (5) is a piecewise function about the crack initiation position.

$$t_{re}(x) = \begin{cases} \frac{P}{\sigma_b^B \cdot l} + \frac{R(x)-G(x)}{2}, x \leq x_0 \\ \frac{P}{\sigma_b^W \cdot l}, x \geq x_0 \end{cases} \quad (6)$$

When the groove function *G(x)*, fatigue load *P*, and the mechanical parameters of the base metal and the deposited metal are known, the geometric dimension *R(x)* of the double V-shaped butt joint can be well-determined according to the EFBC realizing condition $N=N_B$.

Table 1 Main chemical composition and mechanical properties

Material	σ_y (MPa)	σ_u (MPa)	C (%)	Mn (%)	S-N formula
Q690	720	788	0.14	0.13	$\sigma_a^{11.1} \cdot N_f = 6.38 \times 10^{33}$
Welded metal	450	647	0.06	6.39	$\sigma_a^{11.1} \cdot N_f = 7.16 \times 10^{32}$

3 Equal fatigue bearing capacity design of V-shaped butt-welded joints

High strength low alloy steel Q690 was chosen as the experimental material to carry out the equal fatigue bearing capacity design. The main chemical composition and mechanical properties of the material are shown in Table 1. While σ_y and σ_u respectively refer to the yield strength and tensile strength of the material and S-N formula refers to the stress-life fitting formula of the material. The geometric parameters of the test plate and groove are shown in Figure 4.

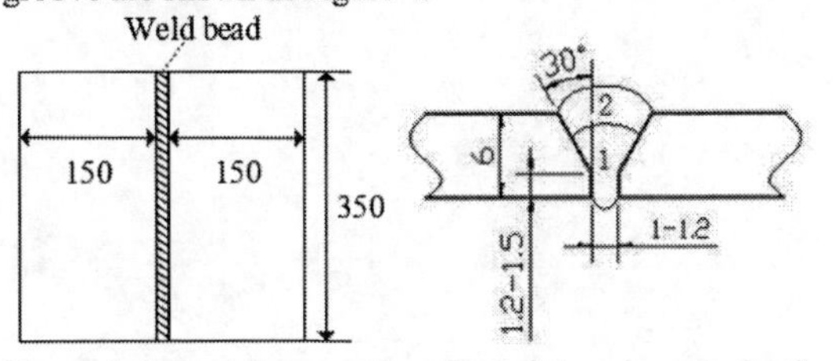

Figure 4. Geometric parameters of test plate and groove (mm)

The fatigue specimen of V-shaped butt welded joint shown in Figure 5 is designed according to ASTM E466-07 [36].

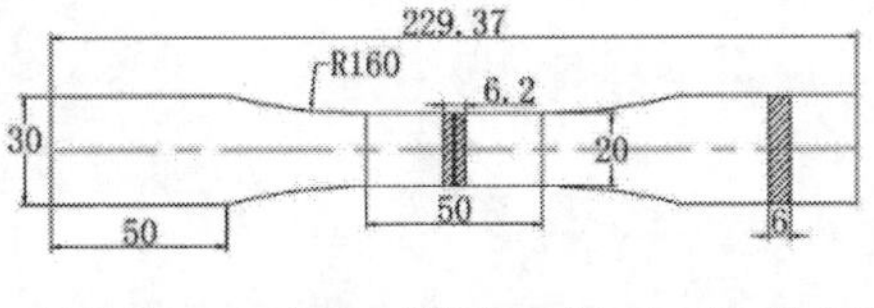

Figure 5. Fatigue specimen of V-shaped butt welded joint (mm)

10^4 stress cycles is used as the design goal of the V-shaped butt-welded joints, while the fatigue strengths of the base metal and the deposited metal are 484.3 MPa and 397.7 MPa, respectively, according to their S-N formulas. Substituting the known material and geometric parameters into equation (5), and then perform arc optimization on the resulting design curve. The final shape parameters of V-shaped butt-welded joints are shown in Figure 6:

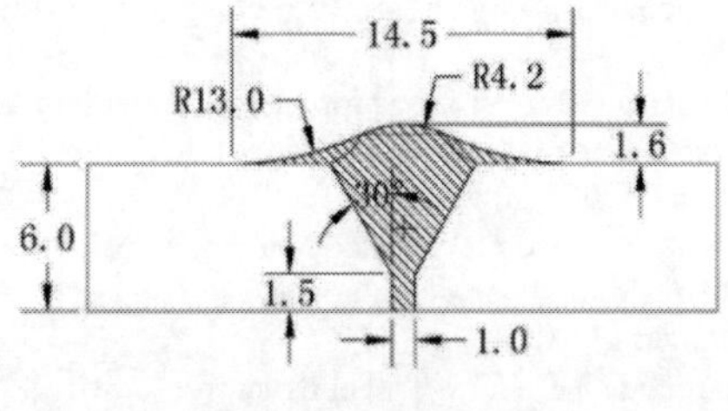

Figure 6. Designed V-shaped butt-welded joints (mm)

Due to the use of material's nominal stress-life formula in the design process, the V-shaped butt-welded joint designed by this method is only applicable to experimental conditions when all hypotheses of the nominal stress-fatigue life formula are satisfied.

4 Conclusion

With the aim of improving the fatigue load carrying capacity of under-matched V-shaped butt welded joints through structural geometric design. Base on nominal stress method, this paper proposed a fatigue life estimation method of V-shaped butt welded joints.

And then, the fatigue life of base metal under the same fatigue load was used as the design goal of the V-shaped butt welded joints and an equal fatigue bearing capacity design method for V-shaped butt welds was thus formed.

Finally, high strength low alloy steel Q690 was chosen as the experimental material to carry out the equal fatigue bearing capacity design, which showed that the proposed design method was theoretically feasible.

References

1. Cao X, Wanjara P, Huang J, et al. 2011. *Mater. Des.* 32(6):3399-3413.
2. Sterjovski Z, Dunne D, Ambrose S. 2004. *Int J Press Vessel Pip* 81(6):465–470.
3. Min D, Xin-hua T, Feng-gui L, Shun Y. 2011. *Int J Adv Manuf Technol.* 55(5–8):527–533.
4. Kim M-C, Park S-G, Lee K-H, Ho Kim S, Lee B-S. 2014. *Curr Nanosci.* 10(1):151–153.
5. Xu W H, Lin S B, Fan C L, et al. 2014. *Int. J. Adv. Manuf. Technol.* 75(9-12):1439-1446.
6. Viano D, Ahmed N, Schumann G. 2000. *Sci Technol Weld Join.* 5(1):26–34.
7. Zhang C, Yang J, Hu X, Lu P, Zhao M. 2012. *Mater Sci Eng A.* 546:169–179.
8. Mohandas T, Madhusudan Reddy G, Satish Kumar B. 1999. *J Mater Process Technol.* 88(1):284–294.
9. Prasad K, Dwivedi D. 2008. *Int J Adv Manuf Technol.* 36(5–6):475–483.
10. Umekuni A and Masubuchi K. 1997 *Weld. J.* 76(7): 256s.
11. Loureiro and A. J 2002 *J. Mater. Process. Tech.* 128(1-3), 240-249.
12. Rakin, M., Medjo, B., Gubeljak, N.and Sedmak, A 2013 *Eng. Fract. Mech.* 109, 221- 235.
13. Zhu, L. and Tao, X. Y. 2016 *Int. J. Pres. Ves. Pip.* 139, 61-68.
14. Eurocode 3-Design of steel structures, Part 1. 8 Joints, EN 1993-1-8.
15. American Bureau of Shipping. Commentary on the guide for the fatigue assessment of offshore structures; January 2004–updated 2010.
16. BS5400, P. 10, 1980, Code of Practice for Fatigue. Steel, Concrete and Composite Bridges. British Standard Institution.
17. Eurocode 3-Design of steel structures, Part 1. 8 Joints, EN 1993-1-8.
18. Zamzami I A, Susmel L. 2016. *Int. J. Fatigue.* 101.
19. Hobbacher A. 2015. *Springer.*
20. Chandran, K. R. 2017. *Int. J. Fatigue.* 102: 261-269.
21. Chandran, K. R. 2015. *Scripta Mater.* 2015; 107: 115-118.
22. Meggiolaro, M. A., & Castro, J. T. P. 2004. *Int. J. Fatigue.* 26(5): 463-476.
23. Brennan, F. P. 1994. *Int. J. Fatigue.* 16(5): 351-356.
24. Paris, P. C., Gomez, M. P., & Anderson, W. E. 1961.*Trend in Engineering.* 13.
25. Paris, P. C. 1963. *Trans.asme Ser.d.* 85(4): 528-533.
26. Suresh, S. 1998. *Cambridge university press.*
27. Castillo, Enrique, and A. Fernández-Canteli. 2009. *Springer.*
28. Kabir A S H, Cao X, Gholipour J, et al. 2012. *Metall. Mater. Trans. A-Phys. Metall. Mater. Sci.* 43(11):4171-4184.
29. Shen J J, Liu H J, Cui F. 2010. *Mater. Des.* 31(8):3937-3942.
30. Zhao Y Y, Zhang Y S, Hu W. E. 2013. *Sci. Technol. Weld. Join.* 18(7):581-590.
31. Li P, Dong H, Xia Y, et al. 2018. *J. Manuf. Process.* 33:54-63.
32. Mao Y, Ke L, Chen Y, et al. 2018. *J. Mater. Sci. Technol.* 34(1).
33. Chen J, Wei Y, Zhan X, et al. 2017. *Int. J. Adv. Manuf. Technol.* 92(2):1-10.
34. Magnabosco I, Ferro P, Bonollo F, et al. 2006. *Mater. Sci. Eng. A-Struct. Mater. Prop. Microstruct. Process.* 424(1-2): 163-173.
35. Liu H, Fujii H, Maeda M, et al. 2003. J. Mater. Sci. Lett. 22(6): 441-444.
36. Standard A. 2007. *Standard practice for conducting force controlled constant amplitude axial fatigue tests of metallic materials.* 3.

https://doi.org/10.1051/matecconf/201822101010

Material Characterization of Test Contact Pin

Praveen kumar Ramamoorthy[1], Florian Fischer[2], Han Guan[1] and Murad Hudda[1]

[1] *Infineon Technologies, 168 Kallang Way, Singapore- 349253*
[2] *Department of Mechanical Engineering, Technical University of Munich, Munich- 80333, Germany*

Abstract. Test Contact Pin is used for testing the functionality and quality of the micro-devices at extreme temperatures before shipping to customers. The contact pin transfers current & signal to the device tested through dynamic contacting where the material resistance of the pin and the contact resistance between the pin and the device plays a significant role in influencing the effectiveness of the test process. Available material specifications for these pins in the market are for ambient temperature only. The test contact pin was characterized across temperatures onto matte Sn leadframe for its contact resistance and force with respect to number of touchdowns. Contact resistance and pin force are the key variables used to understand the mechanism of the process which is also used for determining the lifespan of the contact pin. An automated and sophisticated contact tester tool (CTT) was used to characterize the test contact pin on the mattes Tin leadframe across temperatures (-43°C, 25°C and 150°C) and at fixed pin deflection. Based on the results, it was observed that the contact resistance was higher at higher temperatures. Further data analysis revealed that this phenomenon was due to influence of various factors such as temperature, leadframe material type and the material migration of Sn from leadframe to the test contact pin tip.

1 Introduction

Testing is a dynamic process in which the micro-devices are checked for their functionality and quality by pumping current and signal to the devices before shipment to the customer. This is possible by dynamic material contact between the lead of the package and the contact pin (Figure 1).

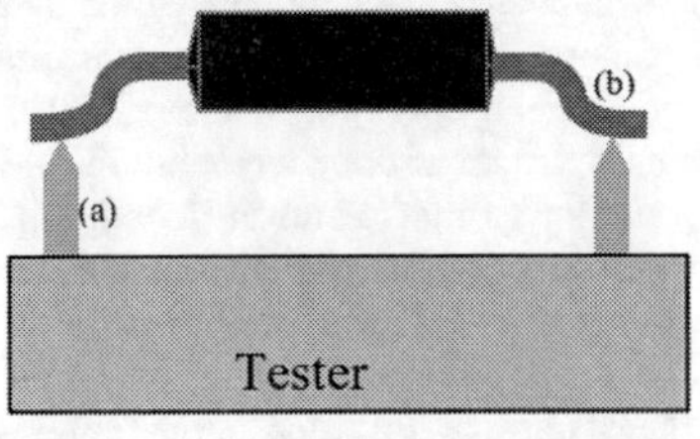

Figure1. Device under Test. (a) contact pin (b) device lead

At micro level real surfaces are rough. During contact between the lead and the pin conducting paths are formed through localized metallic contacts [1]. As the force increases, the number and the area of these small metal–metal contact spots or a-spots will increase (Figure 2).

a-spots are the only conducting paths for the transfer of electrical current and the real contact area is only a fraction of the apparent contact area as illustrated in Figure 2. The relationship between the applied normal load L, hardness of the metal, H, and the apparent contact area, A_a, is given by the following expression:

$$L = \mathrm{x}HA_a. \quad (1)$$

Where H, the hardness is the measure of the ability of a metal to resist deformation due to point loading; x is the pressure factor and is equal to 1 in most practical contact systems [2].

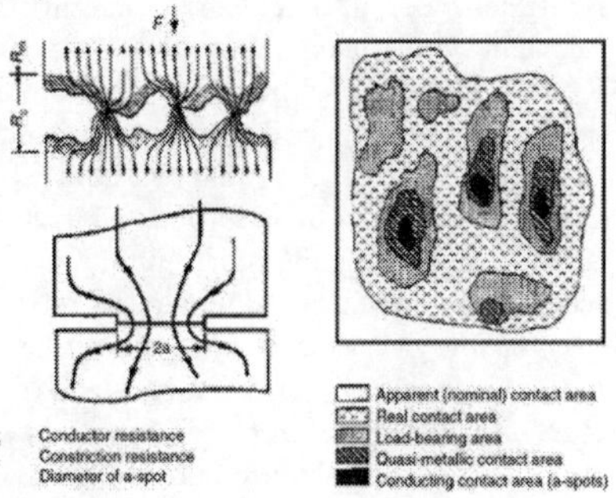

Figure 2. Schematic of current constriction and real contact area.

a-spots are the constricted contact interface through which current passes. The electrical resistance of the contact at *a*-spots is called *constriction resistance* and is related to the basic properties of metals such as hardness and electrical resistivity. Constriction resistance for a single *a*-spot can be expressed as

$$R_s = (\mathrm{r1} + \mathrm{r2})/4a; \quad (2)$$

where r1 and r2 are resistivities of the contacting metals, and *a* is the radius of the metal-to-metal contact area [3]. If two same metals are contacting, then the constriction resistance becomes

$$R_s = r/2a. \quad (3)$$

Usually the metals are contaminated or an oxide layer formed above it due to reaction with oxygen. Thus, the passage of electric current may be affected by thin oxide, sulphide, and other inorganic films present on the metal surfaces. Hence, the total contact resistance of a joint is a sum of the constriction resistance (R_s) and the resistance of the film (Rf)

$$R_c = R_s + Rf \quad (4)$$

$$Rf = s/\pi a^2; \quad (5)$$

where s is the resistance per area of the film. Both tunnelling and fritting are considered operative mechanisms for the current transfer across the film. In most practical applications, the contribution of these films to the total contact resistance is of minor importance because the contact spots are usually created by the mechanical rupture of surface films. The contact resistance is the most important and universal characteristic of all electrical contacts [4].

The contact resistance is the most important and universal characteristic of all electrical contacts and is always taken into account as an integral part of the overall circuit resistance of a device. Therefore, although it is significantly smaller as compared with the overall circuit resist¬ance, the changes in the contact resistance can cause significant malfunctions of the device.

This is because the contact resistance can vary significantly with the changes in the real contact area, contact pressure variations, resistive film nonuniformity, and other factors. This results in large voltage increases, thus making the fine adjustment or good operation of devices difficult. For instance, the instability and high values of contact resistance are especially noticeable in bulk DC potentiometers, whose resistive members are relatively thick and have a high specific resistance.

There are many parameters that can be used to assess the operating efficiency of electrical contacts. Among these parameters, perhaps the most important are electric (the transition voltage drop, commutation noise, erosion resistance), tribological (the wear resistance and friction coefficient) and chemical (corrosion resistance).

Another important factor affecting the contact behavior is the presence of various films (such as oxides, contaminants, and reaction products). The current passes through the "*a*-spots" that are smaller than the real contact spots (Figure 5.1). Since the electrical current lines are constricted to allow them to pass through the *a*-spots, the electrical resistance increases. This increase is defined as the constriction resistance. Contaminant films on the mating surfaces increase the resistance of *a*-spots. The total resistance due to constriction and contaminant films is termed the *contact resistance*.

The fundamental monograph of Holm[1] considered most of the contact phenomena important for mechanical and electrical engineers. Timsit's review updated this topic within the framework of a general Holm's approach describing the factors of either electrical or thermal origin affecting the contact resistance. In the following sections, this approach will be followed with the addition of more recent research data.

In the case of isotropic roughness topographies, the *a*-spots are assumed to be circular or noncircular when the roughness has a directional characteristic (e.g., in rolled-metal sheets or extruded rods). The current passes through the "*a*-spots" that are smaller than the real contact spots.

Since the electrical current lines are constricted to allow them to pass through the *a*-spots, the electrical resistance increases. This increase is defined as the constriction resistance. Contaminant films on the mating surfaces increase the resistance of *a*-spots. The total resistance due to constriction and contaminant films is termed the *contact resistance*.

The fundamental monograph of Holm[1] considered most of the contact phenomena important for mechanical and electrical engineers. Timsit's review updated this topic within the framework of a general Holm's approach describing the factors of either electrical or thermal origin affecting the contact resistance.

In the following sections, this approach will be followed with the addition of more recent research data. In the case of isotropic roughness topographies, the *a*-spots are assumed to be circular or noncircular when the roughness has a directional characteristic.

2 Experimental Section

An industry standard test contact pin was characterized for its force and contact resistance (Cres) with respect to number of touchdowns using a sophisticated tool i.e., contact tester tool. The arrangement of the tool is such that the pin to be characterized is mounted on the Z-drive which moves vertically up and down. The leadframe on which it is to be characterized is mounted onto a XY movable chuck (Figure 3). The entire mechanism is inside a oven which has an operating range of -70°C to 180°C. Experiments were run at three temperatures -43°C, 25°C and 150°C similar to that of testing conditions at production. The deflection of the pin was kept constant and the reactive force and resistance were measured with the help of the tool. Fresh pin was subjected to 1 million touchdowns on the leadframe for each temperature and the experiment was repeated 3 times for repeatability check.

Figure 3. Contact pin under contact with the Leadframe

3 Results & Discussion

3.1 Ambient Run

Test contact pin that was run at ambient exhibited the trend of force and contact resistance (Cres) measurements as shown in the graph below (Figure 5). It was observed that the contact resistance increased from 21mOhms to 78 mOhms after 1 million touchdowns. Examination of the pin tip after each and every run with leadframe revealed the wearing out of the pin and the migration of Sn material from the leadframe to the pin. The reactive force of the pin was observed to be dropping from 38gf to 30gf within the same number of touchdowns.

The increase in contact resistance may be due to multiple factors such as tin migration from the lead to the pin, electron migration at the pin tip etc. The drop in pin force with respect to number of touchdowns maybe due to increased fatigue of the pin, wear and tear of the pin tip which was observed to be upto 50ums in height. Abrasive wear, also known as *abrasion*, is one of the most common forms of wear [5]. The key aspect of abrasive wear is its association with the cutting or ploughing of the surface by harder particles or asperities (Figure 4). These cutting points may either be embedded in the counterface, or lost within the contact zone.

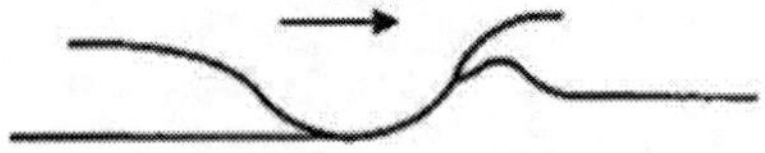

Figure 4. Abrasive Wear Mechanism

Abrasive wear is defined by the ASTM G 40–83 terminology standard as a wear due to hard particles or hard protuberances forced against and moving along a solid surface. Abrasive wear, also known as *abrasion*, is one of the most common forms of wear.

The key aspect of abrasive wear is its association with the cutting or plowing of the surface by harder particles or asperities. These cutting points may either be embedded in the counterface, or lost within the contact zone. The former case is commonly called two-body abrasion; the latter case is three-body abrasion. Sometimes, the abrasion may be classified in terms of the force acting on abrasive material. If the force is so large that the abrasive particles are crushed and the working surface experiences significant cutting and deformation, then *high-stress abrasion* takes place. This form of abrasion is observed, for example, in high-load bearings, where hard particles are trapped between mating surfaces. A particular case of high-stress abrasion is *gouging abrasion*.

This term is used for abrasion by large hard lumps that remove material from the worn surface in large fragments. Yet, the situations when the term may be applied are described ambiguously. When the pressure on the abrading particles or asperities is relatively low and the particles remain unbroken, the abrasion is called *low-stress abrasion*.

Abrasion displays scratches, gouges, and scoring marks on the worn surface, and the debris produced by abrasion frequently takes on the appearance of fine cutting chips similar to those produced during machining, although at a much finer scale. Most of the models associated with abrasive wear incorporate geometric asperity descriptions, so that wear rates turn out to be quite dependent on the shape and apex angles of the abrasive points moving along the surface.

The sources of the abrasive solids are numerous, and the nature of the abrasive wear in a given tribosystem will depend to some extent on the manner in which the abrasives enter the tribo-system: whether they are present in the original microstructure as hard phases, enter the system as contaminants from outside, or are generated as debris from the contact surfaces as they wear.

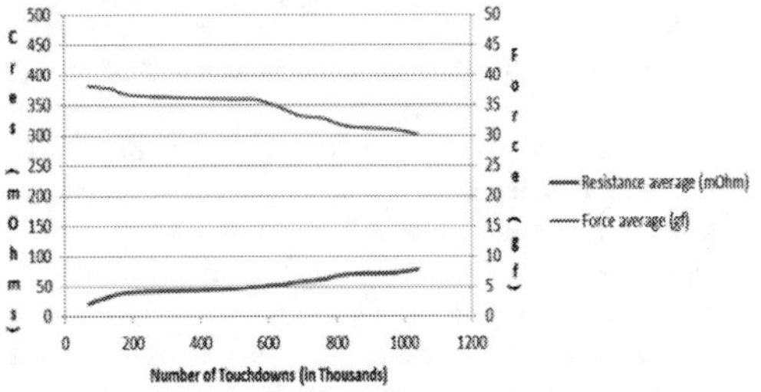

Figure 5. Force and Contact Resistance graph at Ambient temperature

3.2 Hot Run

Figure 6 shows the force and contact resistance measurements at hot (150°C). Drop in force of the pin at 1 million touchdowns was observed to be 20% higher than that at ambient. For resistance it was threefold increase when compared with that of ambient. It's well established that higher temperature accelerates the wear and tear of the contact pin and offers resistance to the flow of electrons thereby increasing the contact resistance [6].

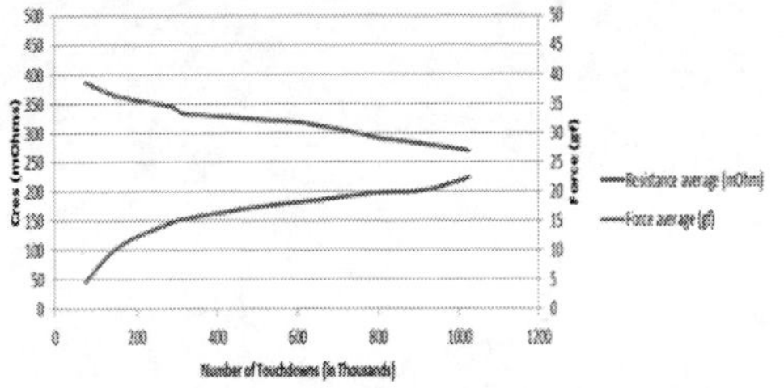

Figure 6. Force and Contact Resistance graph at Hot temperature

3.3 Cold Run

Figure 7 shows the force and Contact resistance measurements at cold (-43°C). Based on the data, it was observed that the force and contact resistance measurements at 1 million touchdowns lie in-between hot and ambient. The reason behind this phenomenon is due to the tin surface of the lead getting harder at lower temperatures [7].

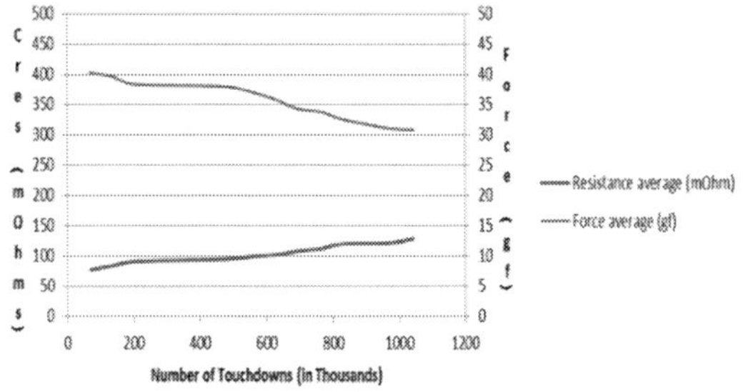

Figure 7. Force and Contact Resistance graph at Cold temperature

4 Conclusion

A detailed characterization of the test contact pin across temperatures was performed. The data exhibited that the contact pin force reduces with respect to number of touchdowns while the contact resistance increases. Based on data analysis and observation pin tip abrasive wear out due to contacting with the leadframe was found out to be the reason behind this phenomenon. It was concluded that the pin tip wear and tear is a function of temperature and the hardness of the contacting surface thereby affecting pin force and contact resistance. The quantified results were used as a reference for troubleshooting at production and also for the selection of prospective pins with better performance.

References

1. Holm, R., *Electrical Contacts*, Springer, New York, 1979.
2. Braunovic, M., Myshkin, N. K., Konchits, V. V. *Electrical Contacts: Fundamentals, Applications and Technology,* Taylor & Francis Group, LLC, pp. 25 – 30, 2006.
3. Timsit, R. S., Electrical contact resistance: Fundamental principles, In *Electrical Contacts: Principles and Applications*, Slade, P. G., Ed., Marcel Dekker, New York, pp. 1–88, 1999.
4. Janga, Y. H., Barberb, J. R. "Effect of contact statistics on electrical contact resistance". *Journal of Applied Physics*, Vol. 94, No, 11 (2003), pp. 7215-7222.
5. Bock, E. M., Whitley, J. H. "Fretting corrosion in electrical contacts". *Twentieth Annual Holm Seminar on Electrical Contacts*, Oct. 1974, pp. 29-31.
6. Tummala, R. R. *Microelectronics Packaging Handbook*: Technology Drivers, Part 1, Springer Science, (2001), pp. 14-17.
7. Beelen-Hendrikx., C.C.M. " Lead-free semiconductor packaging," *Electronics Packaging Technology Conf,* Singapore, Dec. 2002, pp. 11-19.

Dichroism Infrared-reflective Thermochromic Ink of Mobile Terminals' Recognition

Baoning Sha and Shengyao Zhao

Shandong University, Institute of polymer research, Jinan, Shandong, China

Abstract. The decoration of thermochromic ink has applications ranging from engineering devices to liquid sensors. However, when it comes to accurate computer identification, the characteristic as dichroism is not fully used. Thus, we have recently proposed a manufacture method of dichroism infrared-reflective thermochromic ink. (DITI). By rubbing to heat DITI coated layer, this method enables us to finely and separately create dichroism manuscript after the handwriting is complete. Moreover, the microcapsules have cheerful compatibility. We have decorated conventional microcapsules with bola-type copolymer which allow us to fill them with Infrared (IR) reflective perssad. We consider this method promising to distinguish dichroism printing or writing on a tanglesome surface much quicker, for the IR reflective low-colour-density layer will be easily locked by an IR camera or a mobile phone camera to adjust specific characters of interest. We anticipate that this DITI ink will find use in markable temperature measurement, fluid machinery, anti-counterfeiting techniques and daily printing or writing.

1 Introduction

Thermochromic materials have significantly extended the applications of decoration, optical sensors and electronics. In recent years, thermochromic ink has been applied to various fields in life science [1,2]and engineering such as optical temperature measurement, fluid machinery, Anti-counterfeiting techniques and daily printing or writing. DITI ink within flexible structures could open up new applications in electronics identification by adjusting the Chromogenic molecule with high-IR-reflective perssad.

Recent work has shown that temperature-sensitive ink and pressure-sensitive ink (PSI) can be used in fabricating a dual TSI and PSI array sensor by scientific grade inkjet printer, which showed a possible way in operating TSI[3,4]. However, a quick-direct distinguish of dichroism TSI on a tanglesome surface by a commercial camera still remains challenging. Applying the most conservative estimate to current demographic trends, the population of the ones who use mobile terminals will increase by about three billion people by the year 2050. Thus, a convenient way as distinguishing dichroism TSI could further broaden the usage among ordinary users[5,6]. Considering the complementary metal-oxide-semiconductor (CMOS) used in mobile terminals like smart phone has a wider sensitive wavelength in IR sections, we decide to plant several characteristics on traditional TSI.

Here we describe the design and demonstration of DITI system, where the heated IR reflective low-colour-density layer can be precisely distinguished from the unheated layer, whereupon the quick-direct distinguish by mobile terminals is achievable. Distinct from previous reports, IR perssad is filled in the microcapsules mixed with low-colour-density parts so that DITI ink can be used in the similar way as ordinary ink and the distinguish of heated DITI layer can be driven through CMOS used in ordinary mobile terminals. The dichroism infrared-reflective thermochromic ink system concept involves (1) decorating conventional microcapsules with bola-type copolymer, (2) filling IR reflective perssad, (3) adjusting its typical wavelength in IR sections, (4) distinguishing different type of layers through commercial CMOS.

2 Characteristics of DITI and bola-type copolymer

2.1 High fraction of coverage in high-colour-density layer

Reversible thermochromic organic materials usually consist of at least three components: a colour former, colour developer and solvent[2,7]. Common colour formers are electron-donating, leuco dyes,. As these compounds change colour when pH is altered, they are in reality, halochromic. Frequently used examples include crystal violet lactone and thymolphthalein. Their reaction with the second component, the electron-accepting developer, defines the position of the longest wavelength absorption which causes colouration/decolouration[3,8]. For better understanding, we create a microcapsule simulation with low colour density, which is shown in figure 1.

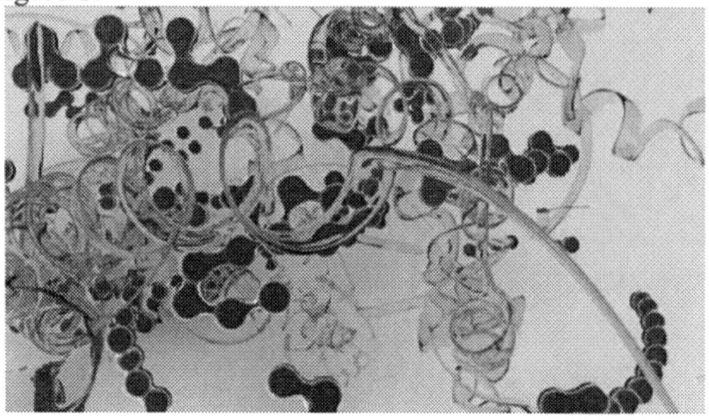

Figure 1. Model simulation made with Cinema 4D

The thermochromic pigment particles are analysed by using a Hitachi su70 field emission SEM. As only the particles locate at the very top of a dry screen-printed sample can be observed, more particles were made visible when the topmost layer of the binder was selectively etched by weakly ionised highly dissociated oxygen plasma. According to the previous study, plasma was created in a glass Pyrex tube with an inductively coupled RF generator operating at a frequency of 27,12 MHz and an output of 200 W. The oxygen pressure was 75 Pa[9]. The procedure is based on selective interaction of activated gaseous particles with the film coating. Selectivity of the etching process was achieved due to different oxidation probability of the polymer binder and TC pigments. The method is frequently used to detect distribution and orientation of particles in polymer matrix. The removal of the highest layers of polymer binder in UV-cured TC printing inks was completed in a couple of minutes. This way the top-laying TC pigment particles become visible on SEM micrographs.

Figure 2. DITI painted layer (without reform)

Although the ink has similar pigment particle size distribution and comparably thick microcapsule shells with stability against oxygen plasma etching, we still need to develop a certain pattern that can protect the capsule from being oxidized, due to the unique electron-accepting developer decorated with IR perssad. The SEM of DITI painted layer (without reform) is schematically shown in figure 2.

Conventional pigments are relatively stable against activated neutral oxygen radicals which are responsible for selective etching in weakly ionised oxygen plasma. The situation is different in systems with DITI ink: under some proportion of oil-type IR reflective perssad, DITI ink and additive can caused an aggregation. When they are destroyed, pigment particles change simultaneously so the method cannot be applied to evaluate the particle size distribution. However, this experiment shows that polymer envelopes are much more stable against selective etching than the binder itself; the binding energy of the molecular species in the polymer protecting capsule is much larger than it is inside the binder. Thus, a more efficient way to protect the binder, which contains with IR reflective perssad is the eye to see this system activated. We develop a prereaction by using DCPyn to protect the system from unwanted reactions with their surroundings. The SEM of DITI painted layer (with DCPyn) is schematically shown in figure 3.

Figure 3. DITI painted layer (with DCPyn)

2.2 Instant-high-temperature colour recovery

Previous work had pointed out that the colourimetric properties of the reversible process were tested up to the temperatures at which a short exposure should cause no long-lasting colour change (up to 140°C, according to the producer data [10,11]). As a condition of our handwriting or painted DITI to remain stable when being used, we measured several copolymers mixed with DITI in Pre-Experiment. The size of the largest particles in DITI inks was 12 μm; this is the so-called grindometer value. The DITI mixture was first added with DCPyn and thenpainted on PVDF membrane. The thickness of the dry layers was measured using a Talysurf (Taylor Talysurf i120). Those measurements finally led us to the found of new utilization in DCPyn, and they also brought us to an unrecorded phenomenon. We found that if the largest pigments could fill the entire thickness three times of dry layer (averagely 20μm), the dry layer can recover from the low-colour-density states.

Reversible colour change occurs via two competing reactions namely that between dye and developer and the other between solvent and developer. The first of these interactions prevails at lower temperatures where the solvent exists in its solid form and gives rise to coloured dye developer complexes; at higher temperature the solvent melts, causing the solvent developer interaction to dominate[12-15]. Thus, dye developer complexes are destroyed which converts the system into its colourless state. Leuco dye developer solvent composites are by far the most important systems to achieve thermochromic properties using organic materials. The temperature at which decolour- isation/colourisation occurs is controlled by the melting temperature of the solvent. We think the Instant-high-temperature colour recovery may be caused by the catalytic system of DCPyn, electron-donating, electron-accepting and PVDF membrane. The longest heating time of the accurate of colour recovery phenomenon is related to the proportion of DCPyn. It also confirmed that large proportion of DCPyn (about 4.5%)will Reduce the number of decolourisation/colour-isation reaction cycle.

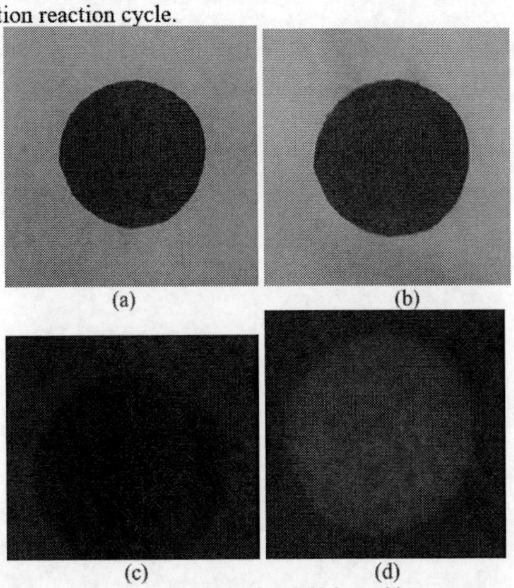

Figure 4. Images of (a) unheated and (b) heated layers captured with SLR camera. Gray scale images captured with band-pass filter of 530 ± 60 nm when the layer is (c)unheated or (d)heated.

3 Properties of written DITI layer

3.1 Sensitivity

Here, the final expected application is mostly used by ordinary mobile terminals. We used RGB colour models to measure the sensitivities of different types of DITI. The emission images were captured by several mobile terminals that contain different type of CMOS and a commercial SLR camera (Canon 760D).

To get the data of its typical wavelength in IR sections and test the stabilization of the scanning recognition system, we attached Infrared filters to the lens of the commercial SLR camera (Canon 760D). As shown in figure 4 the emission images can be separately detected by the Infrared filters. We confirmed that the isolated ink mark of the two states (colourless and coloured) can be recognized by mobile terminals and can be recognized even better if the incident spectrum contains infrared.

4 Conclusion

We have proposed a DITI system. The heated low-colour-density layer can be finely distinguished and adjusted by the scanning (CMOS) system we built. By carefully discussing the setting of ink components, the thermochromic microcapsules is peacefuly stacked on the surface of DITI layer, leading to prevent the unheated DITI from being written or printed at a low-colour-density rank. The characteristics of DITI on IR reflection is a promise of commercial distinguish. Special Emission spectra in IR sections is another key to accomplish the quick-direct distinguish by mobile terminals. We also endued DITI with some tricky characteristics by adding bola-type copolymer(DCPyn), such as colour restoration, high-colour-density structure and isolation from being affected by the IR perssad. These characteristics mentioned above perfectly match the needs in electronics identification and high functioning paper work, which could also be a promise of sustainable connection between electronic office and traditional office.

Acknowledgments

This work was financially supported by Carbon Fiber Engineering Research Center of Shandong Province in China, and by the integration projects of Shandong University. Both are gratefully acknowledged.

References

1. J W Gregory, Porous pressure-sensitive paint for measurement of unsteady pressures in turbomachinery 2004, *AIAA* Pap. 0294.
2. J W Gregory, K J Disotell, D Peng, T J Juliano, J Crafton and N M Komerath, Inverse methods for deblurring pressure-sensitive paint images of rotating Surfaces 2014, *AIAA J.* 52 2045–2061.
3. Maclaren DC and White MA. Dyeedeveloper interactions in the crystal violet lactone-lauryl gallate binary system: implications for thermochromism. *Journal of Materials Chemistry* 2003 13 1695e700.
4. Maclaren DC and White MA. Competition between dyeedeveloper and sol-ventedeveloper interactions in a reversible thermochromic system. *Journal of Materials Chemistry* 2003 13 1701e4.
5. Zhu CF and Wu AB. Studies on the synthesis and thermochromic properties of crystal violet lactone and its reversible thermochromic complexes. *Thermochimica Acta* 2005 425 7e12.
6. Maclaren DC and White MA. Design rules for reversible thermochromic mixtures. *Journal of Materials Chemistry* 2005 40 669e76.
7. Seeboth A, Klukowska A, Ruhmann R and Lötzch D. Thermochromic polymer materials. *Chinese Journal of Polymer Science* 2007 25 123e35.
8. White MA. Thermochromism in commercial products. *Journal of Chemical Education* 1999 76 1201e5.
9. Mozetic M. Controlled oxidation of organic compounds in oxygen plasma. *Vacuum* 2003 71 237e40.
10. Vesel A, Mozetic M, Hladnik A, Dolenc J, Zule J and Milosevic S. Modification of ink-jet paper by oxygen plasma treatment. *Journal of Physics D-Applied Physics* 2007 40 3689e96.

11. Kunaver M, Klanjsek Gunde M, Mozetic M and Hrovat A. The degree of dispersion of pigments in powder coatings. *Dyes and Pigments* 2003 57 235e43.
12. Y. Matsuda, T. Uchida, S. Suzuki, R. Misaki, H. Yamaguchi and T. Niimi. Pressure-sensitive molecular film for investigation of micro gas flows. *Microfluid Nanofluid*. 2011 165–171.
13. Ma C, Xia Y. *Colloids Surf* 1992 66 215.
14. Ma Z., Chen M and Glass J E. Colloids Surf 1996 112.
15. Y. Egami, K. Fujii, T. Takagi, Y. Matsuda, H. Yamaguchi and T. Niimi. Reduction of temperature effects in pressure-sensitive paint measurements. *AIAA J.* 51 2013 1779–1783.

Perceived Affordances in Older People with Dementia: Designing Intuitive Product Interfaces

Li-Hao Chen[1], Yi-Chien Liu[2] and Pei-Jung Cheng[3]

[1] *Department of Applied Arts, Fu Jen Catholic University, No. 510, Zhongzheng Rd., Xinzhuang Dist., New Taipei City, Taiwan*
[2] *Neurology Department, Cardinal Tien Hospital, No. 362, Zhongzheng Rd., Xindian Dist., New Taipei City, Taiwan*
[3] *Department of Media Design, Tatung University, No. 40, Zhongshan N Rd., Taipei City, Taiwan*

Abstract. The objective of this study was to examine the relationship between intuitive interactions and perceived affordances in product interfaces among older people with dementia. Experiments were conducted by examining the cooking time in the user interfaces of microwave ovens, and the participants involved 25 older people with mild dementia. The results showed that a simple arrangement of bottoms was more desirable for presenting visual information, but the usabilities of these interfaces were undesirable.

1. Introduction

Dementia is a neurodegenerative condition among the elderly. In most patients, dementia begins with the atrophy of the hippocampus, which is closely associated with memories. Patients in the early stages of dementia often cannot remember recent events. This study started with improving the intuitiveness of product interface designs can alleviate the problems that older patients with dementia are faced with in operating common everyday products. This enables such patients with dementia to experience a more independent live and decreases the burden of their family members and relevant healthcare personnel, thereby enhancing the quality of life.

Of the various theories related to user–product interactions, the theory of affordance proposed by perceptual psychologist Gibson [1] is dissimilar to theories in the fields of linguistics, cognitive sciences, and human factors engineering. Specifically, it pertains to a design perspective that emphasizes direct correspondence between perception and intuitive use and has become increasingly important in the field of human-machine interactions. The concept has also been applied to virtual interface designs. According to related literature, the perception of affordances, when applied to design, involves two dimensions, namely the observer's motor skills [2, 3] and experiences and culture [4, 5]. The perceived affordances of users are affected by their physical motor skills and by prior experiences and knowledge. Blackler et al. [6] asserted that intuitive interactions are based on a user's past experiences and are both fast and non-conscious. To achieve intuitive usability, designers must understand user characteristics to accurately design elements that promote intuitive interactions. Immediate interactive behavior is produced when user ability matches the affordances of an object [7]. However, the motor and cognitive skills of older people with dementia differ from those of ordinary users. As their cognitive ability deteriorates or fails completely, such patients may rely increasingly on their innate perceptions to interact with the external environment. The perspectives of direct perceptions and intuitive, natural interactions noted by the theory of affordances are highly applicable to the design of user interfaces for older people with dementia and exhibit tremendous development potentials. Therefore, this study explored the intuitive usability regarding the affordances established between products and older people with dementia. In addition, the relationship between intuitive interactions and perceived affordances was examined.

2. Method

The objective of this study was to examine the intuitive usability of user interface elements and shapes for older people with dementia. Experimental testing was performed through the user interfaces of microwave ovens. The participants were older people with mild dementia. The experimental task is to input cooking time. Patients with mild dementia and acceptable communicative and cognitive functions were nominated by their physicians to participate in this study.

2.1. Test interfaces and experimental procedure

We examined current microwave oven models to compile different shapes and arrangements of buttons for inputting cooking time. Nine different test interfaces were created

(Figure 1) using graphics software to draw the basic contours of the buttons. These interfaces were differentiated by whether there was a border around the button, whether the button shape was round or square, and whether the buttons were arranged in one, two, or three columns. The test interfaces were installed on a tablet computer with a 10-inch touch screen. During the testing, these interfaces were displayed one at a time on the tablet. The test interfaces involved a custom program that recorded the amount of time a participant took to complete a task and determined whether the task was completed accurately.

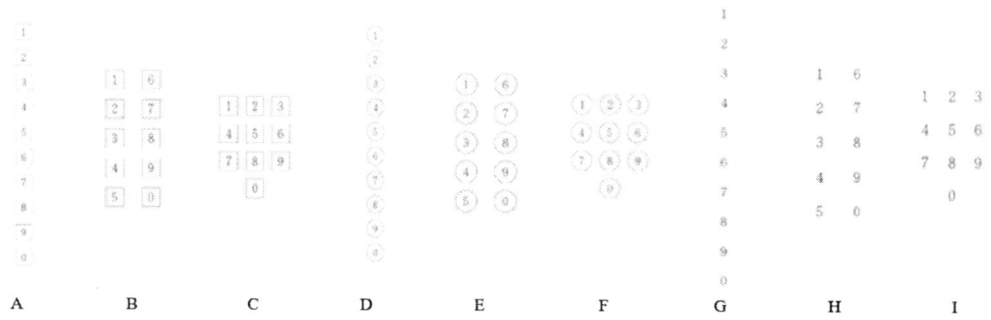

Figure 1. Test interfaces for inputting cooking time.

The custom program recorded two different times. First, the time elapsed before a participant touched the screen on the first attempt (i.e., the initial reaction time) was recorded to analyze whether the test interface features can elicit intuitive use. Second, the time elapsed before a participant completed the task (i.e, the task completion time) was recorded to analyze the usability of the test interface. At the beginning of the test, the participants each sat in front of a 10-inch touch-screen tablet and read the task instructions displayed on the screen. In the first test stage, the participants were asked to input a specified cooking time. If participants did not understand the written instructions, verbal explanations of the task were provided. After the participants confirmed that they understood the task, they manipulated the test interface on the tablet computer. No time limit was set for completing the task. The test interfaces for inputting cooking time were displayed one at a time in a random order on the screen. Video recording equipment was used to capture the entire test process. Some participants forgot the goal of the task or did not understand the test instructions. When this happened, researchers verbally guided the participants to complete the task.

2.2. Experimental task and participants

The experimental task is to input cooking time with no time limits imposed on the participants. The different cooking times were displayed in a random order, and the participants were asked to input the cooking times for each test interface. This test protocol was designed to reduce the learning effect to ensure the results more accurately reflected the intuitive operability of each test interface. After completing each task, the participants pressed a "confirm" button to signal that he or she was finished. The custom program then recorded the time elapsed.

To ensure that the participants did not forget the instructions of the cooking time task, the target cooking time was displayed in each test interface. The tests were performed in a hospital clinic. Hospital neurologists nominated 25 participants who had mild dementia, demonstrated acceptable communicative and cognitive functions, and were experienced in using home appliances. The participants averaged 81.8 years (SD = 7.2 years) of age. The definition of mild dementia was based on a score of 0.5 or 1 in the clinical dementia rating scale. Regarding cognitive skill assessments, patients who scored moderately low in short-term memory function, long-term memory function, and hand-eye coordination were chosen as the participants.

3. Results

3.1. Initial reaction times

For each of the nine test interfaces, the mean initial reaction time of the 25 participants was calculated and compared. As shown in Table 1, the mean reaction times for Interfaces C and F were relatively slow. The post-hoc ANOVA results also showed significant differences between the mean reaction times of Interfaces E, F, and H. Generally, the buttons arranged in a 3x3 configuration elicited lower levels of intuitive operability among the participants. Nonsignificant differences in reaction times were observed between the buttons arranged in one and two columns or between the buttons of different shapes.

3.2. Task completion times

For each of the nine test interfaces, the mean task completion time of the 25 participants were calculated. The average task completion times of the nine test interfaces were determined to be nonsignificantly different (Table 2). This indicated that the user interface designs with different button arrangements or shapes did not differ significantly in their usability levels.

Table 1. Anova results for the initial reaction times of the cooking time task.

Initial Reaction Times in the Cooking Time Task					
Interfaces	Average Time	SD	N	Sig.	p
A	8.12	4.95	25		
B	8.83	6.25	25		
C	9.48	7.76	25		
D	8.94	6.79	25		
E	7.85	5.57	25	*F	0.545
F	10.46	4.34	25	*E、H	
G	9.12	7.06	25		
H	8.02	3.26	25	*F	
I	9.00	5.36	25		

3.3. Observational analysis

Analyzing the observational data of the cooking time task (Table 3) compiled the number of participants who failed to complete the task accurately and the reasons for their failures. The most common reasons for the failing the task were (a) the participant did not understand how to input cooking time and therefore pressed the incorrect buttons (e.g., the participant did not understand that inputting a cooking time of 2:30 required sequentially pressing the buttons 2, 3, and 0); (b) the participant forgot the goal of the task and pressed the incorrect buttons; (c) the participant inadvertently pressed the incorrect buttons because of age-related cognitive decline; and (d) the participant did not successfully complete the task after failed attempts.

Table 2. Anova results for the completion times of the cooking time task.

Completion Times in the Cooking Time Task					
Interfaces	Average Time	SD	N	Sig.	p
A	19.90	12.53	25	A	
B	18.75	12.82	25	B	
C	18.53	13.44	25	C	
D	17.91	11.60	25	D	
E	17.17	12.41	25	E	0.942
F	19.26	9.63	25	F	
G	21.10	18.68	25	G	
H	19.10	8.88	25	H	
I	18.87	11.01	25	I	

Table 3. Observation records of the cooking time task.

Observation / Test Interface	Number of participants who failed to complete the task accurately	Percentage	Reasons
A	6	24%	• Inadvertently pressed the wrong button • Did not understand the task • Did not understand how to input cooking time
B	4	16%	• Did not understand the task • Inadvertently pressed the wrong button • Did not understand how to input cooking time • Did not complete the task
C	2	8%	• Did not understand the task • Inadvertently pressed the wrong button • Did not understand how to input cooking time
D	3	12%	• Did not understand the task • Did not understand how to input cooking time
E	2	8%	• Did not understand the task • Did not understand how to input cooking time
F	6	24%	• Did not understand the task • Inadvertently pressed the wrong button • Did not understand how to input cooking time
G	6	24%	• Did not understand the task • Inadvertently pressed the wrong button • Did not complete the task • Did not understand how to input cooking time
H	2	8%	• Did not understand the task • Did not understand how to input cooking time
I	3	12%	• Inadvertently pressed the wrong button • Did not understand how to input cooking time • Inadvertently pressed the delete button

4. Discussions

The study results showed that the interface with the 3x3 configuration of buttons attained lower levels of intuitive operability, but the mean task completion time using this interface differed nonsignificantly from those of the other interfaces. In addition, Interfaces A, E and H were more likely to entice participants in their initial encounters with the interfaces, but the usabilities of these interfaces were undesirable. This indicated that a simple arrangement of bottoms was more desirable for presenting visual information, thereby facilitating older users to process the information more effectively; hence, their initial reaction times were reduced. However, the different button arrangements and shapes did not result in significant differences between the task completion times. This finding accords with those of Chen and Liu [8]. Therefore, the present study confirmed that although a simple and clear arrangement of interface elements resulted in more intuitive reactions among the users, these interfaces were not necessarily more intuitive to use. Analyzing the observational data showed that the most common reason for user mistakes was that the participants did not understand how to input the cooking time. Blackler et al. [6] asserted that intuitive interactions are based on a user's past experiences and are both fast and non-conscious. Thus, a product with operating styles that match user experiences can enhance the usability of the product.

5. Conclusion

This study examined intuitive interactions between older people with dementia and user interfaces from the perspectives of perceived affordances. The results showed that simple user interfaces are more likely to elicit intuitive reactions among older people with dementia, but this does not necessarily equate to facilitating the intuitive use of such interfaces. The prior experiences of users are crucial to intuitive use; hence, user interfaces should be designed to comply with users' prior experiences. Understanding user characteristics is the key to designing elements that promote intuitive interactions. However, older people with dementia may experience temporary short-term or long-term memory loss because of their disease condition, causing them to forget how to operate once-familiar products. Therefore, designers may experience difficulty accounting for the prior experiences of this user group, and they may need to instead examine the direct relationship between users and product external features noted in the theory of affordances.

In the present study, we tested and analyzed the effective data of 25 older patients with mild dementia. The findings enabled proposing a model describing the perceived affordances between in the intuitive interactions of product interfaces and older people with dementia, despite participant selection being a notable limitation of this study. In addition, the physicians advised us to avoid subjecting the participants to complex interface operating procedures and a high amount of test interfaces. Therefore, we could not adopt comprehensive operating procedures, such as asking the participants to place food in the microwave, turn on the microwave, input cooking time, adjust cooking power, and wait for the food to finish cooking. The experimental tasks were designed to involve abbreviated instructions. Subsequent studies are suggested to increase the number of participants and examine additional tasks and user interfaces, thereby providing research implications for designers to create user interfaces that are intuitively interactive for older people with dementia.

Acknowledgements

This research was supported by the grant from the Ministry of Science and Technology, Taiwan, Grant MOST-106-2410-H-030-057.

References

[1] Gibson J J 1979 *The Ecological Approach to Visual Perception* (Boston: Houghton Mifflin Company)
[2] Gero J S and Kannengiesser U 2012 Representational Affordances in Design, with Examples from Analogy Making and Optimization. *Research in Engineering Design* vol. 23, No.3 pp 35-49
[3] You H C and Chen K 2007 Application of affordance and Semantics in Product design. *Design Studies* vol. 28, No.3 pp 22-38
[4] Cooper A 1995 *About Face The Essentials of User Interface Design* (New York: IDG Books Worldwide)
[5] Krippendorff K 2006 *The Semantic Turn: A new foundation for design* (Boca Raton: Taylor & Francis).
[6] Blacker A, Popovic V, Mahar D, Lawry S, Gudur R, Kraal B and Chamorro- Koc M 2011 Researching Intuitive Interaction. *Proc. IASDR2011 Int. Conf.* pp 1-12
[7] Drewitz U and Brandenburg S 2010 From Design to Experience: Towards a Process Model of User Experience. *Proc. 9th Pan-Pacific Conference on Ergonomics Int. conf.* (Taiwan: Kaohsiung/CRC Press) pp 117-122
[8] Chen L H and Liu Y C 2013 Designing Intuitive Interfaces with Affordances for Older People with Dementia. *Proc. 2013 Orange Beneficence— International Design Conference of Smart Living for Elderly.* (Taiwan: New Taipei City/Fu Jen Catholic University) pp142-149.

Fault Identification in Pipeline System Using Normalized Hilbert Huang Transform and Automatic Selection of Intrinsic Mode Function

Hanafi.M.Yusop[1], M.F.Ghazali[1], M.F.M.Yusof[1], M.A.Pi Remli[1], M. Samykano[1]

[1]*Advanced Structural Integrity of Vibration Research (ASIVR), Faculty of Mechanical Engineering, University Malaysia Pahang'26600 Pekan, Pahang, Malaysia*

Abstract. Pressure transient analysis has been widely used to monitor the condition of pipelines and its assessment in water distribution systems. This is a low-cost nonintrusive technique with the ability to locate uncertainties (leak, pipe fitting, blockage) at a greater distance from the measurement point. In this research, Normalised Hilbert Huang Transform (NHHT) is used as the method to analyse the pressure transient signal. However, this method has difficulty in selecting the suitable intrinsic mode function (IMF) for the advance data analysing. As an alternative, Integrated Kurtosis-based Algorithm for z-filter Technique (Ikaz), which allows automatic selection of intrinsic mode function (IMF) been used to substitute the NHHT limitation in this study. The analysis is conducted on a 67.9-meter Medium High-Density PolyEthylene (MDPE) pipe installed with single artificial leak simulator at a water pressure in the range of 1-4 bar.

1 Introduction

Water management is essential to ensure every drop of water can be distributed to the right places. The loss due to improper management is defined as non-revenue water (NRW). The problem of water leakage in pipeline distribution system is the primary cause of Non-revenue water (NRW).

Signal processing is the most famous method among the researchers due to its ability to detect and localize any disturbance (leak) and blockage in the pipeline system. Pressure Transient analysis is the most recently developed method and attracted researchers due to their cost-effectiveness. Theoretically, transient phenomena are generated due to sudden changes in the fluid propagation filled in the pipeline's system caused by rapid pressure and flow fluctuation in the systems such as rapid closing and opening valve. The wave characteristic represents the pipeline information and conditions from various boundaries such as pipe feature, junction outlet and the presence of the leak. Signal analysis is a way to extract the characteristics, identity, and information of pressure transient signal captured along the pipeline distribution system.

The possibility to analyse the phenomenon in both time and frequency domain has been satisfied by implementation the time-frequency signal processing techniques such as Normalised Hilbert-Huang Transform (NHHT). NHHT analysis has been proposed in many studies due to the ability of this method to detect and capture transient phenomena that occur in the non-stationary signal. In recent years, HHT has been widely used in time-frequency analysis as signal processing method [1], [2]. This method also has difficulty in selecting the suitable IMF for advance data post-processing method known as Normalized Hilbert Transform (NHT). NHHT through Empirical Mode Decomposition (EMD) will decompose the signal into the different monocomponent and symmetric component by the sifting process. The monocomponent signal is termed as an intrinsic mode function (IMF) [3]. The main drawback of this approach is it necessitates to know the prior frequencies and level of original signal that required to be analysed [4]. Therefore, computational mathematical technique and statistical value analysis have been used by the researchers to increase the degree of automation, which eliminates the interaction of the skilled user to select the relevant IMF.

This paper proposes an Integrated Kurtosis Algorithm for Z-notch (Ikaz) filter hybridized with higher order statistical method (Kurtosis) for automatic selection of intrinsic mode function (IMF). The Ikaz method was chosen since this method is adaptive in general and powerfully detects any changes and uncertainties in the measured signal [5]. Unlike the existing statistical analysis such as variance, standard deviation, and kurtosis, Ikaz method capable of indicating both amplitude and frequency difference simultaneously by obtaining the Ikaz representation and Ikaz coefficient, $Z\sigma$. [5] and [6] has performed a detailed analysis using Ikaz on the vibrational signal acquired from rotating machinery part (Bearing) to detect tool wear during the turning process. For more details refer to Nuawi et al 2014 [5].

This paper also focuses on the implementation of Normalized Ikaz coefficient for automatic selection of intrinsic mode function using real pressure transient signal captured along the 67.9-meter medium density polyethylene pipe with single artificial leak simulator attached at a distance of 19.7 meters from the point of analysis. The result for the real pressure transient signal shows that the IMF with the highest value of Normalized Ikaz coefficient is suitable and appropriate for further analysis. Furthermore, the Normalized Ikaz also proved to be suitable as an automatic selection method in choosing better and relevant IMF. Therefore, the degree of automation for Normalised Hilbert Huang Transform (NHHT) has been improved to the detect pipe leakage in live water distribution system using pressure transient signal.

2 Integrated Kurtosis Algorithm for Z-Filter (Ikaz) Technique to Kurtosis Ratio

2.1 Integrated Based Kurtosis Algorithm for z-filter (Ikaz) technique

Ikaz method is developed based on the concept of data scattering to its centroid. The sampling frequency of the raw signal is chosen as 2.56 referring to Nyquist number [15]. Most of the researchers in signal analysis and processing are at ease with this number. To avoid the aliasing effect, the maximum frequency span is chosen using equation 1.

$$F_{\max} = \frac{fs}{2.56} \tag{1}$$

Ikaz decomposes the time domain signal into three level of the frequency range. The x-axis represents low frequency (LF) with a range of 0 – 0.25 of f_{max}, followed by y-axis which represents high frequency (HF) with a range of 0.25 – 0.5 of f_{max} and the third, z-axis represents very high frequency (VF) with range 0.5 of f_{max}. The 0.25 f_{max} and the 0.5 f_{max} been selected as low and high-frequency range limit respectively by referring to the 2nd order of Daubechies concept in signal decomposition process [17]. Ikaz method also contributes to the three-dimensional graphical representation of the measured signal frequency distribution. The variance (σ^2) of each frequency band; σL^2 represent a low-frequency band, σH^2 represent a high-frequency band and σV^2 represent a very-high-frequency band which calculated from equation 2, 3 and 4 to measure the scattering from data distribution.

$$\sigma L^2 = \frac{\sum_{i=1}^{N}(x_i - \mu_L)^2}{n} \tag{2}$$

$$\sigma H^2 = \frac{\sum_{i=1}^{N}(x_i - \mu_H)^2}{n} \tag{3}$$

$$\sigma V^2 = \frac{\sum_{i=1}^{N}(x_i - \mu_V)^2}{n} \tag{4}$$

The Ikaz coefficient, $Z\sigma$ can be simplified in term of variance, σ as in equation 5 [16].

$$Z\sigma = \sqrt{(\sigma L^2)^2 + (\sigma H^2)^2 + (\sigma V^2)^2} \tag{5}$$

2.2 Kurtosis

Kurtosis is a measure of peakness and hence is a good indicator for spikes detection in a non-stationary signal component such as pressure transient. Kurtosis is expressed as;

$$\text{Kurtosis}(x) = \frac{E\{(x-\mu)^4\}}{\sigma^4} \tag{6}$$

where, μ and σ represent the mean and standard deviation of time series signal respectively. The E illustrate the expectation operation. The kurtosis demonstrates the spikiness and peakness of the probability distribution associated with the instantaneous amplitudes of the time-series analysis [18]. Ikaz-kurtosis ratio is expressed in equation 7.

$$ZK\sigma = \frac{(\sqrt{(\sigma L^2)^2 + (\sigma H^2)^2 + (\sigma V^2)^2})(\sigma^4)}{(E\{(x-\mu)^4\})} \tag{7}$$

2.3 Algorithm Automatic Selection of Intrinsic Mode Function (IMF) Based on Pressure Transient Signal

Automatic Selection of Intrinsic Mode Function (IMF) Based on Pressure Transient Signal

1. Perform EMD on a pressure transient signal
2. Find local maxima of the absolute value of IMF (to take advantages of using both upper and lover envelopes) and fix the end values as maxima to ameliorate the end effect.
3. Construct a Spline Envelope (SE) through the maxima.
4. When envelope goes under the data, straight line envelope will be used for that section of the SE.
5. EMD decomposed the signal into so-called level of Intrinsic Mode Function
6. Calculate the Integrated Kurtosis Algorithm for z-filter technique to kurtosis ratio (Ikaz-Kurtosis) coefficient using Eq. 7.
7. Identified the IMF that corresponds to the largest Ikaz-kurtosis ratio coefficient.
8. Perform Hilbert Transform (HT) and Hilbert Spectrum (HS) for IMF level that contain the largest coefficient of Ikaz-Kurtosis ratio.
9. Normalize the data using SE : N-data = Data/SE. This steps can be repeated.
10. Compute IF (FM) and Absolute Value (AV) from Hilbert Transform of N-data.
11. Definition: Error Index = $(AV-1)^2$.
12. Compute Instantaneous Frequency for SE (AM).

3 Methodology

In this work, the response of the signal generated by solenoid valve is acquired using a pressure sensor. During the experiment, the solenoid valve is set to a normal closed condition, and the water hammer phenomena were induced during the opening and closing of the valve for three consecutive times. The experimental data is captured using MatLab Software.

3.1 Experimental Setup

The experimental test was performed in a circular loop pipeline system with 67.90-meter long. The network was constructed using Medium High-Density Polyethylene pipes (MDPE) with an outer diameter of 60 millimeters, the internal diameter of 55 millimeters and 2.6 millimeters mean thickness. An artificial leak simulator was installed at 19.7 meters from point of analysis in the pipeline system. The outlet of the pipe is kept connected to a free surface tank where the water from the pipe ends discharged. This is to prevent sudden expansion phenomena due to the pressure waves, and to minimize the adverse pressure wave, as it will affect the data in the transducer. The calibrated speed of sound from the test rig is 524.3005 m/s.

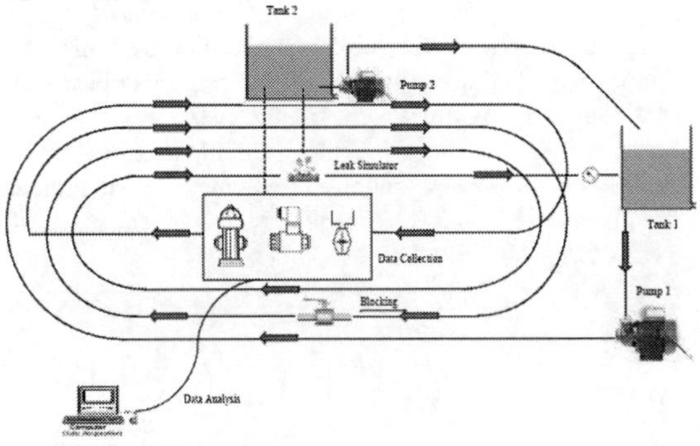

Figure 1: Schematic Diagram of Test Rig design for laboratory and experimental test

The Figure 1 schematic diagram shows the pressure sensor and solenoid valve which been installed 10 meters from the electric pump to avoid the noise during collection the data. This noise is due to the turbulent flow wave characteristic when the water flows out from pump outlet; the flow the approaches to laminar or steady state condition as it travels further from the pump. This phenomenon also generates friction to the wall pipe. Therefore 10-meter-long distance from pump to the sensor is adequate to reduce the noise. This method is commonly practiced in the pipeline leakage detection methods. The idea was adopted from the usage of pressure transient flow of water in the pipeline network. Theoretically, the pressure transient response occurs when a sudden change in the flow of water by closing or opening the valve in the pipeline network. As these phenomena happen inside the pipeline network, a wave propagation of water is expected to be created along the pipeline network, and also the characteristic of wave propagation will be used as a medium to detect leak and pipe feature along the pipeline network. The wave inside the pipeline network will propagate with different signal depends on the size of the leak, distance of leak, the type of pipe feature and distance of pipe feature.

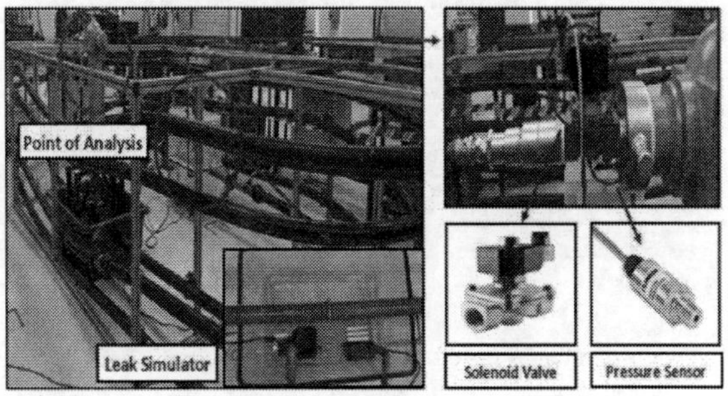

Figure 2: Experimental test rig in laboratory scale

Figure 2 shows the experimental test rig installed in Fluid Mechanic Laboratory at University Malaysia Pahang, Malaysia. The fire hydrant cap was designed and fabricated to fix the solenoid valve and pressure sensor as well as to make the system robust for field test. Both components play an essential role, the solenoid valve is used to generate the water hammer phenomenon and the pressure sensor is used to acquire the signal. In a few cases, the system shut down on its own due to sudden opening or closing valve and malfunction of the pump. This will generate "water hammer" occurred on the pipeline system. This phenomenon commonly happens in all pressure pipe systems, which causes substantial vibration and destruction to the pipeline system [19]. To generates this phenomenon, a solenoid valve (Figure 2) is utilized as a tool to create water hammer along the pipeline system. This situation creates the same phenomenon happens in the real system when there is a pressure disorder occur at the underground pipe.

4 Results and Discussion

The experimentation begins with acquiring the pressure transient signal from test rig using Matlab. The signal acquired retrieves the characteristics of the whole pipeline system itself. The pressure transient waves propagated along the pipeline system in both directions, away and from the burst origin through the speed of sound in water distribution system. In this work, Normalized Hilbert-Huang Transform (NHHT) which contains Empirical Mode Decomposition Method (EMD) data pre-processing method and Normalized Hilbert Transform (NHT) data post-processing method is employed as the tool to analyze the pressure transient signal.

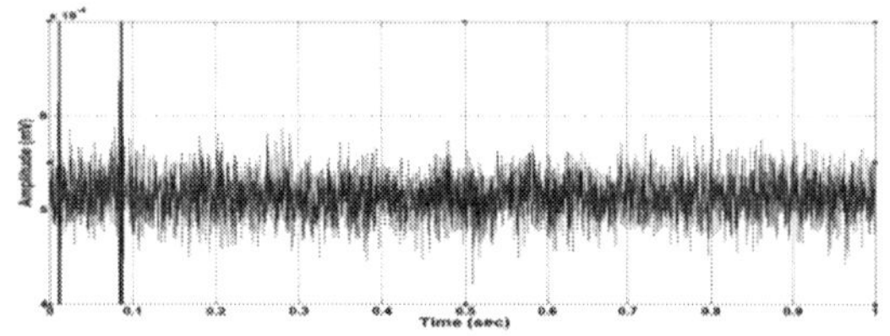

Figure 3: Raw Data for Leak and No-Leak Water Pipeline

Figure 3 illustrates the signal response from the leak and no-leak pipeline system. The first two maximum amplitudes on the figure represent the opening and closing of the solenoid valve which produces the pressure transient signal, which is transmitted along the pipeline system and reflected back due to the disturbance. The original response was recorded at two different levels of pressure: 1 bar and 2 bar. The signals captured at a different level of pressure demonstrated different results. Higher pressure creates a higher frequency of the signal and this causes for the higher amplitude of the signal. The raw signal was then analysed to extract the information contained in the signal. In this research, NHHT was used as the signal analysis tool and Normalized Ikaz was used as a tool to automatically select the right IMF. The analysis using empirical mode decomposition (EMD), decomposed the signal into level 13 of intrinsic mode function (IMF). The first level of IMF is a group of higher frequency signal which is noise signal. The last level was a reserve for the lower frequency signal. Usually, the first and second levels of IMF are not considered for further analysis because these levels contain the noise frequency signal. Meanwhile, IMF level 7 and the rests contain unnecessary response of the network. All these IMF were therefore discarded. The rest, IMF level 3-6 are recombined to produce a signal without noise (Yusopet al 2017). To identify the right IMF automatically without the interaction of the skilled user, normalized Ikaz is used as an advanced statistical method for autonomous selection of the right IMF. The value of normalized Ikaz coefficient computed for each level of IMF values is presented in table 1.

Table 1: Normalized Ikaz Coefficient

IMF	Kurtosis Coefficient	Ikaz Coefficient	Normalized Ikaz Coefficient
1	2429.65	7.10e-09	2.92e-12
2	919.59	3.64e-09	3.96e-12
3	535.22	1.03e-09	1.94e-12
4	63.67	2.87e-09	4.52e-11
5	26.48	3.83e-09	1.44e-10
6	16.92	6.10e-10	3.60e-11
7	9.31	2.06e-10	2.21e-11
8	4.47	1.19e-10	2.67e-11
9	2.65	5.62e-11	2.11e-11
10	2.58	3.83e-11	1.48e-11
11	1.46	9.56e-13	6.50e-13
12	2.14	4.62e-19	2.15e-19
13	2.9	2.74e-12	9.44e-13

Table 1 illustrates the comparison among kurtosis, Ikaz and Normalized Ikaz coefficient for each level of IMF. From the observation, IMF level 1 has the highest kurtosis and Ikaz coefficient. By referring to [20], the first and second level of IMF contains the signal noise. Therefore, in this research, Ikaz and kurtosis coefficient is not suitable to be used for automatic selection of IMF. Besides that, IMF 5 contains the highest Normalized Ikaz coefficient and by referring to [20], this level is in the range of IMF 3-6 that should contain the signal without noise. To ensure IMF 5 has the highest Normalized Ikaz coefficient for consideration as the right IMF for further analysis, IMF 5 was subjected to further analysis using Normalized Hilbert Transform (NHT). The result of NHT is presented in Figure 4 and 5.

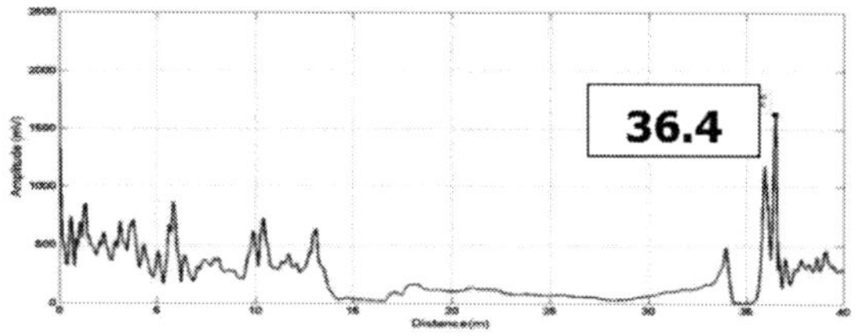

Figure 4: NHT for no-leak pipeline system

Figure 4 shows the result of Normalized Hilbert Transform (NHT) for no-leak pipeline system while Figure 5 for leak pipeline system. During this phase, non-stationary signal, in which frequency value changes at any moments, is more useful to characterize a signal in terms of its instantaneous frequency. Instantaneous frequency describes the frequency that locally fits the signal [20].

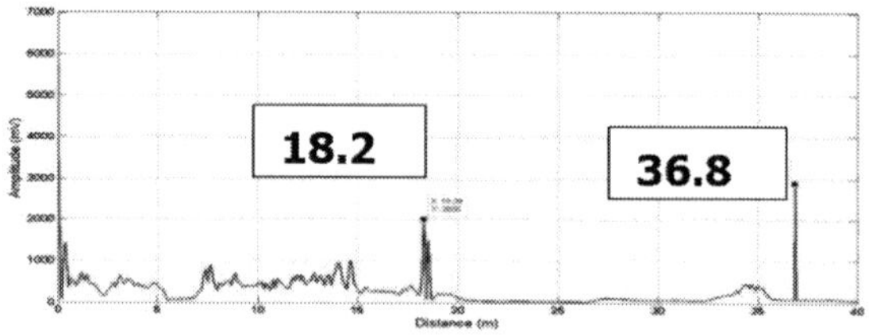

Figure 5: NHT for leak pipeline system

From the observation, it is evident that the tallest spikes and amplitude occurs at a distance 36.46 meters. By comparing to the test rig blueprint, this position is identified to be the outlet for pipes. Figure 5 shows that the tallest spikes and amplitude are at the distance of 18.29 meter and 36.88 meters. By comparing to the test rig blueprint, at distance of 18.29 meter identified to be the leak, while at the distance of 36.88 meters identified to be the outlet pipe. The experimental investigations and analysis were repeated with different pressure and the obtained findings are tabulated in Table 2.

Table 2: Comparison position of pipeline features between measured and experimental position

Water Pressure	Signal Response	Pipe Feature	Measured Position (m)	IMF contain Maximum Normalized Ikaz Coefficient	Experimental Position (m)	Error %
2 bar	Leak Data	Leak	19.7	5	18.29	7.6
		Outlet	39.8		36.88	7.3
	No Leak Data	Outlet	39.8	4	36.46	8.3
4 bar	Leak Data	Leak	19.7	5	19.24	2.3
		Outlet	39.8		37.08	6.8
	No Leak Data	Outlet	39.8	4	36.24	8.9

Table 2 illustrates the comparison of the position of pipe features between measure position and experimental position. For leak pipeline system data with 2 bar water pressure, the position of leak shows 7.6 % error and position of outlet show 7.3 % error. Meanwhile, for no-leak pipeline system data present 8.3 % error for outlet position. The final part of NHHT analysis evidently shows the position of pipe feature that is leak and pipe outlet appears at the experimental position close enough compared to each the measured position and theoretical calculation. It also demonstrates that NHHT analysis was able to detect and position the transient event occurrences in the non-stationary pressure transient signal.

5 Conclusion

This paper discusses self-decision method for IMF selection through Normalized Hilbert Huang Transform (NHHT) analysis by using a high-frequency piezoelectric pressure sensor. The results of the analysis demonstrate that the Normalized Ikaz is suitable and recommended as a self-decision method for IMF selection to implement in Hilbert-Huang Transform (NHHT) analysis. The development of automated self-decision of IMF through NHHT has been built and statistically analyzed using Normalized Ikaz. Therefore, this method is proposed and recommended to be implemented. Normalized Ikaz can be efficiently used to help in developing a real-time monitoring system for live water distribution system and can also be used to manage the issue of IMF selection. Therefore an efficient online monitoring system for leak detection can be developed. The results also verify the ability of piezoelectric pressure sensor in quickly detecting the change in response frequency of the pipeline system.

Acknowledgement

University Malaysia Pahang fully supports the facilities and resources for this research. The authors would like to acknowledge the support from the University Malaysia Pahang internal grant RDU170386 and Ministry of Higher Education Malaysia Grant (FRGS/1/2017/TK03/UMP/02/1).
The title is set in bold 14-point Arial, flush left, unjustified. The first letter of the title should be capitalised with the rest in lower case. You should leave 35 mm of space above the title and 6 mm after the title.

References

1. Lai, C., *Unaccounted for water and the economics of leak detection.* Water Supply, 1991. **9**(3): p. 4.
2. Taghvaei, M., S. Beck, and W. Staszewski, *Leak detection in pipelines using cepstrum analysis.* Measurement Science and Technology, 2006. **17**(2): p. 367.
3. Liou, C.P., *Pipeline leak detection by impulse response extraction.* Journal of Fluids Engineering, 1998. **120**(4): p. 833-838.
4. Ghazali, M.F., *Leak detection using instantaneous frequency analysis.* 2012, University of Sheffield.
5. Stephens, M.L., et al. *The detection of pipeline blockages using transients in the field.* in *South Australian Regional Conf.* 2002.
6. Huang, N.E., et al. *The empirical mode decomposition and the Hilbert spectrum for nonlinear and non-stationary time series analysis.* in *Proceedings of the Royal Society of London A: Mathematical, Physical and Engineering Sciences.* 1998. The Royal Society.
7. Shi, C.-x. and Q.-f. Luo, *Hilbert-Huang transform and wavelet analysis of time history signal.* Acta Seismologica Sinica, 2003. **16**(4): p. 422-429.
8. Bin, G., et al., *Early fault diagnosis of rotating machinery based on wavelet packets—Empirical mode decomposition feature extraction and neural network.* Mechanical Systems and Signal Processing, 2012. **27**: p. 696-711.
9. Huang, N.E., *Hilbert-Huang transform and its applications.* Vol. 16. 2014: World Scientific.
10. Manjula, M. and A. Sarma, *Comparison of empirical mode decomposition and wavelet based classification of power quality events.* Energy Procedia, 2012. **14**: p. 1156-1162.
11. Ricci, R. and P. Pennacchi, *Diagnostics of gear faults based on EMD and automatic selection of intrinsic mode functions.* Mechanical Systems and Signal Processing, 2011. **25**(3): p. 821-838.
12. Maji, U., M. Mitra, and S. Pal, *Automatic Detection of Atrial Fibrillation Using Empirical Mode Decomposition and Statistical Approach.* Procedia Technology, 2013. **10**: p. 45-52.

https://doi.org/10.1051/matecconf/201822102002

13. Kedadouche, M., M. Thomas, and A. Tahan, *Monitoring machines by using a hybrid method combining MED, EMD, and TKEO*. Advances in Acoustics and Vibration, 2014. **2014**.
14. de Souza, D.B., J. Chanussot, and A.-C. Favre. *On selecting relevant intrinsic mode functions in empirical mode decomposition: An energy-based approach*. in *2014 IEEE International Conference on Acoustics, Speech and Signal Processing (ICASSP)*. 2014. IEEE.
15. Nuawi, M.Z., et al., *Development of integrated kurtosis-based algorithm for z-filter technique*. Journal of applied sciences, 2008. **8**(8): p. 1541-1547.
16. Rizal, M., et al., *A Comparative Study of I-kaz Based Signal Analysis Techniques: Application to Detect Tool Wear during Turning Process*. Jurnal Teknologi, 2013. **66**(3).
17. Daubechies, I., *Ten lectures on wavelets*. Vol. 61. 1992: SIAM.
18. Wang, Y., et al., *Spectral kurtosis for fault detection, diagnosis and prognostics of rotating machines: A review with applications*. Mechanical Systems and Signal Processing, 2016. **66**: p. 679-698.
19. Amin, M., et al. *Leak detection in medium density polyethylene (MDPE) pipe using pressure transient method*. in *IOP Conference Series: Materials Science and Engineering*. 2015. IOP Publishing.
20. Ghazali, M., et al., *Comparative study of instantaneous frequency based methods for leak detection in pipeline networks*. Mechanical Systems and Signal Processing, 2012. **29**: p. 187-200.

Reliability Implementation and Cost Trade-off of Product during the Development Process

Wei Wang[1], Yuan Li [1], Qi Zhang [1], Weijia Feng [1], Huichao Liu [1], Pengyan Liu [1]

[1] *School of Reliability and Systems Engineering, Beihang University, Beijing, China*

Abstract. Modern product needs to meet the reliability requirements during the development process. The reliability in this paper refers to an integral view of a product's reliability, maintainability, supportability, testability, safety and environmental adaptability. However, during the product development process, the two problems are how to evaluate the implementation and how to determine the work input costs of reliability. This paper proposes a method to evaluate the degree of reliability implementation. And it researches the schemes and targets decision-making method based on trade-off analysis. Through establishing and solving trade-off optimization model, the results can help decision makers find the optimal parameters program and cost goals.

1 Introduction

The developed reliability engineering integrates the separated engineering work such as reliability engineering, testability engineering and safety engineering and considers the interactions among each engineering work, making the product quality into a higher level [1]. During the development process, it is necessary to periodically evaluate the degree of completion of the reliability work aiming to ensure the product reliability meet the requirements. And the basic way of evaluation is expert review. Chen evaluates reliability of armored product by establishing evaluation index system based on RST-QFD-Bayes [2]. When choosing the development scheme for the product, determining the costs required of the reliability work and making trade-offs are critical to ensure the realization of reliability of the product more economical. The CAIV method (cost as an independent variable) proposed by the US military is a more advanced cost management method nowadays and the trade-off method part is used for reference in this paper [3-5].

2 Reliability implementation evaluation

The purposes of the reliability implementation evaluation are evaluating the degree of completion of the reliability work and providing the reliability implementation model for trade-off analysis.

2.1 Reliability implementation system

The reliability implementation system is established as Figure 1 by considering the various factors that affect the reliability implementation and the relationships among them based on the evaluation index system construction principle [6]. In order to establish the system, the reliability requirements of product and the reliability work specified in national military standards are analyzed firstly. Then the factors affecting the realization of reliability, maintainability, supportability, testability, safety and environmental adaptability are extracted and merged. Finally, the correlation among the factors is considered and the structure of reliability implementation system is determined.

2.2 Reliability implementation evaluation based on AHP

The analytic hierarchy process (AHP) which has the special advantage in multi-criteria evaluation is very suitable for the evaluation of this paper [6]. The evaluation process of reliability implementation based on AHP is as follows. First, establish the evaluation indexes of reliability implementation. Second, determine the weights of the first-level assessment index and the second-level assessment index. Third, calculate the evaluation value of the first-level assessment index based on the scores of the experts. Forth, calculate the evaluation value for reliability implementation degree.

2.2.1 Evaluation indexes for reliability implementation

The evaluation indexes are established based on the factors of reliability implementation system. The evaluation indexes are shown in Figure 2. The evaluation indexes include two layers. The first-level assessment index includes the six assessment indexes: organization, product and facilities, management and controlling, reliability design and analysis, reliability test and assessment, reliability attainment and improvement, which are denoted as X_1, X_2, X_3, X_4, X_5, X_6. In the second-level assessment index, organization includes the three second-level assessment indexes: staffing, agency, post, which are denoted as R_{11}, R_{12}, R_{13} respectively. Product and facilities includes the three second-level assessment indexes: facilities and environmental conditions, product allocation, product maintenance, which are denoted as R_{21}, R_{22}, R_{23} respectively. Management and controlling includes the eight second-level assessment indexes: plan, procedure controlling, review management, closed-loop controlling and management, information management, hierarchical management, state management, material controlling and management, which are denoted as R_{31}, R_{32}, R_{33}, R_{34}, R_{35}, R_{36}, R_{37}, R_{38} respectively. Reliability design and analysis includes the six second-level assessment indexes: reliability design guidelines, reliability model, reliability distribution and forecast, reliability failure and hazard analysis, design and iteration, integrated design, which are denoted as R_{41}, R_{42}, R_{43}, R_{44}, R_{45}, R_{46} respectively. Reliability test and assessment includes the four second-level assessment indexes: engineering test, synthetic test, simulation, comprehensive evaluation, which are denoted as R_{51}, R_{52}, R_{53}, R_{54} respectively. Reliability attainment and continuous improvement includes the three second-level assessment indexes: qualification, out-field performance, continuous improvement, which are denoted as R_{61}, R_{62}, R_{63} respectively.

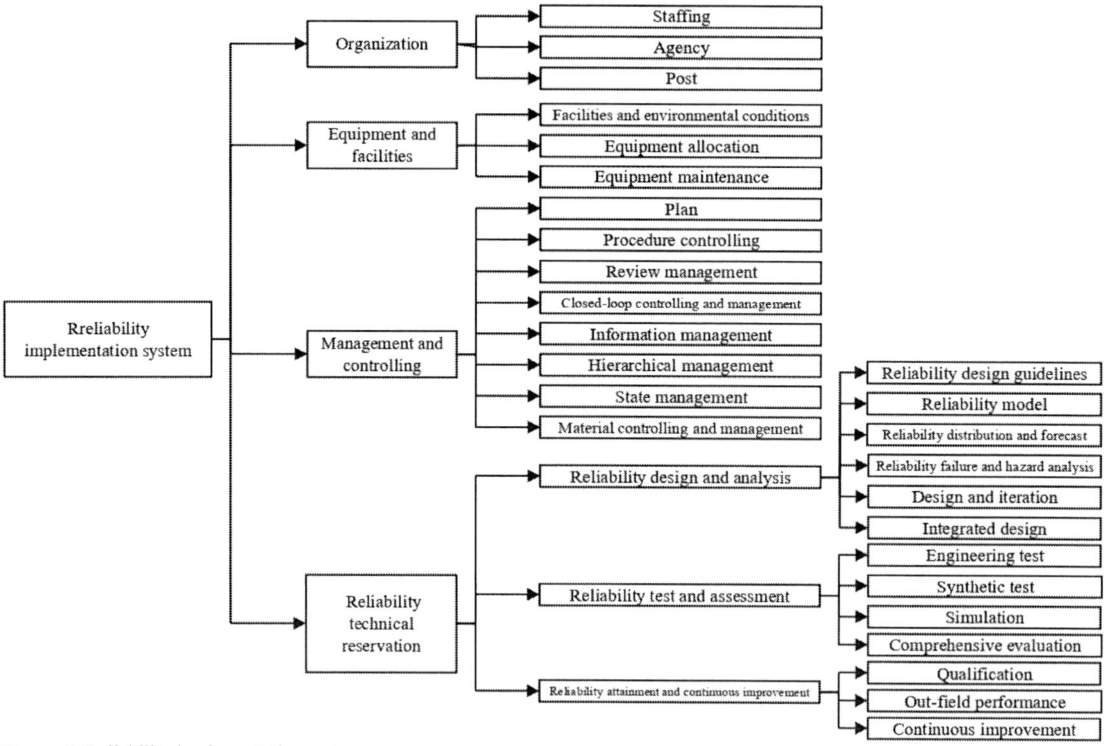

Figure 1. Reliability implementation system.

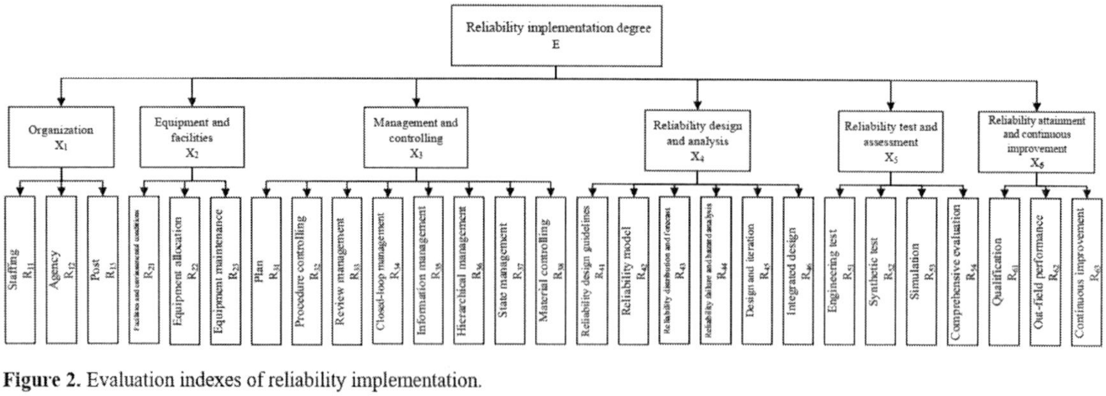

Figure 2. Evaluation indexes of reliability implementation.

2.2.2 Weight Calculation of the Assessment Index

This paper uses the AHP method to establish a weight set, which classifies various related factors in a complex problem into an interrelated and orderly hierarchy. Based on certain objective judgments, the first-level and the second-level assessment indexes are compared and analyzed. In the weight calculation of the two layers of assessment indexes, some experts are invited to construct the discriminant matrix by comparing the index among each other. 1~9 are used as comparison scales among evaluation factors. By solving these discriminant matrices, it can be calculated that the six weights of the first-level assessment index for reliability implementation degree which are denoted as $\omega_i(i = 1,2,3,4,5,6)$ and the weights of the second-level assessment index for the first-level assessment indexes which are denoted as $\omega_{ij}(i = 1,2,3,4,5,6; j = 1,2\cdots)$.

2.2.3 Calculation of evaluation model

The evaluation value for reliability implementation degree is calculated by values and weights of every evaluation index, which is expressed as:

$$E = \sum_{i=1}^{6} X_i\omega_i \quad (1)$$

where $\omega_i(i = 1,2,3,4,5,6)$ is the weight of the first-level assessment index; $X_i(i = 1,2,3,4,5,6)$ is the value of the first-level assessment index, which is expressed as:

$$X_i = \sum_{j=1}^{n_i} R_{ij}\omega_{ij} \quad (2)$$

where $\omega_{ij}(i = 1,2,3,4,5,6; j = 1,2\cdots)$ is the weight of the second-level assessment index; $R_{ij}(i = 1,2,3,4,5,6; j = 1,2\cdots)$ is the value of the second-level assessment index, which is scored by experts during the evaluation.

In the actual evaluation, the second-level assessment index need to be decomposed into the third-level or even fourth-level index to evaluate the values. In the subsequent related studies, we will give more details. The value of the second-level assessment index will reflect the completion degree in the form of percentage, which means from 0 to 100%.

2.2.4 Result analysis

The reliability implementation degree can be obtained through calculating the weights and evaluating. the values of the indexes. The reliability implementation degree is a value between 0 and 100% can be rated five levels. The corresponding level and significance of each value are shown in Table 1.

Table 1. Reliability implementation degree and level correspondence table.

E	Level	Significance
[0,30%)	I	Reliability work is basically not implemented and the product cannot meet the requirements.
[30%,55%)	II	Reliability work centered on typical projects are implemented and the product cannot meet the requirements.
[55%,75%]	III	Most reliability work are implemented and the product partially meet the requirements.
(75%,90%]	IV	All reliability work is implemented and the product basically meet the requirements.
(90%,100%]	V	Reliability work is fully implemented and the product meet the requirements.

3 Trade-off analysis between reliability implementation and cost

The process of reliability implementation and cost trade-off analysis can be described as follows: according to certain evaluation criteria, the evaluation value of each alternative scheme can be calculated by establishing and solving trade-off model, then the alternative scheme with the highest evaluation value can be selected as the optimal decision. A more detailed process is shown in Figure 3.

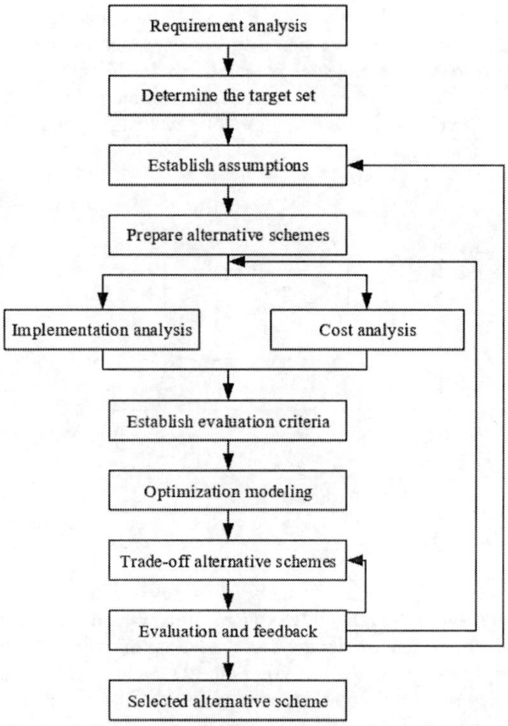

Figure 3. Process of reliability implementation and cost trade-off analysis.

In order to establish a link between reliability implementation and cost for trade-off, the first-level assessment index of reliability implementation evaluation is used as a decision variable to participate in trade-offs as parameters that connect reliability implementation with cost. The optimal solution consisting of the decision variable can be obtained by establishing and solving the trade-off model.

3.1 Reliability implementation model and cost model

Since equation (1) can measure the degree of reliability implementation, it can be used as the reliability implementation model for trade-off. In this section, it is written as equation (3):

$$\mathrm{E} = f(X_1, X_2, X_3, X_4, X_5, X_6) \quad (3)$$

The cost model can be established by multiple linear or nonlinear regression using the evaluation results of reliability implementation and costs of some similar product in the past [7]. The cost model is as follows:

$$\mathrm{C} = g(X_1, X_2, X_3, X_4, X_5, X_6) \quad (4)$$

where C is the cost of the reliability work.

3.2 Trade-off analysis

Not only the indexes of reliability implementation model have a certain range of values, but also the cost of reliability work must be subject to certain constraints. Therefore, when making trade-off analysis, it is necessary to specify the value range, which is also called the trade-off space which is shown in Figure 4. The trade-off space can be described as a multi-dimensional closed space. It represents a feasible set of solutions. In this paper, the trade-off space refers to the maximum value range of the first-level assessment index of evaluation and the maximum cost limit.

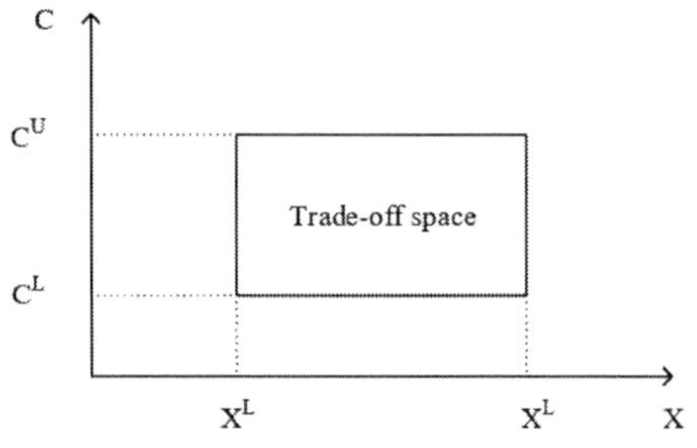

Figure 4. Trade-off space of cost and index of reliability implementation.

Goal of trade-off analysis is to obtain the best decision of the reliability implementation degree and cost, which means minimum cost and maximum reliability implementation degree are both the goals. This goal could be expressed as a function composed of E and C, as shown in equation (5):

$$\mathrm{K} = \mathrm{F}(E, C) \quad (5)$$

where K is the evaluation value of the alternative schemes; F is the evaluation criteria.

Evaluation criteria means the trade-off principles of reliability implementation and cost. The most commonly used evaluation criteria are maximizing the reliability implementation, minimizing the cost of reliability work, maximizing the reliability implementation cost ratio. The evaluation criteria are chosen based on the resources, economic, political and so on. In this paper, a trade-off model is given as equation (6).

$$\begin{aligned} &\max \mathrm{K} = \mathrm{F}(E, C) \\ &\text{s.t. } \mathrm{E} = f(X_1, X_2, X_3, X_4, X_5, X_6) \geq E_0 \\ &\quad \mathrm{C} = g(X_1, X_2, X_3, X_4, X_5, X_6) \leq C_0 \\ &\quad X_i^L \leq X_i \leq X_i^U (i = 1,2,3,4,5,6) \end{aligned} \quad (6)$$

where X_i^L and X_i^U means the lower and upper limits of the decision variable X_i.

4 Conclusion

This paper proposes a reliability implementation system based on the research of the reliability work of product during the development process. This system includes the factors that affect the reliability implementation and is used as the evaluation object to establish an AHP-based evaluation method that can be applied to the periodic review of product development process and to guide the reliability work. In addition, the paper also proposes a method of trade-off the reliability implementation degree with the cost of reliability work, which can guide the input and implementation of reliability and increase the economic efficiency of reliability work during the development process. One of the great advantages of the evaluation method and trade-off method is that reliability management problem can be solved well in the absence of actual understanding and data in the development process.

Further study will consider the constraints of risk control to perfect the reliability implementation model and the cost model and to make the solution more instructive.

References

1. Zhao Qing-li, "An Analysis and Discussion on the Implementation of Common Quality Characteristic Standards," Electronic Quality. 44-48(2015)
2. Chen Chunliang and Liu Yan, "Research on Establishing Evaluation Index System of New Type Armored Product Common Quality Characters Based on RST-QFD-Bayes," Computer Measurement & Control. 2438-2442(2015)
3. J. Brady, "Systems engineering and cost as an independent variable," Systems Engineering, vol. 4, no. 4, pp. 233-241(2001)
4. Q.S. Guo and L. Zhang, "Research summary of weapons and product systems effectiveness evaluation methods," Computer Simulation, vol. 30, no. 8, pp. 1-4(2013)
5. Karen Tyson, John Hiller, David Hunter. Assessing Cost-Reduction Initiatives in a Changing Defense Acquisition Environment-A Handbook. Alexandria, Virginia: Institute for Defense Analysis(1998)
6. L. Breiman, "Hinging hyperplanes for regression, classification, and function approximation," IEEE Transactions on Information Theory, vol. 39, no. 3, pp. 999–1013(1993)

7. T.A. Stukel, "Generalized logistic models," Mathematical & Computer Modelling, vol. 83, no. 402, pp. 426-431(1988)

A TRL Assessment Method Considering Reliability Requirements

Wei Wang[1], Qi Zhang[1], Yuan Li[1], Weijia Feng[1], Wei Zhang[1] and Huichao Liu[1]

[1] School of Reliability and Systems Engineering, Beihang University, Beijing, PR China

Abstract. In addition to considering the functional performance characteristics requirements of the product, reliability requirements should be also considered in the current product design. However, the reliability requirements of products have not been fully considered in the assessment method of TRL. This may lead the result of TRL assessment distortion and cannot help developers identify project technology risks accurately. Aiming at this, this paper improves the traditional TRL assessment method, and proposes a new TRL assessment method that considers reliability requirements. First of all, we expound the method of CTE recognition. Secondly, the TRL definition which incorporated with the reliability requirements has been proposed. Thirdly, the criteria of TRL has been put forward. Then the assessment method of the TRL has been given. Finally, the influence of reliability requirements on TRL has been described.

1 Introduction

Technology Readiness Levels(TRLs) are a systematic metric/measurement system that support assessments of the maturity of a particular technology and the consistent comparison of maturity between different types of technology [1]. There are several advantages by assessing technology readiness levels. First of all, it is possible to assess the state of technology development more objective. Moreover, TRL assessment can also help developers to identify project technology risks more scientifically and prevent the resulting adverse consequences.

When people develop and produce the product, they consider both the functional performance characteristics requirements and the reliability requirements. The reliability in this article includes reliability, maintainability, supportability, testability, safety, and environmental worthiness. However, in the assessment of the TRL, the current formulation of TRL criteria always considers the technical status, technical carrier, verification environment, process design, manufacturing process, equipment and facilities, cost management and so on [2-3]. They do not fully consider reliability requirements [2]. As a result, the reliability of the system entities integrated by the technologies to be evaluated may not be ensured. It will make the result of TRL assessment distortion and cannot help developers accurately identify project technology risks. Because of this, the system faults will increase easily and its service life will decrease and the actual needs of the product will be not satisfied.

According to the problems above, this paper first expounds the method of Critical Technology Element (CTE) recognition. Secondly, we adjust TRL definition which is incorporated with the reliability requirements. Thirdly, on the basis of the TRL definition above, the relevant contents of the reliability elements are added and the TRL assessment method was proposed. Finally, the influence of reliability on the assessment results of TRL is described.

2 CTE recognition method

2.1 Build TBS

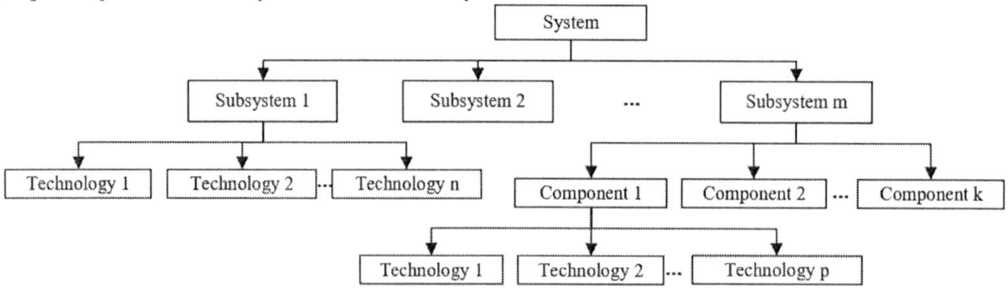

Figure 1. The TBS of the system.

When constructing the Technology Breakdown Structure(TBS), the Work Breakdown Structure(WBS) should be determined firstly. Based on the above WBS, people should identify the technologies used to implement products, establish the relevance of products, and finally determine the TBS [4]. The schematic of TBS is shown in Figure 1.

2.2 Identify CTE

On the basis of the TBS, the technology should be screened and then the CTE can be identified. The CTE should have both the importance and the high risk [5].

Importance means that the technology has a significant impact on the product's functional performance, development progress, life cycle cost, or it plays a significant role in promoting the independent innovation and overall level of science and technology.

High risk refers that the technology must meet the requirements specified in the agreement in the actual use environment.

After forming a list of CTE, experts should be organized to review them to finalize the CTE.

3 TRL definition

The definition of TRL always presents the degree which the CTE maturity degree meets the project's expected goals. It is usually used in the process of TRL assessment to determine the level of technological maturity of CTE.

In this paper, we supplement the traditional TRL definition [1,2,4,6]. The definition of TRL in this paper adds the factors related to the reliability requirements based on the definition of the original TRL definition, so that the designers can consider the reliability requirements when they determine the TRL of the CTE in initial judgement stage of TRL assessment through the definition of TRL shown in this paper. According to this, designers can effectively reduce the distortion of the assessment results caused by the failure to consider the reliability during the TRL assessment of the CTE. It can make the designers control the project's technical risks more scientifically.

The 9-level division of TRL adopted in this paper is defined in the table1.

4 The TRL assessment method

4.1 The criteria of the TRL

When measuring the TRL of aviation products, people should consider either the design and verification of technologies or the circumstances of manufacturing and management. Therefore, when designers establish a TRL criteria, the evaluation elements are refined to three dimensions: technology, manufacturing, and management. The TRL criteria of CTE can be seen in figure 2.

(1) Technology

This dimension mainly includes technical design, reliability design and analysis, technical test and evaluation, reliability test and evaluation, technical compliance and improvement, reliability compliance and improvement.

Technical design refers to the completion status of the research and design work that should be completed in

Table 1. The definition of TRL

Level	Definition
1	Basic principles observed and formal reported.
2	Propose technology concept and application assumption. According to this, establish preliminary life profile and mission profile.
3	Finish the concept and the feasibility verification of the requirements in the basic functions, performance and the reliability of the application assumption.
4	Verify the basic functions, performance and the reliability of the intended system in laboratory environment using the breadboard.
5	Verify the basic functions, performance and the reliability of the intended system in simulated environment using the brass board.
6	Verify the basic functions, performance and the reliability of the real system in simulated environment which can sufficiently represent the actual operating environment using the system or subsystem.
7	Fully demonstrate and verify the basic functions, performance and the reliability of the real system in typical application environment which can sufficiently represent the actual operating environment using the prototype of the system.
8	Tests in the simulated use environment with the actual system as a carrier show that the actual system can meet the functional, performance, and reliability requirements by the specifications.
9	The actual system is used as a carrier to verify the actual use of the task and all functions, performance, and reliability can meet the requirements.

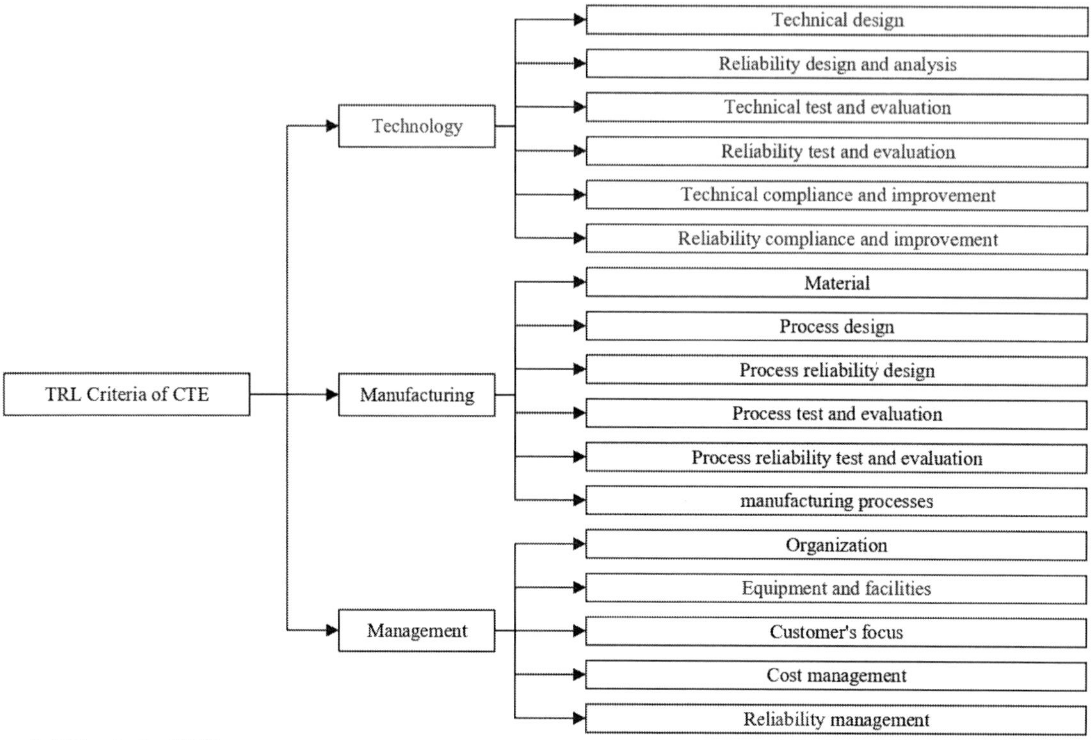

Figure 2. TRL criteria of CTE.

technical research and development, that is, the coverage of the technical design work, the quality of the work and the progress of the work ("The completion of the work" in epilogue all includes the above three points).

Reliability design and analysis refers to the completion status of the work that should be done in the design and analysis of the reliability when researching and developing the technology.

Technical test and evaluation refers to the contents (including the test and evaluation object, the test environment, the test and evaluation results) of the test and evaluation of the technical research and development results and the completion status of the work.

Reliability test and evaluation refers to the contents (including test and evaluation objects, test environment, test and evaluation results) of the test and evaluation of reliability design and analysis results and the completion status of the work during the technical research and development.

Technical compliance and improvement refers to the completion status of technical compliance and continuous improvement work that should be completed during technical research and development.

Reliability compliance and improvement refers to the completion status of the work that should be completed when the reliability is achieved and continuously improved during technical research and development.

(2) Manufacturing

This dimension mainly refers to the materials, process design, process reliability design and analysis, process testing and evaluation, process reliability test and evaluation, and manufacturing processes which are involved in the production phase.

Material refers to the extent to which the material meets manufacturing requirements and its availability.

Process design refers to the completion status of research and design work on the processes involved in the manufacturing phase that should be completed during trial production.

Process reliability design and analysis refers to the completion status of research design work for process reliability that should be completed during trial production.

Process test and evaluation refers to the content of testing and evaluation of the processes involved in the manufacturing stage and the completion status of the work that should be completed during trial production.

Process reliability test and evaluation refer to the process reliability test and evaluation of the process reliability that should be completed during trial production and the completion status of the work.

Manufacturing process refers to the reflection status of the manufacturing process of the product on the design intent (reproducibility and economic affordability) of the critical characteristics.

(3) Management

This dimension mainly refers to organization, equipment and facilities, customer's focus, cost management, and reliability management.

Organization refers to the number of people whose work are involved in the function, performance, and reliability during the process of technology maturity, the satisfaction status of the workers' technical level and the degree of the organization setting perfection.

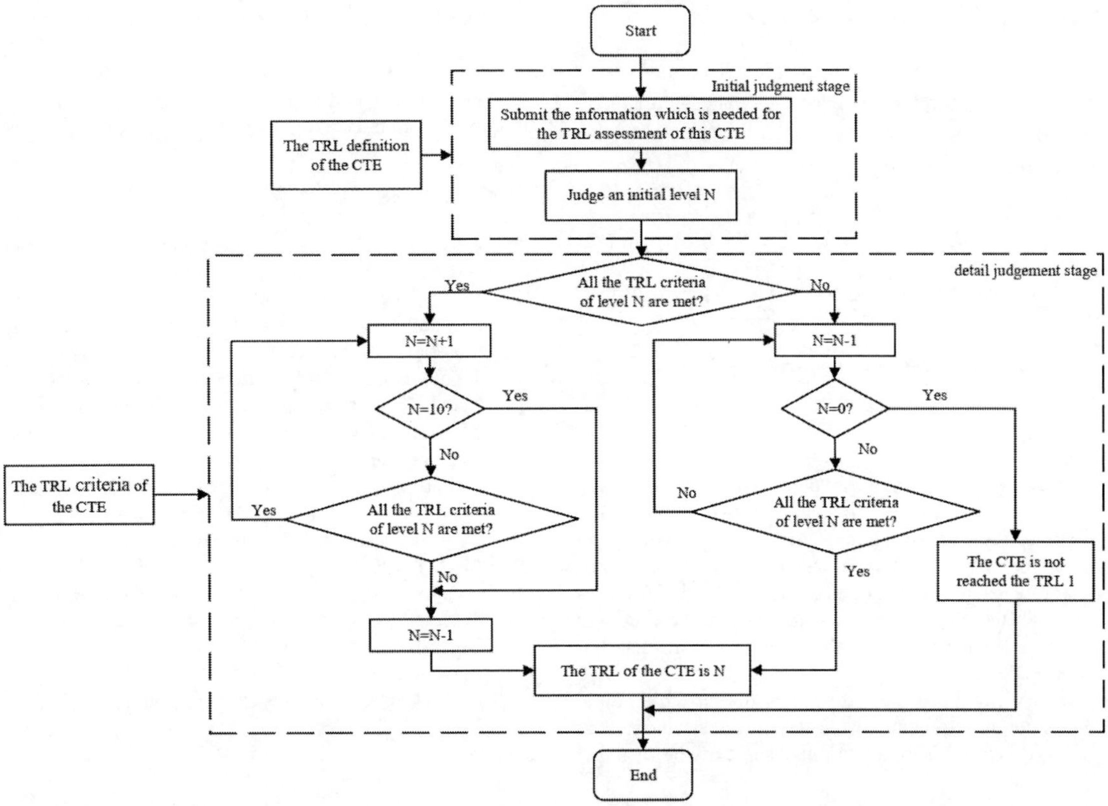

Figure 3. The judgement process of each CTE.

Equipment and facilities refer to the level of development and application of various equipment which are used in various work of functions, performance, and reliability during the technology maturity process and the construction progress of each facility.

Customer focus is the ability to clearly identify and remember user needs and he ability to target customers [6].

Cost management refers to the completion status of management work of the products' life cycle cost.

Reliability management refers to the completion status of reliability management work during the technological maturity process.

On the basis of the TRL assessment system, designer should formulate the criteria for the level division of each underlying element that is in accordance with the overall hierarchy according to the actual situation of the system or the project. These rules for the level division of each underlying elements will serve as a basis for evaluating technology maturity.

4.2 The TRL assessment method

The TRL assessment process of each CTE includes two stages: initial judgment stage and detailed judgment stage.

In the initial judgment stage, experts judge an initial level based on the TRL definition. After that, the assessment enters the detail judgement stage.

In the detailed judgment stage, experts will compare the parameters of the technology with the criteria of the initial level. If all the criteria of the initial level are met, developer need to assess the criteria of the higher level until the CTE meets all the criteria of the level N and does not completely satisfy the criteria in N+1 level. As a result, the TRL of the CTE of is N. On the contrary, if the criteria of the initial level are not completely satisfied, experts should determine whether all the criteria in the lower level is satisfied until the criteria of the level are completely satisfied or when the criteria for the level 1 are not completely satisfied. Then the TRL the CTE can be determined.

The judgement process of each CTE is shown in figure 3.

5 Influence of reliability requirements on TRL assessment

In this paper, the relevant contents of the reliability requirements are added in the definition and the criteria of the original TRL. The content to be considered in the TRL assessment are increased. Because of it, the severity of the TRL assessment are also increased. Therefore, the TRL which are proposed in this paper will be lower than the original TRL in most cases. At the same time, because TRL provides an important reference for technology development and transition, the increased severity of assessment of TRL can effectively reduce technology risks.

It can be assumed that there are a underlying elements in the TRL criteria. Among these underlying elements, there are b underlying elements related to reliability requirements.(a>b) For this reason, there are a-b non-universal quality characteristic elements exist.

It can be assumed that the weight of the ith underlying element in all underlying element of the TRL criteria is ω_i. Besides, $\omega_1,\ldots,\omega_b$ refers to underlying elements related to reliability requirements; $\omega_{b+1},\ldots,\omega_a$ refers to other underlying elements.

In this way

$$\omega_1+\ldots+\omega_b+\omega_{b+1}\ldots+\omega_a = 1 \quad (1)$$

The impact of reliability on the TRL can be expressed as

$$Effect = \omega_1+\ldots+\omega_b \quad (2)$$

6 Conclusion

Traditional TRL definition and its criteria do not fully consider reliability requirements. This article first describes the CTE recognition method. Moreover, based on the maturity of traditional technologies, the factors of reliability requirements are added to form a definition of TRL. Furthermore, based on the original technology readiness level criteria, the relevant contents of reliability requirements are added and the assessment method of TRL is proposed. Finally, the influence of the factors that add the reliability requirements on the assessment method of TRL is described. Through the analysis, it can be seen that the TRL proposed in this paper can increase the severity of the assessment of TRL, which reduces the risk of the project.

The work we should do in the future is to further refine the TRL criteria and measure the influence of reliability requirements on the assessment of TRL more reasonably.

References

1. J. Mankins, Technology readiness levels White Paper, NASA, (1995)
2. PLA General armament Department, Classification and definition of the technology readiness levels for materiel:GJB 7688-2012, (2012)
3. J.Xu,Y.Liang,Y.Qi,Research on the Method of Technology Readiness Levels Evaluation in the Material Development, Science and Technology Management Research,2(2016)
4. United States Department of Defense,Technology Readiness Assessment (TRA) Guidance, (2011)
5. Y.Wu, Technology Readiness and Its Assessment Method, (2012)
6. Assistant Secretary for Research and Engineering. Technology Readiness Assessment Guidance. USA Department of Defense,2011
7. X.Zhang, Theory and Application of Maturity Model in Defense Material System Engineering, (2013)

MATEC Web of Conferences **221**, 02005 (2018) https://doi.org/10.1051/matecconf/201822102005
ICDME 2018

Systematic Innovation for Manufacturing Quality Improvement

Shu Luing Nikalus Swee[1], Guat Guan Toh[1], Mum Wai Yip[1] and See Chew Tai[1]

[1] *Tunku Abdul Rahman University College, Malaysia*

Abstract. This paper provides a detailed analysis on the systematic innovation process in improving the quality control of latex gloves production. The systematic innovation tool such as TRIZ is applied in this case study. Function analysis, cause and effect chain analysis, physical contradiction, By-separation model and 40 Inventive Principles are applied in order to derive some feasible and low cost solutions to alleviate the problem. Findings revealed that the rejected (leaking) gloves on the production line will be manually monitored by a checker during the air blowing test and will be discarded by the same checker instantly. The main root cause is that the quality control worker is not able to concentrate all the times to detect the torn gloves, mainly is due to the fast speed production line and other distractions. The problem is solved by applying function analysis, physical contradiction, by-separation tool and Inventive Principles to generate low cost but elegant solutions within the defined scope of several constraints and without making the production line more complex. Therefore, it can be concluded that TRIZ is a systematic and innovative problem solving methodology.

1 Introduction

Manufacturing lines cover a set of serial processes. Raw materials are channeled into refining operations to produce an end-product that is suitable for consumer market [1]. It also includes inspecting, labelling, and packing the final product before the product is released. Machines, materials and workforce influence the reliability and capacity of the production operation. Therefore, every process is crucial for seamless production [2]. The production of latex gloves involves a process of coagulant dipping, in a continuous chain dipping line. High speed continuous dipping line are commonly used for surgical and examination gloves production. During the coagulant dipping process, clean porcelain formers are first dipped into a coagulant solution and then into a natural rubber latex compound. The next stage involves the bead rolling of the gloves by rotating brushes after a short drying. Then, leach the wet gel on the former in hot water. Corn starch is then applied to the outer surface of the formers via a slurry dip before the gloves are subjected to a drying and vulcanization process in an oven (tumble dryer). Subsequently, the dried gloves are manually stripped from the porcelain formers, tested and packed. [3-4] Leakage and tensile testing is part of the quality control measures. Latex glove durability is measured in tensile properties (strength and elongation) and barrier integrity (water leakage test). Leakage is identified most commonly in the finger parts of the glove [4-5].

TRIZ is a Russian acronym that is equivalent to "Theory of Inventive Problem Solving" in English. TRIZ methodology was introduced by Genrich Altshuller and his team in 1940's [6]. From 200,000 patents, he discovered and categorized his study of 40,000 patents according to innovative patterns of design and the inventive principles in these innovative solutions. Findings revealed that solutions and patterns of technical evolution were recurring across industries and sciences, and innovations used scientific effects outside the field where they were developed. Therefore, Genrich Altshuller created 40 inventive principles [6-13]. TRIZ comprises of several crucial tools such as engineering systems analysis, function analysis, cause and effect chain analysis, trimming, engineering contradiction, by-separation, substance-field model, Trends of Engineering System Evolutions, ARIZ, and etc. TRIZ uses 40 inventive principles and 39 parameters to help inventors to derive many solutions [6-13]. Besides that, there are also 4 basic concepts of TRIZ such as systems approach, contradiction, ideality and resources. Technical contradiction is the key concept in TRIZ. A problem is solved only is the technical contradiction is identified and eradicated [14-15].

2 Problem Statement

On-site inspection revealed that the quality control worker is not able to focus all the times to detect and to discard the leaking gloves during the air blowing test which may fail the AQL, or Acceptable Quality Level. There are a few constraints to be taken into account such as the speed of the production line cannot be altered, dipping process can be altered, power of air blower cannot be toned down and there is no budget for sophisticated solutions such as robotic arms. The following are the details on how TRIZ is applied in this case study.

3 Systematic Innovation Tools (TRIZ)

TRIZ flow of processes is shown in Figure. 1. First, define the engineering system of the initial problem, and this is followed by function analysis, cause and effect chain analysis, and Physical contradiction. Finally, By-Separation model is applied to derive specific inventive principles and specific solution(s) [6-7].

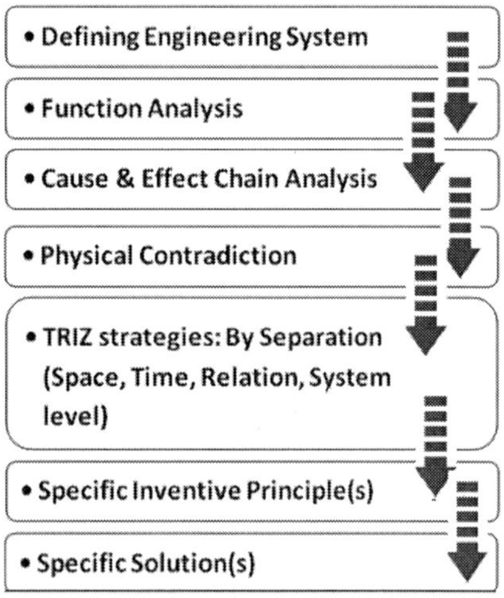

Figure 1. TRIZ Processes

3.1. Engineering System Definition

An Engineering System comprises of several components that are interacted among each other. All these components are commonly recognized as system components (subsystems) that are listed in Table 1. Along with subsystems, there are also connections between engineering system and external entities called Supersystems. Supersystems are not the members of the Engineering System; however, they can influence the Engineering System. [7-8].

Table 1. Engineering System

Sub /System Components	Formers(mold), coagulant tank, latex tank, rinser, dipping track, oven, bead roller, motor, air blower
Supersystems	air particles/dust, workers, Humidity, Thermal

3.2. Function Analysis

Function analysis analyses the contacts/interactions between two or more systems/subsystems (Engineering System components) which are listed in Figure 2. These interactions are called functions. Functions are interactions between two components, i.e., a subject and an object in which the subject acts upon and modifies a parameter(s) of the object [6]. For instance, 'a robotic arm holds a bottle. Two main types of functions are useful function and harmful function. As for useful function, it has "normal", "insufficient", and "excessive" functions. [6-7].

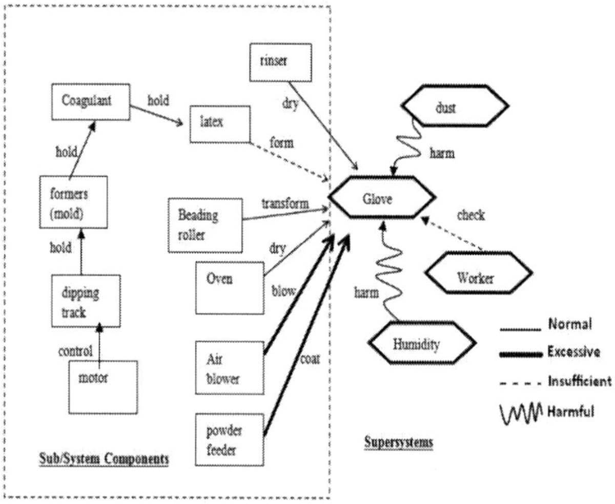

Figure 2. Function Model

3.3. Cause and Effect Chain Analysis (CECA)

Cause and Effect Chain Analysis(CECA) helps identify the right root cause(s) in relation to the problem in Figure 3. If the wrong root cause is derived, it leads to ineffective solution. We prompt for causes continuously for the problem from high level causes to low level causes by asking "the question "why?" [6]. From the CECA, the root cause identified is that checker is not able to concentrate all the times to detect and pull out the leaking gloves in the event of the high speed production line. Nevertheless, the function model shows that there exists other problems which may affect the gloves quality, but these problems are not crucial.

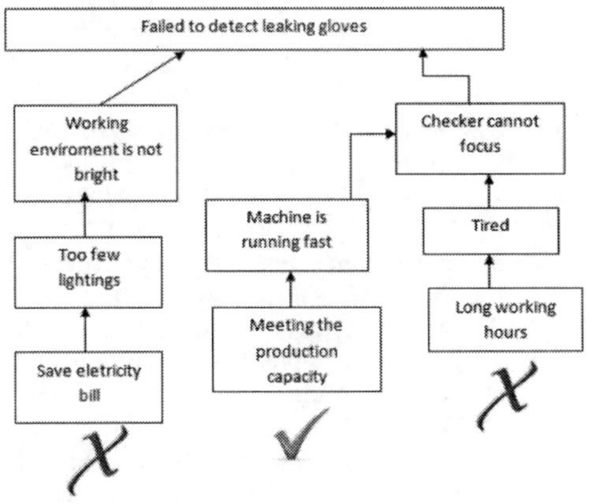

Figure 3. Cause and Effect Chain Analysis Diagram

3.4. Physical Contradiction

Physical Contradiction refers to the existence of a contradiction in an engineering system involving a single parameter (e.g. speed, weight, strength, temperature, length, volume, etc.) which has a contradiction at two different values. [6-7]. For example, Sandblasting abrasive needs to be present to remove material stuck to an object and must not be present when sandblasting operation is completed. The essence of the Physical Contradiction is to find the controlling characteristic. The Physical Contradiction can be solved through methods such By-Separation in (Space, Time, Relation, System level) and By–Satisfaction, and Bypass. In this case study, we employ By-Separation model. Based on the case study, the following Physical Contradiction is proposed.

-The speed of the production line needs to be fast for high productivity.
-The speed of the production line needs to be slow for checker to detect leaking gloves easily.

3.5. By Separation principles (Space, Time, Relation, System level)

By separation principles are used commonly to generate new ideas by resolving some key paradoxes of industrial innovations. To investigate for the type of the Separation whether in Space, in Time, or in Relation, the following scenarios should be established [6]:

- •Separation in Space – **Where** is condition Z required? **Where** is condition –Z not required?
- •Separation in Time – **When** is condition Z required? **When** is **condition** –Z not required?
- •Separation in Relation – I need condition Z **IF…** ? I need condition -Z **IF…**?
- •Separation in System level – The Physical Contradiction's controlling parameter has a value at the system level but has an opposite value at the component level or a controlling parameter exists at the system level but not at the component level.

Referring to the above scenarios, the speed of the production line needs to be fast in relation to the high productivity, and the speed of the production line needs to be slow in relation to the effectiveness of quality check. Hence, Separation in Relation model is deemed suitable to reflect the problem. There are many recommendations of Inventive principles in Separation in Relation[6], only Inventive Principle #32 Color Changes and #28 Mechanics Substitution are deemed suitable for the problem after thorough discussion with production engineers.

4 TRIZ Solutions and Discussion

Referring to the proposed inventive principles above, Inventive Principle #28 Mechanics Substitution suggests replacing a mechanical means with a sensory means such as optical, or acoustic means. #Inventive Principle #32 Color Changes suggests to change the color of an object or its external environment, or to change the transparency of the object, or to use colored additives (luminescent elements) to improve the observability of the object [6]. Based on the suggested Inventive principles, production engineers can install infrared sensors (IR) that emit and detect IR radiation to sense its ambience. IR transmitters and IR receivers are installed at the level of the finger parts of the glove in which the glove is sandwiched between the IR sensors. When the glove is tested with air blowing test, leaking finger parts will allow the IR receiver to detect the IR radiation from IR transmitters. Therefore, IR receiver can trigger color LED to light. Color LED attracts attention from the checker to discard the torn glove effectively [16-17]. In addition to that, production engineer may also consider other useful Inventive Principles even though the Inventive Principles are not recommended through the systematic approaches above. For instance, Inventive Principle #29, Pneumatic and hydraulics, suggests to use gas or liquid parts of an object instead of solid parts, e.g. transition from mechanical to pneumatic drive [6]. Therefore, the discarding of leaking glove can be automated by installing a simple vacuum suction cup to perform the task [18].

5 Conclusion

The manufacturers should be able to process faster, reliable with lower cost for any types of products [19-20]. In this case study, the problem was confined by applying TRIZ tools particularly the Physical Contradiction and By-Separation tools. TRIZ helps engineers generate more practicable ideas which may derive multiple low cost and elegant solutions. TRIZ needs field specialists and TRIZ consultant to work together to derive effective solutions. It can be concluded that TRIZ is a systematic and an innovative problem solving methodology.

References

1. Sarder, MD Baniamin, Liles, Don H, Ali M Yousuf 2005 Design & Implementation of production line

inapparel manufacturing industry *IIE*, Annual Conference. Proceedings, pp. 1-6 (2005)
2. U. Michael, G. Van and U. Elisabeth, Improving production line performance, IIE Solutions; 32, 11 ProQuest, p. 36 (2000)
3. Jose Paul Meleth, *An Introduction to Latex Gloves*, Lap Lambert Academic Publishing (2012)
4. Ploytip Jirasukprasert, Jose Arturo Garza-Reyes, Horacio Soriano-Meier, Luis Rocha-Lona, A Case Study of Defects Reduction in a Rubber Gloves Manufacturing Process by Applying Six Sigma Principles and DMAIC Problem Solving Methodology, Proceedings of the International Conference on Industrial Engineering and Operation Management Istanbul, Turkey, July 3 – 6, (2012)
5. A. Douglas, T. R. Simon and M. Goddard, Barrier durability of latex and vinyl medical gloves in clinical settings, American Industrial Hygiene Association Journal, 58, 9; ProQuestp. 672 (1997)
6. T.S. Yeoh, T.J. Yeoh, and C.L. Song, *TRIZ-Systematic Innovation in Manufacturing*, Selangor: Firstfruits (2009)
7. A.A. Guin, A.V. Kudryavtsev, V. Yu. Boubentsov and A. Seredinsky, *Theory of Inventive Problem Solving*, First Fruits (2009)
8. D. Mann, *Hands-On Systematic Innovation*. Belgium: CREAX Press (2003)
9. D. Mann, *Hands-On Systematic Innovation for Business and Management*, Devon: Lazarus Press (2007)
10. E. M. Smith, From Russia with TRIZ, Mechanical Engineering, 125, 3; Academic Research Library, p. D18 (2003)
11. G.S. Altshuller, *The Innovation Algorithm, TRIZ Systematic Innovation and Technical Creativity*, Worcester, MA: Technical Innovation Centre (2000)
12. G.S. Altshuller, *40 Principles, TRIZ Keys to Technical Innovation*, Worcester, MA: Technical Innovation Centre (2002)
13. J. Terninko, A. Zusman, and B. Zlotin, Systematic Innovation: An Introduction to TRIZ, New York, NY: CRC Press (2010)
14. Bowyer, D., Evaluation of the effectiveness of TRIZ concepts in non-technical problem-solving utilizing a problem solving guide. ProQuest (2008)
15. Ilevbare, I. M., Probert, D., & Phaal, R., A review of TRIZ, and its benefits and challenges in practice. Technovation, 33, pp. 30–37 (2013)
16. M.Gordon, Collison avoidance and detection: Part 1, Poptronics; 1, 12; ProQuest, p. 53 (2000)
17. S. Piet 2003 Electro-optical sensors: How good are they?, Naval Forces; 24, 1, Research Library, p. 13 (2003)
18. Zalud, Todd, Using suction cups to design a vacuum system, Machine Design, 70, 22 p. *114 (1998)*
19. Andy Longford, Chip packaging challenges ...a roadmap based overview, Microelectronics International, 22/2, pp. 17–20 (2005)
20. Lagner, P., Advantest Handling challenges of small & fine pitch devices, Advantest Technical Forum (2004)

Design and Construction of Sago Rasper (As Business Process Re-Engineering of Sago Flour)

Taufiq Immawan[1], Dicky Bayuadi Saputro[1] and Winda Nur Cahyo[1]

[1] *Industrial Engineering Study Program, Faculty of Industrial Technology Universitas Islam Indonesia Jalan Kaliurang Km. 14.5 Phone (0274) 895287 Sleman Yogyakarta 55584*

Abstract. The forest area of Sago in Indonesia is about 1.25 million hectares with a seedling density of 1,480 per hectare. Only about 40 percent of the area is productive with the starch productivity of 7 tons per hectare per year, because of the number of sago crops that are harvestable but not harvested so that they are damaged. This happens because the business and the raw material sources are at a great distance. Therefore, there is a need for business process re-engineering of sago to be more efficient and effective. By redesigning the sago rasper to fit into the truck and can be laid out with other tools. The method used in this research was parameter measurement technique of Quality Function Deployment (QFD), validity and reliability test, morphological chart then designed by using SolidWork software. There were 5 attributes used in QFD: safe, high productivity, strong, durable, and price. Eventually, virtual design of sago extraction tool in accordance with customer requirements and expert machining was designed and constructed. The benefits of the research for Sago flour business process are machine optimization, reduced labor, shorter production time and higher productivity.

1 Introduction

Sago is an alternative food substitute for rice. In Indonesia, rice has been an essential staple food for the society, the most common food ingredients as a source of carbohydrates although there are some regions in Indonesia that make alternative food as staple food [1]. The demand for rice in Indonesia populated 212 million people is 30 million tons per year and it will continue to increase as the population grows. If the average population growth is 1.8% per year, then the population of Indonesia in 2010 is estimated 238.4 million and in 2015 is 253.6 million. Considering the potential condition of national rice production, it is estimated that by 2015 the rice supply will experience a deficit of 5.64 million tons [2].

Indonesia has the largest sago plantation area in the world of which about 48% of the total world sago area or approximately 994,000 ha are in Papua and West Papua provinces. It is estimated that the potential of dry starch of sago in the Indonesian easternmost province is about 4.75 million tons every year, but the actual production is only about 200 thousand tons. This means that 4.55 million tons of dry starch is wasted because they have not been harvested [3]. Based on the problem, there is a great business opportunity for sago flour processing in Indonesia. In Indonesia, sago is generally eaten as traditional food such as papeda, consumed as the staple food. Besides, it is also consumed as complementary food such as plate sago, sinoli, bagea and others [4]. Additionally, sago may be used as raw material for various industries such as food industry, adhesive industry, cosmetics and other industries [5]. Business Process Engineering is one thing that can be done to improve the effectiveness and efficiency [6]. To undertake Business Process Engineering required information on the entire process used by the company [7]. By conducting business process engineering, it is expected that there will be a significant and drastic change in the company performance. The purpose of using the concept of Business Process Reengineering is that the processing system will be more computerized so that the presentation of information run more quickly, accurately, effectively and efficiently, and the improvement in the operational management system as an effort to survive and develop among other competitors can be expected.

The result of business process re-engineering is a proposed business process. Increasing the capacity of sago processing at the farm level can certainly be improved by improving the processing techniques at each stage, especially at the stage of pith destruction and extraction because these two stages are the most time and energy consuming of all processing stages. One of the alternatives is to create sago processing tools which are interconnected at each stage so that the process starting from the pith destruction to starch drying can be done more efficiently and effectively in terms of time and energy to increase productivity. The real demand of the consumer needs to be identified to redesign the tool. The process design is performed in response to consumer needs by using the QFD method [8]. To serve this purpose, QFD has applied in many area, such as: supplier selection in supply chain [9]; developing scenario for industrial development [10]; and infrastructure [11].

Based on the result of the field survey, the rasper machine used in the sago flour processing industry in Klaten is made of wood cylinders and nail spikes, driven by 5.5 HP gasoline engine. The disadvantage of the machine is the construction of the machine directly contact with the product so that it might be easily rusty resulted in the quality of the rasped starch. There is the other problem for the industry that is the lack of raw material supplies caused by the far distance between the industry and the source of raw materials. The far distance causes the high cost of sending raw materials. The traditional business processes which are semi-mechanical results in wasteful expenses, time, and workers.

Business process re-engineering on sago flour processing has not been conducted previously. Several previous studies have been conducted on the prototypes of sago rasper, sago starch extractor, and sago starch dryer. This research combines the processing tools in one set aims to improve the efficiency of the business process which is later expected to increase the productivity in the business process of sago flour. This study focuses on the design and construction of mobile sago rasper to be operated with other sago processing tools such as the extractor, precipitator, and dryer without reducing the effectiveness of the rasper. It is expected to improve sago processing production system through the transfer of sago processing technology with production tools such as sago rasper, extractor, and dryer previously constructed.

2 Research Method

The focus of this study was to design a more efficient sago rasper design by using Quality Function Deployment (QFD) and Morphological Chart methods. The object of this research was a sago processing manufacture, sago rasper section, with the subject of sampling was the experts in sago rasper in Daleman, Tulung, Klaten, Central Java and the experts in the engineering at Indotech Cipta Mandiri workshop Jl. Taraman Raya, Sinduharjo, Ngaglik, Sleman, Yogyakarta

3 Result and Discussion

3.1 HOQ results

The relationship between technical requirements and customer requirements, the relationship between technical requirements, importance rating, direction of improvement and importance values are described in the form of HOQ matrices as shown in Figure 1.

3.2 Morphological Chart

Having obtained technical requirements, the goal was defined by creating a morphological chart with several goal options. After that, the expert chose an option that met customer and technical requirements. The morphological charts are presented in Table 1.

3.3 Design of the Sago Rasper

The design shown in figure 2 was referred to the QFD so that the design had been built on customer requirements, technical requirements and morphological charts. Thus, this design was expected to meet consumer desires.

3.4 How it Works

The device was operated manually by inserting blocks or pieces of sago that had been separated from the bark of the stem into the hopper and dropped to the rasper cylinder through the hopper to be shredded. At the top of the hopper there was a booster for the sago pith. This booster served to provide a thrust force on the slices of the sago pith so that the pieces of the pith are pushed toward the rasper cylinder. The pressure on the pith will accelerate the rasping process. This driver is very easy to use and very helpful for the operator. The sago spikes bar consists of 6 main parts, namely: 1. Main Frame, 2. Motor drive (3-phase electric motors), 3. Transmission system using a belt pulley, 4. Hopper, 5. Rasping result Channel (unloading), and 6. Rasper Cylinder. A propulsion mechanism had been created and located at the top of the hopper so as to improve the safety of the use of the tool so as to reduce the accident rate at the time of the rasping process.

3.5 Advantages

The research was about designing sago rasper with high mobility by being loaded up to the truck to be laid out with other tools without reducing the value of the tool usage. The following are advantages of the machine toward the sago business process:

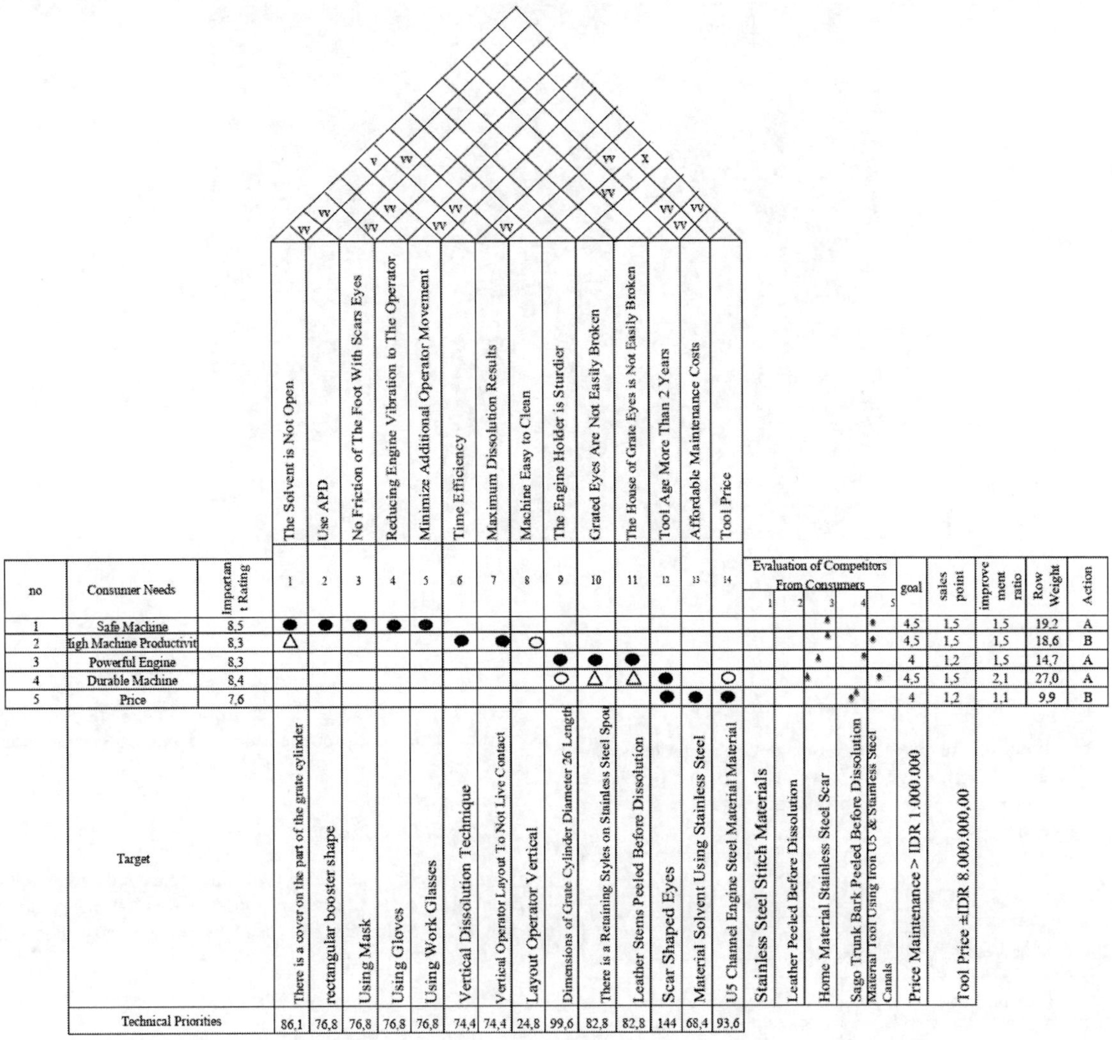

Figure 1. The House of Quality

Table 1 Morphological Chart

NO	FUNCTION	MEANS
1	Adding Cover On Rasper Cylinder	AVAILABLE
2	BOOSTER ON HOPPER	RECTANGLE
3	MASK	YES
4	GLOVES	YES
5	WORK GLASSES	YES
6	REDESIGN GRATE TECHNIQUE	DESIGNED WITH VERTICAL LAYOUT
7	Minimize Direct Contact Between Operator And Machines	DESIGNED WITH VERTICAL LAYOUT
8	REDESIGN OPERATOR LAYOUT	DESIGNED WITH VERTICAL LAYOUT
9	Redesign Re Dimensions Length And Cylinder Diameter	DIAMETER 26CM LENGTH 40CM
10	ADD RETAINING PARTITION ON DISCHARGING FUNNEL	YES, BY USING STAINLESS MATERIAL
11	Separate Sago Pith Before Grated	YES
12	RASPER SPIKES	NEEDLE
13	MACHINE MATERIAL	STAINLESS STEEL
14	MACHINE CANTILEVER MATERIAL	CHANNEL IRON U5
15	SPIKES	STAINLESS STEEL
16	SEPARATE SAGO SKIN FROM PITH BEFORE GRATED	YES
17	SPIKES CASE MATERIAL	STAINLESS STEEL
18	SEPARATE SAGO SKIN FROM PITH BEFORE GRATED	YES
19	MACHINE MATERIAL SELECTION AND THE FRAME USING CHANNEL IRON U5 AND STANLESS STEEL	
20	AFFORDABLE MAINTENANCE COST	> ± RP 1.000.000,00
21	AFFORDABLE TOOL PRICE	±RP 8.000.000,00

https://doi.org/10.1051/matecconf/201822102006

Figure 2. Proposed Design of Sago Rasper.

3.5.1 Reducing Manpower

Based on the result of the reengineering process, the number of labour can be reduced. The existing process is shown in Figure 3 and the proposed process in shown in Figure 4.

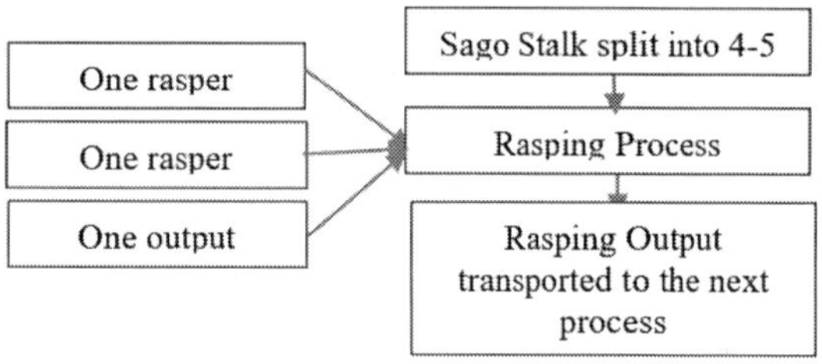

Figure 3. Flowchart of the existing Rasping process in Sago Flour Industry in Klaten.

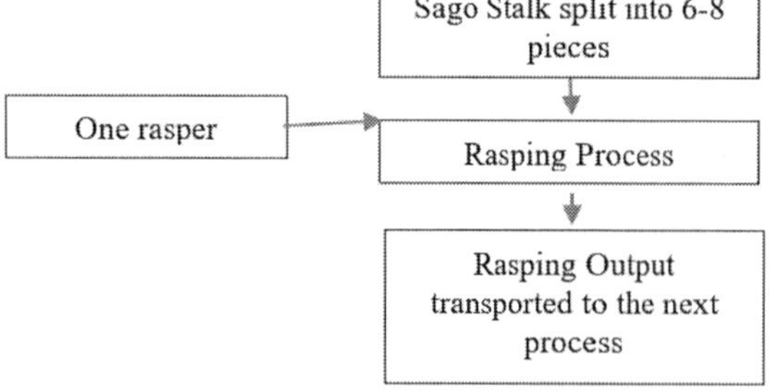

Figure 4. Flowchart Rasping process using the proposed design

It can be compared from Figure 3 and Figure 4 that there is a reduction in labour from three workers to one worker. Thus, the proposed sago-rasping process is more efficient.

3.5.2 Tool Optimization

By optimizing the design of the improved rasper, the capacity of rasping process obtained was 649.38 kg/hour, compared to the machine used by the industry that was only 416.67 kg/hour.

4 Conclusion

The conclusions drawn after data processing and discussion are as follows: There are five attributes used in QFD namely safe, high productivity, strong, durable, and price. Technical requirements and goals for designing and developing sago extraction tools are obtained from the experts. The virtual design of sago extractor fulfils customer requirements and the expert expectation. It provides a positive impact on the business process of sago flour that is reducing manpower and optimizing tools.

References

[1] Sakiynah, N., Tigor, A.R., dan Setyawan, H. 2013. Design of Sago Processing Plant. *Jurnal Teknik Pomits*. Vol. 2, No. 1, (2013) ISSN: 2337-3539.

[2] Iswahyudi. 2016. Evaluation of Food Security to Resolve Food Crisis in North Aceh. Agrosamudra, *Jurnal Penelitian*. Vol. 3 No. 1 Jan – Juni 2016 : 39-46.

[3] Aceng, K. 2012. The Development of Sago Industry in Papua to Support Food Security and Diversity. *Prosiding Insinas*.

[4] Harsanto, P.B. 1986. Sago Cultivation and Processing. Kanisius. Yogyakarta.

[5] Haryanto, B. dan Pangloli, P. 1992. Sago Potency and Usage, *Kanisius*. Yogyakarta.

[6] Mochyidin, A. Hartanto, D. M. Devara, R. & Rantetana, M. 2011. Business Process Reengineering in Marketing, Logistic, and Accounting. Mochyidin, ainun, et al / Journal of Applied Finance and Accounting 4(1) 39-50.

[7] Opit, F. P. 2012. Business Process Modelling of Procurement Division, J@TI Undip, Vol VII, 3 September 2012.

[8] Aji, E.R. dan Yuliawati, E. 2016. Desk Study Lamp Product Development using Kano and Quality Function Deployment (QFD). *Journal of Research and Technology*. Vol. 2. No. 2. Desember 2016.

[9] Francisco Rodrigues Lima-Junior, Luiz Cesar Ribeiro Carpinetti, A multicriteria approach based on fuzzy QFD for choosing criteria for supplier selection, Computers & Industrial Engineering, Volume 101, 2016, Pages 269-285, ISSN 0360-8352,

[10] Meelis Eldermann, Andres Siirde, Julija Gusca, QFD framework for selection of industry development scenarios, Energy Procedia, Volume 128, 2017, Pages 230-233, ISSN 1876-6102,

[11] Aman A. Bolar, Solomon Tesfamariam, Rehan Sadiq, Framework for prioritizing infrastructure user expectations using Quality Function Deployment (QFD), International Journal of Sustainable Built Environment, Volume 6, Issue 1, 2017, Pages 16-29, ISSN 2212-6090.

Combining Time and Physical Workload Analysis on Cold Press Working Group for Operator Management in Manufacturing Company

Muhammad Ragil Suryoputro[1], Tresna Candra Ginanjar[1], Amarria Dila Sari[1]
[1]*Industrial Engineering Deparment, Faculty of Industrial Technology, Islamic University of Indonesia, Jl. Kaliurang km 14.5, Yogyakarta Indonesia*

Abstract. Efficiency improvement in manufacturing company is essential to ensure the cycle of continuous improvement. Prior decision was reducing 6 operators in the cold press team work into 5 workers. Thus significant changes, of course, would resulted in the change of the workload experienced by each operator. The research objective in this study is validating from the productivity of the reduction conducted by measuring the workload with combining Full Time Equivalent (FTE) and Cardiovascular Load Percentage (%CVL). The result states that there are differences in measurement results workload from both methods. Based on the FTE method, all Cold Press operator's workload has an index in average of 1.4128 with all was above 1.28 indicating the "Overload" condition. On the other hand, based on the physical workload method obtained that % CVL in average was 8.272% with all operator was below 30% which indicates the workload of all Cold Press operators were in a not experiencing fatigue category.
Keywords: Workload, FTE, %CVL, productivity, Physical Workload.

1 Introduction

XYZ Manufacturing company is a manufacturing company that are producing musical instrument that located in the main production area in Indonesia. Based on previous pilot study for this manufacturing company, in order to improve productivity on a production line, the Cold Press working group control labour efficiency by reducing the number of operators from the original 6 operators into 5 operators. Determination of the amount is based on a pilot workload analysis. In its production activities that were divided into several working groups that have different kinds of production depending on the production process, the type of musical instrument, and what kind of cabinet in the process. For each working group in XYZ Company provides a Value Stream Mapping and Industrial Engineering project, where each working group is targeted to perform labour productivity increases of up to 15% for the cabinet or products assembly and 30% for the cabinet or non-assembly products, supported by the application of Kaizen in each working group. The Cold Press working group is one group of non-assembly work in XYZ Company, which has run the project. In order to increase the productivity of the Cold Press working group production control labour efficiency by reducing the number of operators from the original 6 carriers to 5 operators.

Every worker in a company must have a workload. The workload can be defined as the price or value of the achievement on the target activity [1]. If this value is too high from the job performed, the fatigue level is also high. Fatigue will reduce performance and increase the error rate. Increased faulty work will provide greater opportunities for work accidents [2].

There were researches regarding the workload using various methods in companies, NASA TLX (Task Load Index) for mental workload analysis [3] and [4]; WISN (Workload Indicators of Staffing Needs) for wellbeing of household, medical staff [5], and Namibia [6]; Cardio Vascular Load Method for physical workload [7] and Stopwatch method in standard time measurement for time based workload analysis [8].

With the change of the number of operators in the working group Cold Press, of course, will change the workload perceived by each operator. This study will conduct research on methods of measuring the workload with Full Time Equivalent and Percentage CVL, continued with a comparison of the results of the two methods. The determination of these two methods is based as a link between the type of work and the time of the type of work performed by the operator. In general, most types of work performed by the operator in the Cold Press working group is physical workload types such as lifting the cabinet, push the cabinet, and also cut the cabinet in various types, sizes and weights. Furthermore, every job in the Cold Press working group each cabinet has a cycle time, for example the time required to process a single cabinet. This shows the necessity of understanding the difference in the type of workload measurement results based on the method of Full Time Equivalent (FTE)-time workload measurement and the method of percentage of CVL-physical workload measurement. So the final result of this study is expected to be known to the workload perceived by each operator based on time and physical workload, and also the correlation based on both methods.

2 Methods

2.1. Research subject

In this study sampling was conducted using purposive sampling technique. The study was performed on the woodworking department, especially focused on the Cold Press working group that meets the criteria specified. The criteria used to select the sample is as follows. The first is good mental and physical health during the data retrieval process, continued with male operator targeted. The age range was between 20 and 35 years old. In addition, the operator should understand the production process and the occupational safety and health requirements

2.2 Research Procedure

Workload analysis was conducted using two stages of measurement and one stage of qualitative interpretation. There were three guideline defined to conduct the study, FTE calculation, %CVL calculation and consultation of combination results from both criteria. Procedure in performing workload analysis in this research were divided into three stages:

a. **Full Time Equivavlent (FTE) Stage:** The FTE was the beginning step in conducting the research. FTE was performed to simplify the work measurement by converting hours' workload into the number of workers needed to completing the task/job required [9]. There were three types of the implication on FTE score, Overload (>1.28), Normal (1-1.28) and Underload (>1) [10]. There were five steps to measuring the FTE score: measuring allowance and effective time calculation, adequacy test and data uniformity, adjustment rating factor, normal time and standard time calculation, and full time equivalent measurement.

b. **Cardiovascular Load Percentage (%CVL):** Meanwhile, %CVL was conducted by examining the physical workload using cardiovascular load with four distinctive results, acceptable level (<30%), moderate level (30%-59%), high level (60%-99%) and intolerable high level (100%) [11]. The physical workload measured the working and rest pulse rate by the help of stopwatch. The pulse rate was used to indicate the workload and calculated the %CVL. this stage will be determined allowance consisting of leniency for personal needs, allowance to eliminate the sense of fatique, allowance for inevitable obstacles and work environment.

c. **Consultation of combination results from both criteria:** Both criteria were analysed and consulted with the management to understand the nature of the result, and recommend the further action needed. At this stage will be determined the effectiveness on the measurement of time base workload and physical workload.

3 Results and Discussion

Cold press working group was in the wood working area that performing 27 cabinets and 17 models that would be distributed across sections. The job description involved in this group were veneer cut, panel saw, chipping, glue spreader, press machine, bench saw, and key bed bone assembly. There were 5 operators with age range from 22-33 years old as the respondents.

This study conducted based on company's data on June to August 2017. Operators work hours were from 07.00 until 16.00 on working days (5 days a week, Monday to Friday).

3.1 Full Time Equivalent

The Full Time Equivalent (FTE) was calculated based on the job description performed for each operator. The job description then would be measured for each working element attached. After the working hours were accumulated for each operator and each working element, this cycle time was then multiplied by the rating factor and allowance to get the normal time and standard time accordingly. As the example, the panel saw for Side Board B1 model, the working time was 0.39 minutes. The calculation of rating factor and allowance were displayed based on the table 1 and table 2 respectively.

Table 1. Rating Factor

Factors	Category	Score	Accumulation
Skill	Superskill (A2)	0.13	
Effort	Good (C1)	0.05	
Conditi on	Fair (E)	-0.03	
Consist ency	Good (C)	0.01	0.16
Rating Factor		1.16	

Table 2. Allowance

Factors	Category	Percentage
Work performed	Light	7.5 %
Work Attitude	Stand on both feet	1.5 %
Work Movement	Swing restricted on hammers	1.0 %
Eye fatigue	Disjointed views	1.0%
Temperature	Normal	5.0 %
Atmosphere	Enough	2.0 %
Environment	Repeated circle per 5-10 seconds	2.0 %
Personal Allowance	Male	1.0 %

Total Allowance	21.0 %
Total Efficiency Factor	79.0 %

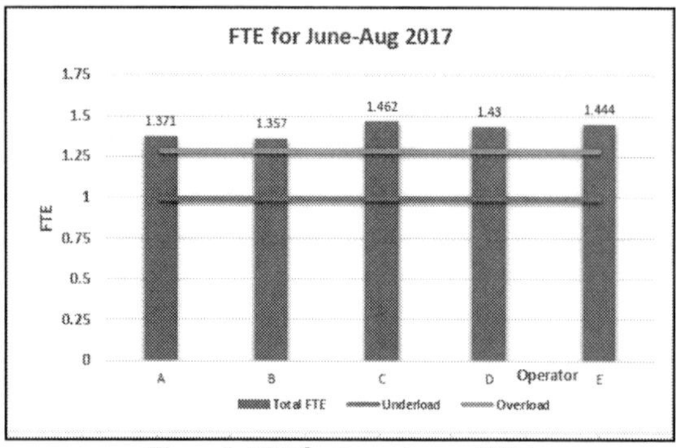

Figure 1. FTE for June-August 2017

Based on the calculation using the rating factor, the normal time for the Panel Saw of the Side Board B1 (average time was 0.39 minutes) would be 0.4524 minutes. In addition, the standard time would be calculated as 0.58 minutes. Based on the frequency observed was 32 times a day, the FTE calculated was 0.0493. Continued by all element and all operator, the complete list for the FTE for all operator was captured in the Figure 1.

The figure 1 shows that all the operator were burdened with overload works due to the FTE all above 1.28. This problem could be solved by adding working hours (overtime) or adding operator (this was unlikely chosen by the company). High FTE index gained by every Cold Press operator was influenced by several factors including production planning and allowance granted. In this study allowance was given quite high at 21%. There were several categories in the allowance could be improved so as to reduce the percentage of allowance. One of the high factors was the environmental factor with the percentage of 5%. This was due to the temperature of the working environment are included in the normal category but still make the operator feel stifling in occasionally. The impact observed was sometimes the operator tried to find a fresh spot for refreshing the body. In addition, atmospheric factors also influenced the allowance in this study. Allowance in the amount of 2% included in the category enough because in the Cold Press working environment, the ventilation was categorized as good but still there was the smell - quite annoying odours such as the smell of Kony Bond glue.

3.2 Cardiovascular Load Percentage

The data required in the measurement of the cardiovascular load percentage (% CVL) as physical workload were data of working pulse, resting pulse rate, and age of each operator. The pulse data obtained by measuring the time of 10 pulse rate both worked and rested on operator using a stopwatch. Based on the working hours implemented by the company, the pulse rates were collected during 5 working pulse (09.00, 10.00, 11.00, 12.00 and 14.00) and 2 resting pulse (09.30 and 12.30). On each operator, the observations/measurements were conducted on this 7 times pulse rate on 4 working days and calculated for the average results. The data for the pulse rate shown in the table 3.

Table 3. Pulse Rate Operator

Operator	Age	Average Time for 10 Pulse		Pulse Rate	
		Rest	Work	Rest	Work
A	33	6.91	6.66	86.81	90.10
B	29	7.65	6.56	78.41	91.43
C	23	7.69	6.95	77.97	86.34
D	23	7.42	6.54	80.88	91.81
E	22	7.49	6.52	80.22	92.06

Cardiovascular strain was an estimation to determine the classification of the workload by increased pulse rate of work from the resting pulse compared with the maximum pulse rate. The following shown figure 2 as the recapitulation of the calculation workload in the Cold Press.

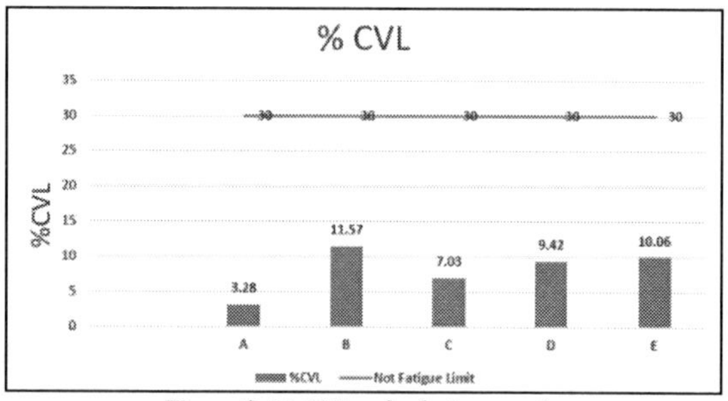

Figure 2. % CVL calculation result

Based on the results of the calculation of the workload for example for the type of work on the operator A aged 33 years with the job type was cutting in the machine panel saw, cut the cabinet key bed P22 or U1J in engine bench saw, and the process of cutting R in the machine chipping the category in the classification of the workload is not experiencing fatigue. This was because the percentage of operator A CVL was 3.28 (% CVL<30 %).

From the workload classification by methods Cardiovascular Load in all sections Cold Press operator had the load of Cardiovascular below 30%. So that the operator workload could be included in the classification did not experience fatigue. The least load of cardiovascular was in a operator that had a lowest percentage of the cardiovascular load (3:28%) with a job description focuses on the cutting process in the machine cabinet Panel Saw. The greatest burden by operator B was equal to 11:57%. the job description was in part the emphasis, gluing and assembling bone Key Bed and U1J P22 models.

3.3 Combination results from both criteria

Based on the results above occurred differences in the results of the two methods of measuring the workload. For the method of Full Time Equivalent all operators at the Cold Press had excessive workloads. This could be caused by several factors such as the length of time of execution of each process per cabinet, the high allowance, many types of cabinet are processed, until the layout is not optimal. While based on the measurement on %CVL workload results obtained that all operator of Cold Press had % CVL below 30%, which indicated that all operators were in a state of not experiencing fatigue. However, there were still some works continued to be performed could pose a potential danger to health and safety. This result was because the workload measurement method used has a different approach. The first method is the measurement of the workload with the method of Full Time Equivalent (FTE) was a measurement of workload by using the approach of working time as a reference. The second method was by the percentage of CVL was measuring the level of physical workloads that require physical energy muscle as a source of energy. From the measurement results of both methods occurred differences in measurement results workload by both methods could be concluded that operator did not experience fatigue physically but operators are in an overload time measured or the operator could still receive excessive workloads FTE based on the calculation method. Based on data from the efficiency of June 2017 there were 5 working days from 15 effective days were overtime work and in July 2017 there were 12 days of overtime of 21 effective working days.

Taking into further account the results of the FTE workload measurement method and the percentage CVL researchers were also trying to find the best alternative between the two methods was based on salary costs incurred. Based on the comparison calculation resulting costs to be paid in the period of September 2017 - November 2017, doing overtime would have 1.5 times more expensive than the cost of the salary increase of the operator. Whereas, if only the company add one more operator, the cost incurred 0.7 times cheaper than by doing overtime. This was merely by the salary and fees only, excluding the long consideration of hire and termination contract, and other cost derived by adding or reducing number of operator.

5 Conclusion

Based on the research, could be concluded that all operator was categorized on FTE as overload, all operator was categorized on %CVL as not experiencing fatigue, and also the correlation between both categories did not state equal comparison. Therefore, it validated on the increased productivity (based on the increased FTE score) even though the physical workload was perceived below the limit (underload) as the research objectives. Regarding the FTE method, the workload experienced by all Cold Press operators were above 1.28 (indication of the "Overload" condition) as 1.371, 1.357, 1.462, 1.43 and 1.444 respectively resulted in average 1.4128. In the contrary, based on % CVL method, the physical workload perceived as non-experiencing fatigue category due to the average score was 8.272% and all score were below 30% (even less than 12%).

References

[1] Hart, S.G. "NASA Task Load Index (NASA-TLX): 20 Years Later", Moffett Field: NASA-Ames Research Center (1990)

[2] Nurmianto, E. Ergonomi: Konsep Dasar dan Aplikasinya. Surabaya: Prima Printing. (2004).

[3] EH. Puspawardhani, MR. Suryoputro, AD. Sari, RD. Kurnia, and H. Purnomo. Mental workload analysis using NASA-TLX method between various level of work in plastic injection division of manufacturing company. Advances in Intelligent Systems and Computing Advances in Safety Management and Human Factors, Orlando (2016)

[4] O. J. Sealetsa and A. Thatcher. Ergonomics issues among sewing machine operators in the textile manufacturing industry in Botswana. Work, 38, pp. 279–289 (2011)

[5] D. Bonfim, A.M Laus, A.E Leal, F.M.T Fugulin, R.R Gaidzinski. Application of the Workload Indicators of Staffing Need method to predict nursing human resources at a Family Health Service. Revista Latino-Americana de Enfermagem, 24 (2016)

[6] P.A McQuid, R.L Kolehmainen-Aitken, and N. Forster. Applying the workload indicators of staffing need (WISN) method in Namibia: challenges and implications for human resources for health policy. Human Resources for Health, 11, 64 (2013)

[7] A.D Sari, M.R. Suryoputro, M.D. Pramaningtyas, P.S Putra and S.B. Maulidyawati. Work Physiology Evaluation of Laundry Workers. IOP Conference Series: Materials Science and Engineering (2016)

[8] H. Purnomo. Workload Analysis For Determining The Number Of Employees At Bank Companies. GEMA Jurnal 201. 2068-2076(2015)

[9] Zimmerman, S., Gruber-Baldini, A. L., Hebel, J. R., Sloane, P. D., & Magaziner, J. (2002). Nursing home facility risk factors for infection and hospitalization: importance of registered nurse turnover, administration, and social factors. Journal of the American Geriatrics Society, 50(12), 1987-1995.

[10] Tridoyo, T., & Sriyanto, S. Analisis Beban Kerja Dengan Metode Full Time Equivalent Untuk Mengoptimalkan Kinerja Karyawan Pada Pt Astra International Tbk-Honda Sales Operation Region Semarang. Industrial Engineering Online Journal, 3(2). (2014)

[11] Ismaila, S. O., Oriolowo, K. T., & Akanbi, O. G. Cardiovascular strain of sawmill workers in South-Western Nigeria. International Journal of Occupational Safety and Ergonomics, 19(4), 607-611. (2013).

A Study for New Balance Sports Products to Analyze and Design

Yu-Che Huang[1], Tai-Shen Huang[1]

[1] *168, Jifeng E. Rd., Wufeng District, Taichung , 41349,Taiwan*

Abstract. In recent years, under the influence of globalization, more and more different needs have created different sports products. In general, they are mainly used for cardio training and muscular endurance training, and their products for human balance training are few. Balance training has always been one of the most important exercises, but most exercise training is often overlooked. In this study, we searched for sports product patents and related literature that improve balance training and determined the direction of innovative design based on the different patent categories. Based on the results of the analysis, we have redesigned a model that can effectively provide innovative sports products with different angular misalignments and different displacements. In this research, we search, analyze and summarize the related patents and further identify the methods of innovative design to accomplish the required technologies for this design. The relevant design of this study is to use SolidWorks to draw 3D models for motion simulation and size checking. Then use the ANSYS11.0 command module to analyze the body structure of the force situation, to determine the strength and safety. Finally, make the original. Finally, this study hopes to provide a design reference for future sports product design engineers through this design process.

1 Introduction

The diminished sense of balance is one of the important indicators of the aging process. Normally healthy adults begin to exhibit initial signs of diminished balance after the age of 60, and then degenerate more rapidly thereafter. The main reason is the degradation of multiple physiological functions, including the nervous system, skeletal joint system and cardiovascular system, all of which cause the degradation of "balance" function. And the degradation of balance will directly affect the daily life of the elderly, if combined with environmental factors (such as poor lighting, slippery ground) are often easy to make the elderly fall. The study found that 30% of elderly people over the age of 65 fell and one quarter of them were severely injured. About 5% of those who received medical examination had Fall twice or more. Research also shows that the chances of falling old people falling again are four times those of those who have never fallen. So preventing falls is very important to the elderly. Preventing falls can start with exercise, improving the home environment and using appropriate walking aids. The purpose of this study is to enhance the promotion of the balance of the elderly through training of balance exercises [1] [2] [3].

In general, after long-term practice, we can rely on the body's reflex and coordination ability to walk easily. However, modern people often get sedentary for a long time and cause muscle soreness for a long time. Even brain response and limbs balance Will be gradually reduced by the fitness equipment to maintain physical fitness, general exercise equipment by monotonous reciprocating exercise forging muscle endurance, and balance training machine can enhance the brain inner ear vestibular, semicircular canal and other organs and the coordination between the muscular Speed, suitable for promoting the brain development of young children, special education for people with intellectual disabilities, video game simulator and indoor surfing training and other purposes. This research by collecting relevant patents and existing products for analysis to find a simple and reliable mechanism with the lowest cost , To achieve the best training effect of the new balance training machine, so that both men and women can easily and safely experience the balance of exercise fun. These guidelines, written in the style of a submission to MATEC Web of Conferences, explain how to prepare your paper using Microsoft Word.

Please submit sources files directly to the conference organizer. If the conference editors chose to provide print-ready PDF documents to the publisher, you have to submit high-resolution PDF file with all fonts embedded (see PDF guidelines) instead of the sources files. In this case, please remember that no final corrections will be made by the publisher. There are quite a lot of patents and products available in balance training machines today. There are no motivation sources for "machine passive" and "motivation for machine". Please submit sources files directly to the conference organizer. If the conference editors chose to provide print-ready PDF documents to the publisher, you have to submit high-resolution PDF file with all fonts embedded (see PDF guidelines) instead of the sources files. In this case, please remember that no final corrections will be made by the publisher. There are quite a lot of patents and products available in balance training machines today. There are no motivation sources for "machine passive" and "motivation for machine".

2 Patent classification and analysis

There are quite a lot of patents and products available in balance training machines today. There are no motivation sources for "machine passive" and "motivation for machine". There are two major categories: machine passive balance training machine, the product itself is not Power-driven, rely on the user to move the body to stabilize the equipment, which has the advantages of small size, low cost, simple structure design is easy, often used to stimulate the development of vestibular vestibular training equipment, but the operation is too monotonous, long-term use of user fun gradually Lost, often can only be placed in the warehouse finally Section Headings.

The second type of active balance training machine has more than one actuator drive mechanism, which can make the pedal perform the deflection, bumpy and swivel as shown in Figure 1, and even the displacement in different angular directions, causing the user's body to disintegrate To do this, the muscle must be moved to maintain the balance of the body. To achieve this goal, an active balance training machine usually has multiple sets of original work units, large volumes and high costs. The following is a list of several active balance trainer patent agencies to introduce, from induction to find innovative design direction.

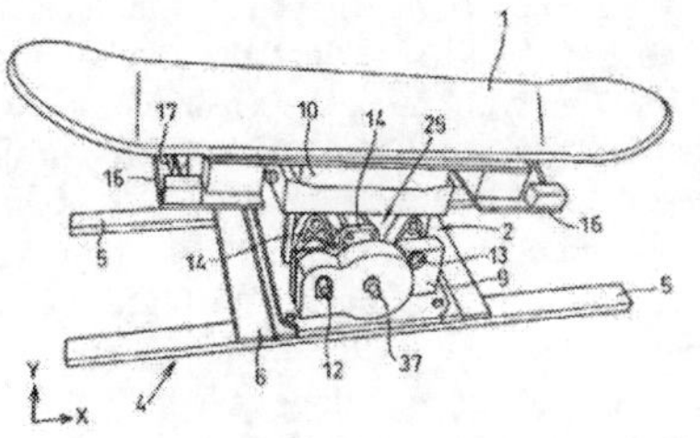

Figure 1. Pedal steering.

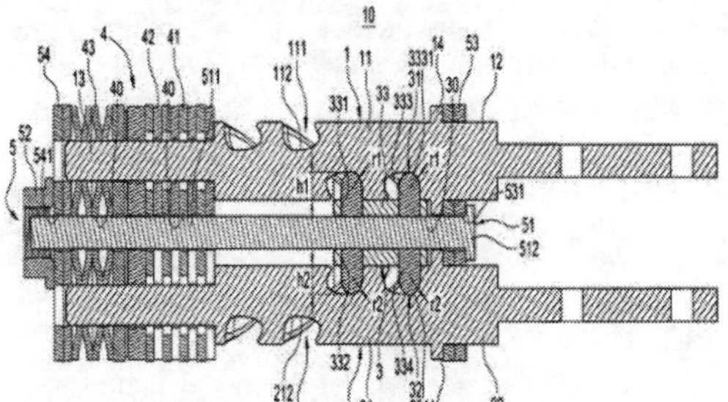

Figure 2. Twin Screw Drive Crank Slider.

Twin-screw-driven balance training machine [4], Figure 2 This patent is driven by the motor leadscrew, inverted crank slider mechanism slider, so that the crank angle changes, the use of two parallel linkage system to drive the pedal , This design can get the bumpy and the rotation of the two directions of the angle change, the disadvantage is unable to produce the user's horizontal displacement of the center of gravity, the use of screw drive can get accurate and stable output, but lost the speed of change, reducing user challenges. Please submit sources files directly to the conference organizer. If the conference editors chose to provide print-ready PDF documents to the publisher, you have to submit high-resolution PDF file with all fonts embedded (see PDF guidelines) instead of the sources files. In this case, please remember that no final corrections will be made by the publisher [4].

As shown in Figure 3, the patent uses two sets of crank connecting pedals. Since the shafts of the two cranks are not parallel, it can generate complex 8-shaped swing and tilt and use a single motor via the bevel gear drives stable rotation of two sets of crankshafts and has a wide range of motion variation. Because of the simplified structure design, the cost is low and the manufacturing is easy. Only the direction of the displacement can not be controlled. Only the speed is adjusted, and the user feels closer to the vibration Body shaping sports equipment [5] [6] [7] [8].

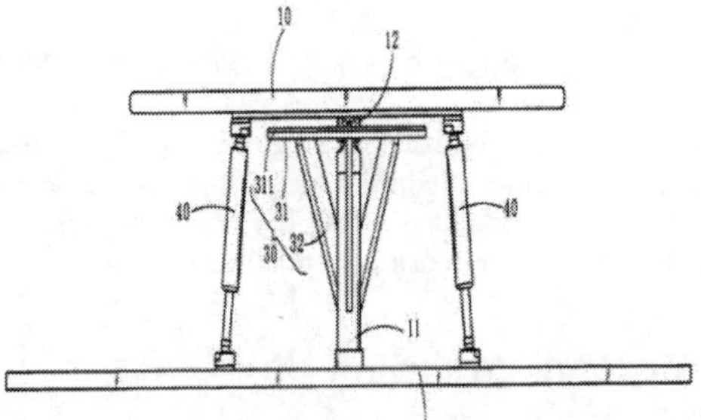

Figure 3. Double crank interlocking.

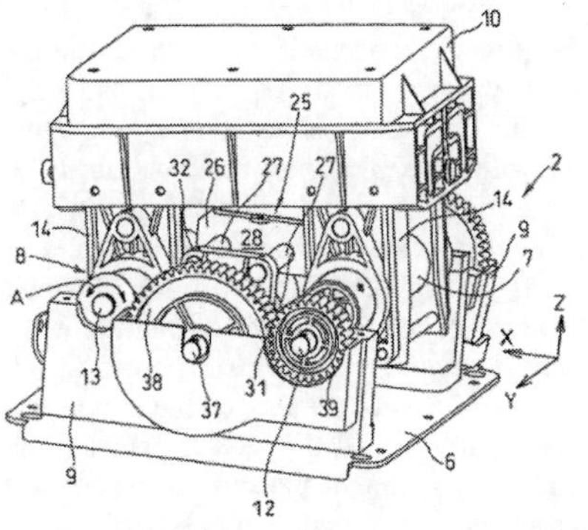

Figure 4. Composite Balance Trainer Driver.

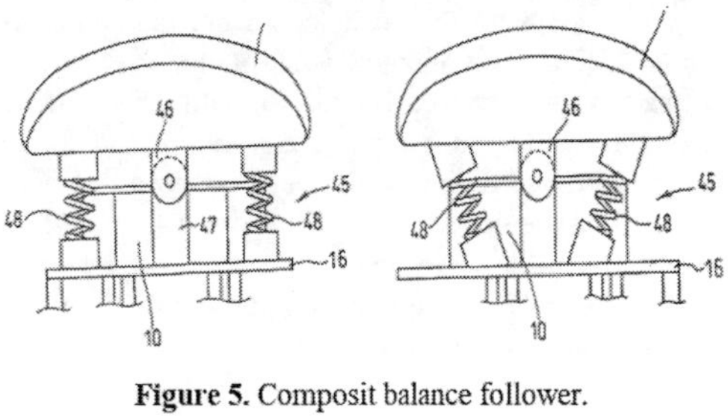

Figure 5. Composit balance follower.

3 The balance of training equipment design conditions

The analysis of available patents and related products available after the hair

Now, the balance training machine designed in this study aims to train the indoor sports equipment that reflects the coordination between the body and the body, and provide the users of any age without any limitation of time and space to exercise and blood easily Cycle effect, so the definition of balance training machine basic design conditions, respectively, as follows:

(a) The design focus must be achieved with the least amount of parts, the ability to achieve the most movement, and the goal of reducing material costs.

(b) Is a surfing simulation training machine, machine running frequency is not high, but the pedal movement range and displacement range should be sufficient.

(c) Balance trainers need to provide better stability, so the lower the level, the better.

(d) It must be closer to real surfing and challenging.

(e) Designed to be safe and comfortable, to simplify and weight the components, and to have sufficient mechanical strength.

According to the above patent classification, "active" and "passive" balance trainers each have their own advantages and disadvantages. If one can design a sport capable of "active" and at the same time has a "passive" small volume for balance training Machine, I believe it will become a new consumer sports equipment.

4 Mechanism design of balance training machine

This time, we hope to achieve the most movement changing ability with the minimum number of machine parts. First of all, we should think about the mechanical structure that has a variety of rotation and displacement capabilities. The spherical joint is a single mechanical structure with yaw, pitch and roll rotation. However, To achieve rapid and accurate positioning, we must rely on an array of complex transmission mechanism to guide, do not meet the design conditions (a) described, so give up this idea; change to use six axes of rotation with two-axis rotation angle cylinder mechanism, partial Angle cylinder is composed of two cylinders combined with each other, shown in Figure 6, and the axis intersect to maintain a fixed angle, fixed one of the cylinder rotation, the other cylinder will show a reciprocating swing, and then use the ring into the swing cylinder to form a surface contact, the pedal Fixed on the ring, and the installation of the bracket to maintain the level to prevent the pedal group with the declination of cylindrical rotation, as long as the declination cylinder fixed axis, you can drive the pedal pitch, yaw, pitch ... reciprocating motion, to a single power input Resulting in biaxial angular displacement target, when the output axis and the drive shaft deflection angle of 10 °, the motor per revolution The plate swings up and down while the output shaft produces 20 ° (± 10 °) pitch, yaw angular displacement.

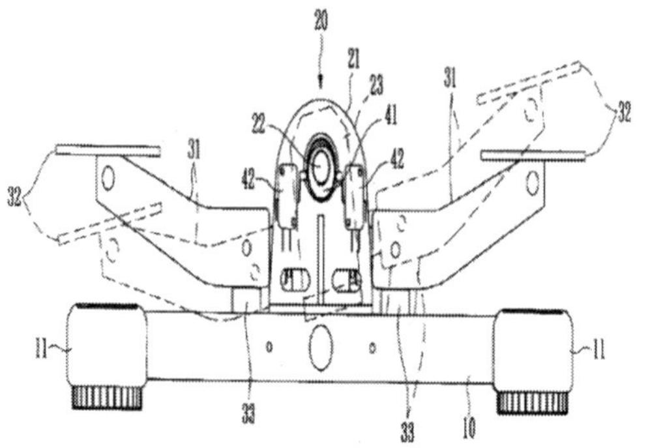

Figure 6. Angular cylinder.

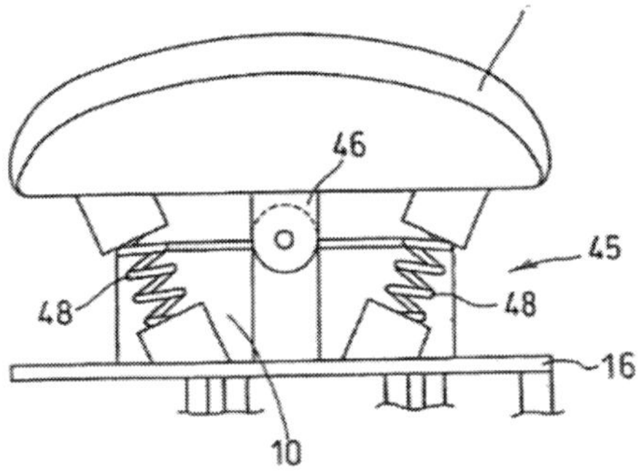

Figure 7. Pedal roll angle displacement front view.

If the pedal continuously changes in the stable angular displacement, it will soon lose the user's fun. Therefore, in combination with the passive balance training machine, the user must overcome the external force of the machine driven by the human body and the additional center of gravity from the passive balance training machine Change, and increase the use of the variability and challenges; Consider surfing the Y direction of the wave is less chance of encountering, so long as the surfer in the event of waves hit the pitch (Pitch), you can still stand on the water, stabilize the lateral (roll) can not fall into the water, so the passive balance training machine swing direction is set to rotate the X-axis, the implementation of the method used to maintain the level of the bracket to remove, to change the angle cylinder drive shaft on both sides, each setting A set of compression springs, similar to the see-saw structure shown in Figure 7, is used to generate the roll angular displacement with a simple design, which meets the design conditions of Article (e) and reduces the size of the body to reduce manufacturing costs.

Horizontal displacement of the pedal includes △ X, △ Y biaxial translation and Z yaw, in order to obtain sufficient movement, use the horizontal position as shown in Figure 8 crank-slider mechanism as the base, the pedal is fixed on the even rod, by adjusting the length of the crank and the spindles, the appropriate pedal horizontal displacement is obtained, and the yaw from the crank is combined with the yaw angular displacement produced by the angling cylinder to produce more variability, You want to be more stable in the pedal above the more challenging.

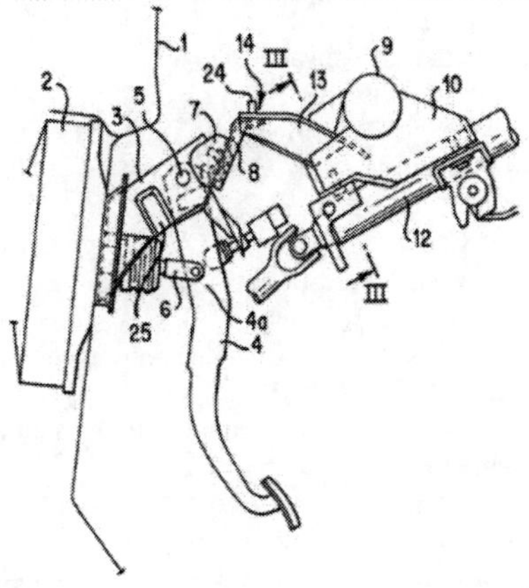

Figure 8. Pedal horizontal displacement mechanism.

4.1 The Establishment of 3D Model and Stress Analysis of Balance Training Machine

According to the design requirements to define the important size of the organization, the use of computer-aided design software AutoCAD 2016 design, draw a detailed part portfolio diagram, and then further use SolidWorks2016 according to the size of 2D drawings to build the 3D model of Figure 9 and Figure 10 parts drawings and portfolio, Test whether the interference between the various components, and consider the ease of assembly, make reasonable modifications, and finally the use of computer graphics software for dynamic simulation of movement and horizontal displacement, you can show the range of components, and then design a sufficient size of the shell cover Internal agencies, to prevent the user due to improper use of the resulting risk of injury.

After completing the 3D model of the balance trainer and confirming that there is no interference on the part of the organization, the stress analysis of the structure is performed to ensure that the strength of the structural members and joints are within the safe range and the entity building can be avoided Finished to find the design strength of the defect. The materials used in the structural part of this study are based on structural steel A36 pipes of 1.2 mm thickness, while bolts are the material of medium carbon steel AISI1045. And simulate the stress analysis as shown in Figure 11, through the 120kg user's weight when operating on the balance trainer. The ANSYS software analysis shows that the maximum stress value is about 149.01MPa, which occurs at the center of the base frame. The bolt material used in this study is AISI1045 medium carbon steel with a yield strength of 530MPa. The structure is made of A36 structural steel. The yield strength of 250MPa, so the bolt and the structure are not damaged, the safety factor of 1.68, the strength of the solid structure within a certain range of safety.

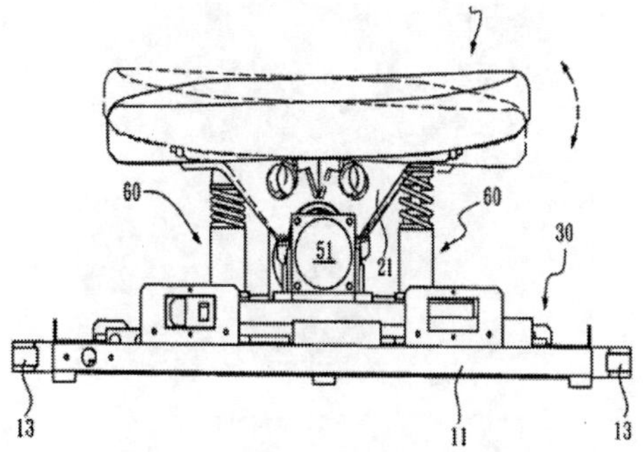

Figure 9. Passive roll rotation system.

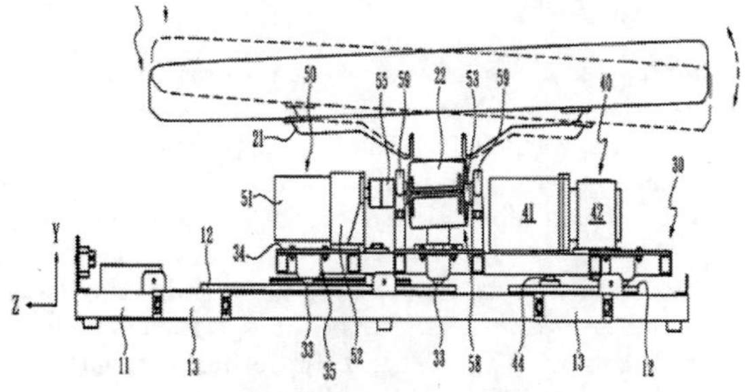

Figure 10. Intermediate base.

Figure 11. Simulation analysis.

4.2 Prototyping

After a series of complete design and analysis, the output drawings of all the parts were processed and manufactured, and some defects were partially improved. The prototype of Figure 12 and Figure 13 was assembled and the operation test was conducted to verify the original design concept. This design also places a sensor under the yaw mechanism and displays the balance of the user through the light on the display of Figure 13 to provide future balance training.

Figure 12. Prototype entity.

Figure 13. Bias correction sensing.

5 Summary

In this study, two DC motors were successfully used as the driving source to achieve the roll, pitch, and yaw three-axis rotation and balance exercise machine that balances the horizontal movement, and saves the parts cost by ingenious design without sacrificing sports performance. One of the directions for future development is to favor large-scale game simulation machines with a wider range of operation and more controllability. The other is the opposite, with light, thin, small size and more cost-saving home-balance training machines.

References

1. S.E. Syu, M.H. Hsu, Kunshan University of Science and Technology Institute of Mechanical Engineering Master thesis, Taiwan (2015)
2. Z. Shen, Jianguo University of Science and Technology Department of Automation Engineering and Electrical and Mechanical Systems Master thesis, Taiwan (2012)
3. Y.L. Chiu, J.N. Chen, Chaoyang University of Science and Technology Industrial Design Institute Master thesis, Taiwan(2014)
4. R.P. Bittikofer, US patent announcement, Patent number: 5593380 (1997)
5. S. Ho, Republic of China patent announcement, Patent number : 310010 (2006)
6. Shimizu stretch one, Republic of China patent announcement, Patent number: 957555 (2006)
7. K. Grondys, I. Kott, A. Wiśniewska-Sałek, Polish Journal of Management Studies, 10(1), 45-53, 2014
8. S. Kot, Promet - Traffic - Traffico, 27 (5), 387-394. 2015

Design of an Automated White Board Profiler

Md. Tanzim Hasan[1], Shahin Parvez[1], Shaylesh Roy[1], Md. Abdullah Sakib[2], and K. M. Khalidur Rahman[2]

[1]*Bangladesh Army University of Science and Technology/School of Mechanical Engineering, Saidpur-5310, Bangladesh*
[2]*Bangladesh Army University of Science and Technology/School of Computer and Electrical Engineering, Saidpur-5310, Bangladesh*

Abstract. This paper presents the design for an automated and modernized white board profiler. The design objectives were to: (a) Introducing the concept for a new method of white board cleaning. (b) Getting rid of time consuming old methods. (c) Introducing the design with a CAD model with circuit analysis. Proving that it is time and labor efficient while being cost efficient to do it by using electricity and mechanical body Also, enhancing the accessibility of the user and increasing efficiency in terms of use. This work is expected to contribute to the overall knowledge of machine design and white board cleaning automation in specific.

1 Introduction

White board is an important aspect of any modern classroom. Teachers, Students all the persons present in a classroom looks up to it to acquire and give knowledge by using it to share information. It's an important part of a classroom. A teacher may give lecture orally. But when it's written or presented on the white board it becomes clear. And it may become clearer when it's graphical as a teacher may draw figures on it to make students understand properly and efficiently. It doesn't need much maintenance. All it needs is a marker to write and a cloth or duster to clean. Although it doesn't take much time to clean it, but it's an orthodox way of cleaning to do it by hand. As, it takes more time and strength to do so, it's easy to clean only one line. But when the board is full it's difficult and tiring to do so, That's why this paper aims to get rid of that way of cleaning and introduce a new and modernized automated way to clean the white board with efficiency and perfection in a much less time. So, it can be ready for the next class or lecture for the students and teachers altogether. It will help in all around class work and save time evidently. It will also get rid of strength consuming and pain elevating wiping that may be done by hand.

2 Conceptual design

The white board profiler was designed as a whole setup and was divided into sections for clear understanding and view. Firstly, we can see the whole setup. The body is envisioned to be made of Synthetic industrial Plastics, [1]. There are rail and gear pathways on top and bottom of the white board. With two obstacles on each sides that acts as a double as a marker holder. And finally the cleaner part that is situated on the rail and gear pathways. It's adjacent and will rub against the white board surface. The whole setup is designed to be wire-less and remote control as a secondary option but it's primary function is to be wired and switched.

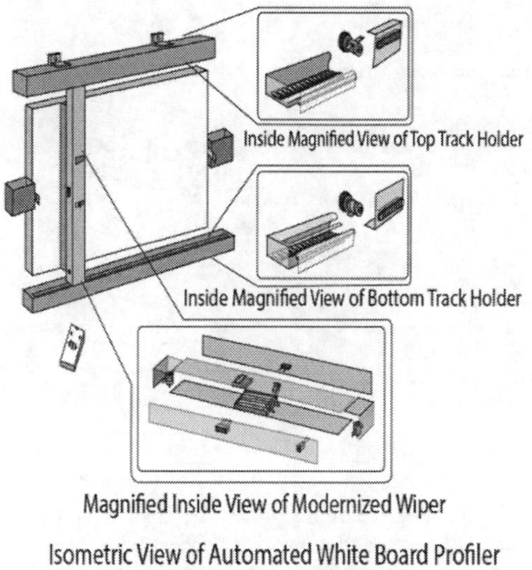

Figure 1. Concept design of the whole setup.

As we can see from Fig. 1, the whole setup is done on wall. The upper and bottom pathways is set up with hinges for support. The obstacles are set up with screw. The modernized wiper part rolls on the pathways with mainly gear and also support wheels for additional support and mobility with less friction.

2.1 Wiper track

The wiper mainly runs on a track path made by gear and wheels. The Track has two parts are: 1. Top Track Holder. 2. Bottom Track Holder. Both of them are mirror of each other and has the same principle.

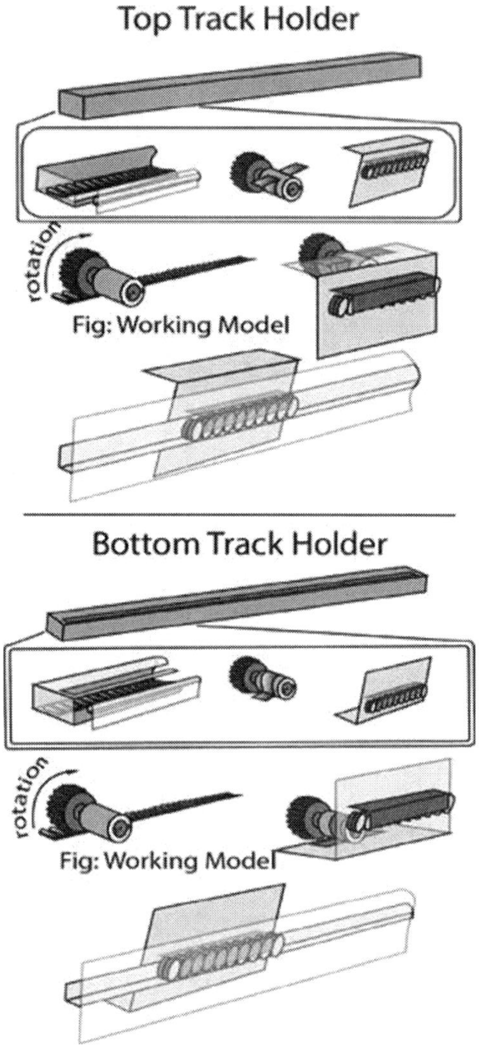

Figure 2. Wiper track.

As we can see from Fig. 2, the track pathways are made with gear and supported wheels. The gears have motor and clam attached with them and it is fixed with the wiper as it moves the wiper. The wiper also has added support wheels which helps in displacement and mobility with also reducing it's friction when needed. The track also has a path for these support wheels. Both the upper and bottom part of the track is distanced from the white board, And has the same mirrored function overall.

2.2 Modernized wiper

Ultimately we come to the main part of this setup that is the modernized wiper. The wiper runs on motors mainly the two that is discussed above. The motor has gears and support wheels are attached to the wiper body to make it more frictionless and mobile. The duster part that cleanes the board is replace-able by simple lock for replacing the duster if and when needed. There are two motors in the model that runs on battery. It also has a remote control receiver and switch inside of it. As, it can be made both wireless or wired, As the controllers are Adaptive and Infrared is used, [2, 3]. It also has a charging port. For charging the re-chargable battery that will be used in the wiper.

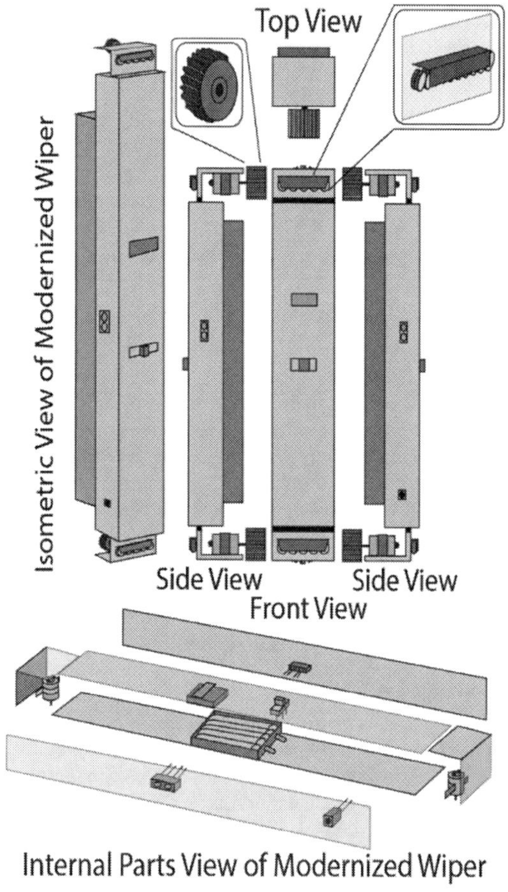

Figure 3. Modernized wiper.

One other important thing that the wiper has is, the obstacle sensor. The obstacle sensor is made to sense the obstacle that was previously discussed and shown in Fig. 1. The sensor senses the obstacle and reverses when it is close to it. All of it is shown in Fig 3. That way it can wipe clean both sides in much less time with more efficiency. All the electrical parts are assembled in a circuit that will be discussed later in the paper.

3 CAD model

The whole setup and it's parts were made into CAD with the help of SolidWorks 2014© software for more realistic approach and design with it's properties and formation. The design will help in understanding how the the model would look in real life.

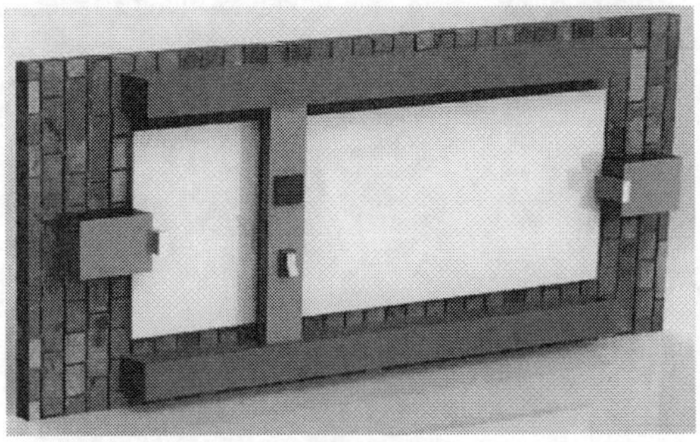

Figure 4. CAD model of the full setup.

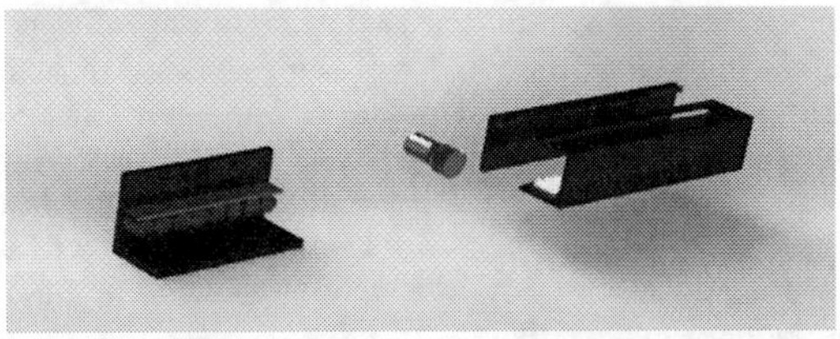

Figure 5. CAD model of the top track holder.

Figure 6. CAD model of the bottom track holder.

Figure 7. CAD model of the modernized wiper.

From Fig. 4 to Fig. 7, These CAD models shows the real life like design of the white board profiler and it's different parts. It is inspired by model based system engineering, [4, 5]. It will help in understanding the design properly and effectively.

4 Circuit design

As the model is run on electricity. May it be ac or dc as the design was made to provide both. That's why a circuit analysis was made to fully understand the circuit works and electrical works. The circuit was made with help of programmable circuit primarily Arduino©. Which helps the censor to find obstacle and report it back to the motors for changing their direction in the opposite side to get the board back.

The circuit was designed with the help of Proteus Design Suite©. The fig. of the circuit is provided below.

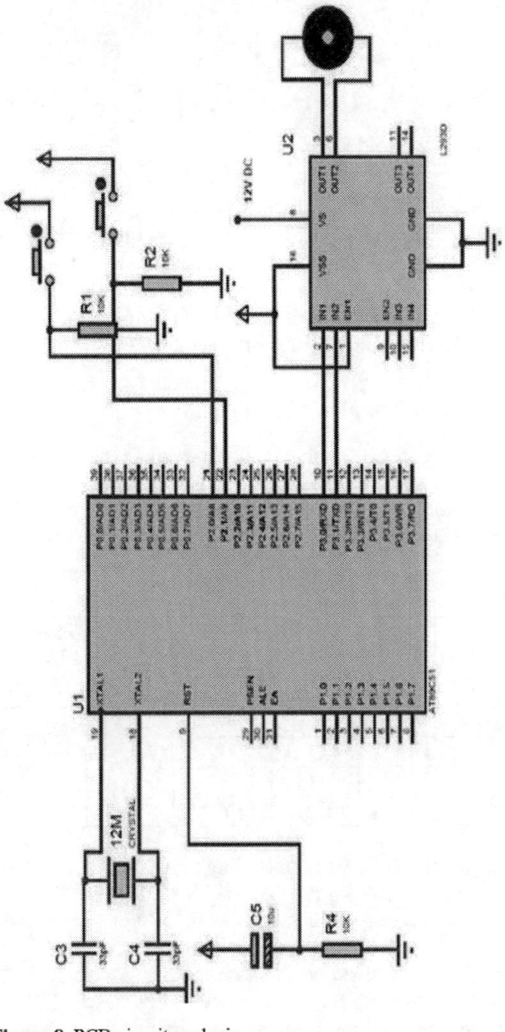

Figure 8. PCB circuit analysis.

As shown in Fig. 8, this circuit analysis describes the use of PCB and its effects on the motor. This circuit is done for the modernized wiper. It helps in turning the wiper back on the track once it has sensed the obstacle by obstacle sensor.

The circuit is an important part of the wiper. It's mainly the backbone that helps in moving the modernized wiper properly. The circuit is mainly to move the motor that holds the gear and sensing the obstacles and reversing the motors. It's an important operation for the modernized wiper. The programming code for the Arduino is also provided below.

```
# define M1 PORTB.F7
# define M2 PORTB.F6
# define B1  PORTB.F0
# define B2  PORTB.F1

void main() {
TRISB=0;
TRISD=0xFF;
PORTB=0;

while(1){
if(B1 == 0) {PortB=0xAA;}
if(B2 == 0) {PortB=0x55;}
}
}
```

Figure 9. Programming code for Arduino.

As shown in Fig. 9, This is the programming code that is put into the Arduino, That oversees the whole operation and controls the PCB. This type of programming is seen in many automated machines. Such as, An Automated Waste Bin Monitor, [6].

4.1 Circuit analysis result

An analysis was made to understand the voltage, current, torque and other properties that is seen in the circuit and the modernized wiper. A table is provided below for proper understanding.

Table 1. Result Analysis.

Input	Voltage (V)	Current (A)	Motor Output (Watt)	Require d Power (Watt)	System Weight (kg)
16	15.2	2.54	38.608	2.99 (Total Force)	≈ 2.00
12	11.4	1.9	21.66		
8	7.62	1.27	9.6774		
6	5.71	0.95	5.4245		

1. Brush Bar = 450 gm including brush.
2. Battery = 34x5 gm.
3. Drag Force = 1380 gm.
4. Approximate Torque 6.52 g.cm.
5. Total Effective Length 240 cm.
6. Total working time 5 sec. Speed = 48 cm/sec (The power required for this motor was driven by comparing to a large motor's reading which was 1000 w 48 V DC).

We can see from the results that the final speed of the board will be 48 cm/sec. The drag force would be 1380 gm. The Approximate Torque 6.52 g.cm. It will be more than enough to clean the board in and under 5 seconds. The device will clean the board in much less time and with efficiency than done by human hands.

5 Cost analysis

Currently the conventional methods of cleaning the whiteboard includes many ways which includes chemicals, cleaning cloth, dusters, wipes etc. They are relatively cheap costing around $5.96 to $212.00. But all of them require handwork and labor. Which eventually is time consuming and painful.

This setup can be made in any industry or at home. The production cost is very low and cost efficient, eventually cheaper than conventional methods. It can be assembled by hand. The whole setup will cost around $30. And will be very time saving while being used. Conventional cleaning requires the use of hand which takes much time. But this automatic setup will do it under 5 seconds.

6 Conclusion

This paper indicates the development and conceptual design for a new technological achievment in the field of classroom and technology. The technology and electrical aspects that are used in in the design for this device is relatively easy and properly used in modern technological areas. This will help in time saving as well as efficiency in modern classrooms. And will help in saving time for the teacher and students alike. Giving more time to improve in many areas and study as well as work. Also having room to improve itself in many areas. Though in this paper the development work is for only one board but it can be developed for more board and more development gradually.

Acknowledgement

This work has been supported in part by the grant provided by the Research Fund of Bangladesh Army University of Science and Technology, Saidpur, Rangpur, Bangladesh.

References

1. G. O. Young, "*Synthetic structure of industrial plastics*," in *Plastics*, 2nd ed. vol. **3**, pp. 15-64. (1964).
2. S. P. Bingulac, "*On the compatibility of adaptive controllers*," in *Proc. 4th Annu. Allerton Conf. Circuits and Systems Theory*, pp. 8-16, (1994).
3. J. U. Duncombe, "*Infrared navigation—Part I: An assessment of feasibility*," *IEEE Trans. Electron Devices*, vol. **ED-11**, pp. 34-39, Jan. (1959).
4. B. P. Douglass, "*Chapter 1 - What Is Model-Based Systems Engineering?*" *in Agile Systems Engineering*, Boston: Morgan Kaufmann, pp. 1–39, (2016).
5. N. Van, K. Gadeyne, and M. Witters, "*Model-based systems engineering of discrete production lines using SysML: An experience report*," *Procedia CIRP*, vol. **60**, pp. 157–162, (2017).
6. Md Ahad Bin Alam, International Journal of Advanced Research in Electrical, Electronics and Instrumentation Engineering. *Automated Waste Bin Monitor*.

Derivation of Life Cycle Cost Model for Selecting Optimum Decisions in Engineering Asset Management

W.N. Cahyo

Department of Industrial Engineering, Faculty of Industrial Technology, Islamic University of Indonesia, Jl. Kaliurang Km. 14.4, Sleman, Yogyakarta, Indonesia

Abstract. It is common that interdepartmental conflict may arise in an organisation to select which decision to be taken, especially in the area of engineering asset management. In this paper, an engineering asset management related case study is discussed and a derivation of a life cycle cost model is proposed to assist finding the optimum decision. Another challenge in this study is a situation that the data and information required to calculate the total cost is unavailable. To deal with this situation, a cost comparison approach is proposed as well as a sensitivity analysis in order to select the optimum decision if the uncontrolled variable in the system nature changes. The result shows that the total cost model derived from the life cycle cost model is capable to assist selecting the optimum decision. A simple spreadsheet based Monte Carlo model is also developed to represent the randomness of the time to failure during the five years of time horizon of the analysis.

1. Introduction

The main purpose of asset management is to balance between cost, performance and risk involved in the asset [1]. It leads to essential functions of asset management in making asset-related important decisions in a corporate level. However, it also may generate disputes between different departments related to asset, such as in making decision to keep or dispose the asset, or decision to keep the scheduled maintenance interval suggested by the manufacturer or change the interval. In making decision related to asset, conflict of interest between departments might happen. In General, the departments involved in this dispute are production (operation), maintenance, purchasing, and finance. It is understandable that the main reason of the conflict is they have their own KPIs to achieve.

To deal with this issue, it is suggested that the related departments should seek a solution that provide global optimum in the corporate level instead of local optimum in the departmental level. Referring to the purpose of asset management, the optimised decision must involve cost, performance, and risk as the main constraint in the decision. To help the decision maker to select what decision to make correct decision, a mathematical model can be suggested. In this paper, a mathematical model is developed to support the process of solving the asset management disputing case. The mathematical model is derived from a pre-developed in Life Cycle Cost model in [2]. The main reason of using the life cycle cost model is based on a consideration that the dispute happens during the life cycle delivery of the asset, so that the life cycle cost can assist to determine which policy may generate lower cost. After the desired mathematical cost model is attained, then the model is verified in the discussed case study to verify whether the model is capable to assist in the process of making a better asset related decision.

2. Life Cycle Cost Model for Asset-Related Decision Making

Life cycle cost model can be implemented in various fields, such as manufacturing [3, 4]; public infrastructure [5] to power plant [6, 7]. In the area of asset management, life cycle cost can be used as a dominant criterion in making decision related to asset [8]. The main concern of asset management is managing the engineering asset through their life cycle to achieve the purpose of the organisation [8, 9]. According to the Institute of Asset Management, there are 6 subject groups in the Conceptual Model of Asset Management. One of the subject groups is the life cycle delivery which includes activities for asset acquisition, asset operation, asset maintenance, and asset disposal. According to [10], all technical system is a physical asset and it goes through different stages during its useful life. Those stages are the stages in the asset life cycle. The activities in the asset life cycle delivery are complex and may consume considerably large amount of financial resource [11, 12]. In this respect, it is important to optimise the cost generated from the asset life cycle delivery as well as maintaining the performance of the asset. The process of optimising the asset performance throughout its life cycles requires the use of life cycle cost analysis and models for such performance optimisation [11, 13-16]. It means that life cycle cost model is very useful to assist management to make any decision related to their asset in the asset management framework [16].

The use of life cycle cost in supporting the management to make decision in the area of engineering asset management has been observed in [16]. In that article, the authors have observed the different use of life cycle cost model in the asset management decision. Briefly, there are five different categories of the use of life cycle cost model is proposed by the authors. One application of the life cycle cost model is to develop a value oriented decision support system in the area of maintenance and replacement or restoration. The example of a complete life cycle cost model that covers all stages in the asset life cycle is presented in [2] and [17]. As thoroughly discussed in [2] that the life cycle cost model has been verified and tailored to different asset management case study case studies and it is capable to serve its purpose. It is also indicated that in [11] the mathematical model of the total cost to select the maintenance policy is derived from a life cycle cost model presented in [2]. Based on this similarities, the life cycle cost model can be employed as a basis model to serve the purpose of this study.

3. Case Study and Analysis

The case study presented in this paper is a case from a government-owned manufacture industry and it is related to a decision for buying a machine spare-part, which is bearing, with a particular specification but different price and quality. There are some argues from different departments in regard to which bearing that should be bought; cheaper one with less reliability or one with more reliability but more expensive. The purchasing department and the finance department suggest buying the lower price bearing to save the expenses but the maintenance department and production department recommend buying the more expensive one because of the reliability reason. Resolving this conflict is the main purpose of this study. This type of asset-related conflict of interest may happen in many different ways in different organisations. It is common that when every department tends to achieve or maintain their Key Performance Indicator (KPI). One of the KPIs of the maintenance department is asset reliability. Asset reliability leads to the reliability of the production system however it may require more cost. Conversely, the finance department is responsible to maintain and push the cost down as one of their KPIs. In this case study, reliability and cost are the conflicting KPI of these departments.

Referring to ISO 55000 that the implementation of asset management should involve balancing cost, risk, and performance. It should be noted that these aspects should be balanced in all asset related process and decision making and mainly to ensure that the asset is capable to meet its function in the process of achieving organisation's competitive strategy [9]. Based on this logic, it can be highlighted that instead of achieving local optimum at the departmental level (either in maintenance, production, purchasing, or finance), it is proposed to achieve global optimum at the corporate level by optimising the combination of cost, risk, and performance. In order to be able to represent this process of analysis, a Life Cycle Cost model can be employed. In developing the life cycle cost model, the risk and performance is also important to be considered. In other words, the life cycle cost model should associate the risk and performance as the cost components.

4. Model Development and Discussion

It is proposed in this paper that to resolve the interdepartmental conflict discussed in the case study, life cycle cost model is used. However, the model is modified and tailored as necessary. It is also to verify that a complete life cycle cost model (e.g. as presented in [2]) can be modified and tailored to provide decision makers with a guidance to attain an optimum asset related decision considering the balance of cost, risk, and performance of the asset. The develop the cost model for the case study in this paper, life cycle cost in [2] is used as the basis model as shown in Equation (1) and (2).

$$\begin{aligned} LCC = \; & \sum_{t=1}^{td}[C_{u,t}(MTBF, MTTR)(m, t+s, t) + F_{o,t} + A_{sys}C_{o,t}m + \sum_{r=1}^{n_{SM,t}}(F_{SM,r,t} + C_{SM,r,t}) + \\ & \sum_{s=1}^{n_{UM,t}}(F_{UM,s,t} + C_{UM,s,t}) + n_{d,t}.F_{SL,t} + T_{d,t}.C_{S,t} + (n_{HR,t}.L_t) + \sum_{l=1}^{n_{p,t}}(\frac{t_{p,l,t}}{365}.L_t) + [(n_{R,t}.F_{R,t}) + \\ & \sum_{p=1}^{n_{R,t}}(n_{NHR,p,t}.\frac{t_{NHR,p,t}}{365}.L_t)] + [(n_{RO,t}.F_{RO,t}) + \sum_{q=1}^{n_{RO,t}}(n_{o,q,t}.C_{o,q,t})] + F_{i,t} + (n_{p,t}.C_{p,t}) + \\ & (n_{c,t}.C_{i,t}) + (\frac{n_{i,t}+n_{c,t}}{365}.C_{inv,t}) - S_{a,t}(m, t+s, t)](1+\pi)^{t-1}(1+r)^{-t} \end{aligned} \quad (1)$$

$$LCC_t = C_{A,t} + F_{o,t} + C_{o,t} + C_{M,t} + C_{SL,t} + C_{HR,t} + C_{PI,t} + C_{S,t} \quad (2)$$

where

LCC_t	: Total Life Cycle Cost at time t
$C_{A,t}$	: acquisition cost at time t
$F_{o,t}$	: fixed cost of operating at time t
$C_{o,t}$	: annual operating cost at time t
$C_{M,t}$	: maintenance cost at time t
$C_{SL,t}$	: stoppage loss at time t
$C_{HR,t}$	: human resource provisioning cost at time t
$C_{PI,t}$	: purchasing and inventory cost at time t
$C_{S,t}$	: salvage value at time t

The life cycle cost in equation consists of eight cost component as shown in equation 2. An investigation has been done to identify what cost components used related to the case study. From the investigation, it is determined that the cost components retained in the model are maintenance cost, stoppage loss, and purchasing and inventory cost as shown in Table 1.

Table 1. The cost component for the new model

No	Cost components	Equation
1	Maintenance cost	$\sum_{r=1}^{n_{SM}}(F_{SM,r} + C_{SM,r}) + \sum_{s=1}^{n_{UM}}(F_{UM,s} + C_{UM,s})$
2	Stoppage loss	$n_d.F_{SL} + T_d.C_S$
3	Purchasing and inventory cost	$F_i + (n_p.C_p) + (n_c.C_i) + (\frac{n_i+n_c}{365}.C_{inv})$

Where :

F_{SM} : fixed cost of scheduled maintenance
C_{SM} : total variable cost for every scheduled maintenance performed
F_{UM} : fixed cost of unscheduled maintenance
C_{UM} : total variable cost for every unscheduled maintenance performed
n_d : number of stoppage occurrences
F_{SL} : Fixed cost of a unit's stoppage
T_d : amount of time the units fail
C_S : opportunity loss per measured time
F_i : Fixed purchasing and inventory cost
n_p : number of purchases
C_p : purchasing cost
n_i : number of initial inventories
n_c : number of components purchased
C_i : cost of a component
C_{inv} : inventory cost

The main reason of retaining the three cost component is because those components present the process of balancing risk, performance, and cost at a particular time frame as mentioned in [1]. Although it is possible to retain other cost components such as annual operating cost, and human resources provisioning cost, it is assumed that other cost components will not significantly affect the decision since those cost component has similar amount regardless what decision alternatives proposed. To fit with the case, the cost component in the Table 1 still needs to be tailored. The complete model suggested to assist making decision for this case study is shown at equation 3.

$$TC = \sum_{r=1}^{n_{SM}}(F_{SM,r} + C_{SM,r}) + \sum_{s=1}^{n_{UM}}(F_{UM,s} + C_{UM,s}) + n_d.F_{SL} + T_d.C_S + F_i + (n_p.C_p) + (n_c.C_i) + (\frac{n_i+n_c}{365}.C_{inv}) \quad (3)$$

Since the company implements breakdown maintenance strategy, the notation for preventive maintenance cost could be eliminated. Also, it is assumed that the required time to replace the any failed bearing is similar regardless its brand. It leads to a circumstance that the breakdown maintenance cost and the stoppage loss can be combined into total cost for replacement (C_R). The information from the company indicates that the fixed cost of purchasing and the inventory cost is relatively small and insignificantly contribute to the total cost, so that for this cost component the significant cost is from the price of the bearing. Unfortunately, the data and information required to complete the total cost equation is unavailable. To deal with this situation, a ratio cost model and a sensitivity analysis as in [18] is proposed. The cost promoted as the basis cost for the cost comparison is the cost of the normal bearing (more expensive bearing and noted as p). A spreadsheet Monte Carlo model is developed to represent the randomness of the time to failure and to analyse the sensitivity of the result. The replacement cost is assumed vary from 1p (replacement cost is the same amount as the component price) to 5p (replacement cost is five times higher than the price of the component) and the lower component price is also various from 0.3p to 0.6p. The time to failure is assumed as Weibull distributed and random. The total cost for each combination is shown in Table 2.

Table 2. The total cost for each combination

$C_{i,2}/C_{i,1}$ / C_r/C_i	p	0.3p	0.4p	0.5p	0.6p
1p	**12**	9.1	9.8	10.5	11.2
2p	**18**	16.1	16.8	17.5	18.2
3p	**24**	23.1	23.8	24.5	25.2
4p	**30**	30.1	30.8	31.5	32.2
5p	**36**	37.1	37.8	38.5	39.2

As aforementioned that the purpose for this study is to select the optimum asset related decision based on the minimum system cost. In this case study, the significant variables affect the total cost is replacement cost and the price of the component. From the Table 2, it can be concluded that if the replacement cost is as much as the cost of the normal component (1p), as long as the price of the replacement component is less or equal to 0.6p, the cheaper component is selected. If the replacement cost increases to 2p, the price of the cheaper component should decrease to 0.5p to be feasible. In a situation when the replacement cost increases to 3p, the cheaper component cost more than 0.4p is not feasible. If the replacement cost reaches 4p, the policy to select normal component is selected because it generates lower system cost.

5. Conclusion

A derivation of a life cycle cost model is developed and proposed in this model to resolve an interdepartmental case related conflict. The result shows that the model is capable to support the process of finding the optimum decision for the case presented in the case study. It is also support a hypothesis that the LCC model used as the basis of the derived model is sufficient to support asset-related decision making and a further study in different area and case to strengthen this statement is recommended. To deal with the situation when is incomplete data and information to complete the total cost equation, a cost ratio and sensitivity analysis is proposed. However, it is suggested that the organisation needs to review the process of purchasing to prevent this conflict as well as considering to implement asset management system in their organisation. A challenge that may arise for the future study is to develop a more robust model that is able to take into account uncertainty in the whole life cycle.

References

[1] ISO, "ISO 55000 : 2014," in *Asset management — Overview, principles and terminology*, ed. Switzerland, 2014.

[2] W. N. Cahyo, "A modelling approach for maintenance resource-provisioning policies," Doctor of Philosophy, School of Mechanicall, Material, and Mechatronics, University of Wollongong, Wollongong - Australia, 2015.

[3] M. Gram and W. E. Schroeder, "Evaluating The Life Cycle Cost of Plant Assets - A Multidimentional View-," *Serbian Journal of Management*, vol. 7, p. 12, 2012.

[4] M. Sheikhalishahi and S. A. Torabi, "Maintenance supplier selection considering life cycle costs and risks: a fuzzy goal programming approach," *International Journal of Production Research*, vol. 52, p. 16, 2014.

[5] J. O. Almeida, P. F. Teixeira, and R. M. Delgado, "Life cycle cost optimisation in highway concrete bridges management," *Structure and Infrastructure Engineering*, vol. 11, p. 15, 2015.

[6] N. I. Sinisuka and H. Nugraha, "Life cycle cost analysis on the operation of power generation," *Journal of Quality in Maintenance Engineering*, vol. 19, p. 20, 2013.

[7] A. Lesmerises and D. Crowley, "Effect of Different Workscope Strategies on Wind Turbine Gearbox Life Cycle Repair Costs," *International Journal of Prognostics and Health Management*, vol. 4, p. 7, 2013.

[8] Asset_Management_Council. (2009) Asset Management. *Asset Management Council and MESA Newsletter*.

[9] K. O. El-Akruti, "The strategic role of engineering asset management in capital intensive organisations," Dissertation/Thesis, UoW, 2012.

[10] W. J. Fabrycky and B. S. Blanchard, *Life-cycle cost and economic analysis*. Englewood Cliffs, N.J: Prentice Hall, 1991.

[11] W. Cahyo, "Integrated Maintenance and Inventory Optimization Model for Offshore Assets," presented at the Asia Pacific Industrial Engineering And Management System Conference 2014, Jeju, South Korea, 2014.

[12] W. Cahyo, K. El-Akruti, R. Dwight, and T. Zhang, "Managing maintenance resources for better asset utilisation," *Australian Journal of Multi-Disciplinary Engineering*, vol. 11, 2015.

[13] B. S. Blanchard and W. J. Fabrycky, *Systems engineering and analysis*. Upper Saddle River, N.J: Pearson, 2011.

[14] R. Dwight, "Frameworks for measuring the performance of the maintenance system in a capital intensive organisation," Doctoral Dissertation/Thesis, University of Wollongong, 1999.

[15] B. S. Dhillon, *Life Cycle Costing for Engineers*. Ottawa: Taylor and Francis Group, LLC, 2010.

[16] K. El-Akruti, R. Dwight, T. Zhang, and M. Al-Marsumi, "The Role of Life Cycle Cost in Engineering Asset Management," presented at the 8th World Congress on Engineering Asset Management & 3rd International Conference on Utility Management & Safety, Hong Kong, 2013.

[17] C. E. Ebeling, *An introduction to reliability and maintainability engineering*. Long Grove, Ill: Waveland Press, 2010.

[18] D. J. Pannell, "Sensitivity analysis of normative economic models: Theoretical framework and practical strategies," *Agricultural Economics*, vol. 16, 1997.

Design for Supplier Performance Assessment Model of Information Technology Service Provider in Small Medium Enterprises in Indonesia

ElisaKusrini[1], Alfanugrah Hi Usman[2]
[1]*Department of Industrial Engineering & Magister of Industrial Engineering, Indonesia Islamic University, Jl Kaliurang Km 14,5Yogyakarta 55584 , Indonesia*
2 Magister of Informatics Engineering , Indonesia Islamic University, Jl Kaliurang Km 14,5Yogyakarta 55584 , Indonesia

Abstract. This study aims to design a supplier performance assessment model for information technology services provider in small medium enterprises (SME) in Indonesia. Supplier performance assessment is important to improve supplier performance and maintain long-term relationships between SME (buyer) and their suppliers. This supplier Assessment models are developed based on the review literature and interviews with procurement staff in SMEs. Performance assessment criteria are identified and tested for validity by providing a questionnaire to procurement section in SMEs. In order to obtain importance weight of each criteria toward performance score, a questionnaire to weighed the criteria are distributed to procurement staff and analyzed the answers using the Analytical Hierarchy Process (AHP) method. The final performance score is calculated based on the standardized value of each criteria multiplied by its weight. Based on the implementation of this model in SMEs, it is known that the proposed model is suitable to assess the supplier's performance and greatly help the SMEs in managing the performance of their suppliers.

1. Introduction

Today, companies tend to outsource their goods and services and focus on their core business. this resulted in the company increasingly dependent on suppliers. The company's performance is highly rely on supplier performance [1]. Based on [2], it was found that the contribution of procurement and supply chain cost is 30% to 70% of cost of good sold in some industries such as chemical and automotive. In managing vendors, many organizations reduce the number of suppliers and prefer to manage relationships with strategic suppliers in the long term [3]. Unfortunately, the company's dependence is not followed by a system of supplier performance assessment. Until now, many companies don't conduct the evaluation and monitoring of supplier performance regularly [4].

There are various types of vendors, one of the types is a vendor that specialized in information technology (IT) provider. Various services that can be provided by IT vendors include: system integration, system inspection, system management, internet management, software development, software verification, software maintenance, hardware maintenance, hardware operation, management of facilities, internet services. Information technology (IT) is very important for a company because through IT, companies can easily achieve competitive advantage and gain profit. The use of IT has evolved into 'commodities' where IT can be said to be a must-have for any company [5]. Considering the importance of IT and its vendors within a company, then the company needs to manage the performance of IT vendors and companies need to perform supplier performance assessment.

Supplier performance assessment has been widely investigated by many researchers in previous study. They used different criteria and different criteria and methods to develop supplier assessment model and up to now there is no agreement what criteria can be used for vendor performance assessment. The focus of previous researchers more on assessing for vendor selection, while for monitoring and assessment of vendor performance is still very limited. Therefore, this research will propose a model for assessment and monitoring vendor performance especially vendor IT in small medium enterprises in Indonesia.

2. Literature review

There are many studies on the measurement of vendor selection criteria but very few for examining the model in measuring and monitoring supplier performance. Supplier performance assessment is a technique to measure the actual performance of suppliers and assesment based on predetermined criteria and give suppliers value based on the measurement results. This becomes important in maintaining relationships with suppliers and for consideration of the future relationship. While the vendor selection is a process of getting vendors with various predetermined criteria [6]. Some research in vendor selection system is done on a large industry and the public procurement of goods and services but on SMEs and for special items of IT is still very limited. The assessment methods are varied and no agreement what method is most appropriate. Here are some researches on vendor selection methods by previous researchers with different methods.

The study of IT vendor selection criteria in SMEs is done by [7] which examines the criteria of vendor selection in SMEs Taiwan by using multivariate analysis. This study yields a conclusion of criteria that can be used in vendor selection. A hybrid model consist of five steps to support the vendor selection is proposed by [8] and this research using Multi criteria decision making (MCDM) approach. Analityc Network Process (ANP) approach is employed to obtain relative weighting of criteria and modified TOPSIS (Techniques for order preference by similarity to ideal solution) for vendor evaluation process. The proposed model also provides a framework for designing and improving adequate criteria and reducing risks in choosing unsuitable vendors. A study conducted by [9] aimed at developing criteria and providing a model for testing the relationships and impacts related to the selection of supplier by using Fuzzy DEMATEL method. The study was conducted using literature study and review some related articles and found 43 important criteria in selecting a vendor. The criteria were chosen according to the importance of using Delphi fuzzy method and obtained 14 important criteria and then made a dematel questionnaire and distributed among 11 experts and members in Islamic Azad University of Yadz. The Result of this study is that stable financial criteria have the greatest impact in optimal supplier selection and security is the lowest. A study focused on developing a decision support system for supplier selection strategies based on the AHP method using case study in automotive industry in Pakistan is conducted by [10]. A sensitivity analysis to check the strength of the vendor also has been done in this research. Another study carried out by [11] to propose voting methods using AHP to determine the final rankings of vendors. This method called Voting AHP (VAHP). The results of the study literature mentioned above will be used develop the criteria of supplier performance assessment models specifically for IT suppliers in SME Indonesia.

3. Research Methodology

The steps of this study consist of 6 stages as follows: identify criteria of supplier performance assessment, criteria screening , determining the weights of criteria using AHP, determining the rating scale, formulation of supplier assessment model, implementation of the model.

4. Result and Discussion

4.1. Identify criteria of supplier performance assessment

Based on literature review, it is found that some criteria are important and often used in vendor assessment. In this study, the model divided into two categories, that is model for Software & Hardware supplier assessment and model for Sistem Development supplier assessment. Table 1 shows criteria and its sources.

Table 1. Criteria and its Sources

Source (Articel Number)	Criteria	
	Software and Hardware	**System Development**
[6]	Quality, On time delivery, Low price	Quality, Timely development, Low cost, Problem-solving skill
[12]	Quality, On time delivery, Low price, Financial stability, Flexibility of payment and delivery time	Quality, Timely development, Low cost, Financial stability, Problem solving skill
[8]	Quality, On time delivery, Low price, Communication, Document completeness, Responsibility	Quality, Timely development, Low cost, Communication, Development tools and document completeness.
[14]	Quality, On time delivery, Low cost, Document completeness, Location, Reputation	Quality, Timely development, Low cost, Development tools and document completeness, Ability to manage projects, Location, Reputation
[11]	Quality, On time delivery, Financial stability,, Document completeness, Responsibility, Reputation, Honest and maintenance of confidentiality of business	Quality, Timely development, Financial stability,, Development tools document completeness, Ability to manage projects, Reputation, Honest and maintenance of confidentiality of business
[20]	Quality, On time delivery, Lawsuits with clients, Document completeness	Quality, Timely development, Lawsuits with clients, Development tools and document completeness
[19]	Quality, Low price, Document completeness, Honest and maintenance of confidentiality of business	Quality, Low cost, Development tools and document completeness, Honest and maintenance of confidentiality of business
[7]	Quality, Financial	Quality, Financial

	stability, Communication, Lawsuits with clients, Document completeness, Flexibility of payment and delivery time, Honest and maintenance of confidentiality of business	stability, Communication, Lowsuits with clients, Development tools and document completeness, Flexibility of contracts in relation to the deadline, Ability to manage projects, Honest and maintenance of confidentiality of business
[21]	Quality, On time delivery, Low price, Communication	Quality, Timely development, Low cost, Communication, Problem solving skill
[22]	Low price, Responsibility, Location	Low cost, Location
[16]	Quality, On time delivery, Low price, Communication, Location, Reputation	Quality, Timely development, Low cost, Communication, Problem solving skill, Location, Reputation
[13]	Quality, On time delivery, Low price, Lawsuits with clients, Flexibility of payment and delivery time, Location, Reputation, Honest and maintenance of confidentiality of business	Quality, Timely development, Low cost, Lawsuits with clients, Flexibility of contracts in relation to the deadline, Location, Reputation, Honest and maintenance of confidentiality of business

In this study, the criteria of previous research were modified and adapted to the needs of IT procurement in SME. the proposed criteria are based on the above review literature and interviews with the procurement staff in the SME, as given in Table 2.

4.2. Criteria Screening

At this stage, the questionnaires are distributed to 12 IT procurement experts and practitioners which aim to determining the importance level of criteria. Only criteria with a value greater than 3 will be used as performance measurement criteria in this study. Base on the screening result, The criteria financial stability, no legal problems are removed because the average value is less than 3.

4.3. Determining the weights of criteria using AHP

The importance weigh of criteria is determined using AHP methods by distributing AHP Questionnaire to 5 expert. The important weight Resulted from of the AHP analysis are given in Table 2.

4.4. Determining the rating scale

The rating scale of the assessment is ranging from 1-5, from poor to Very good.

4.5. Formulation of supplier assessment model

Vendor performance assessment models is formulate based on selected criteria and it's weight. The vendor's performance score is the multiplication of performance value with its importance weight. The Supplier performance assessment model for Sofware&Hardware Category is presented in Table 2

Table 2. Supplier performance assessment model for Sofware&Hardware Category

No	Criteria	Weighting	Assessment scale	Description
1	Quality	0,126	5	Very user friendly and there is no complaint at all and Defect free and no complaints at all
			4	User friendly and There is a defect in the hardware but no Non Conformity Report (NCR)
			3	Not user friendly but still perform and There is a defect in the hardware and Non Conformity Report (NCR) is released
			2	Very not user friendly but vendor willing to improve and More than 1 Non Conformity Report (NCR) are released but supplier is responsible
			1	Not user friendly and More than 1 Non Conformity Report (NCR) are released and supplier is not responsible
2	On time delivery	0,078	5	All of software and hardware is delivered on time
			4	Some can be delivered on time and others are delayed less than 3 days
			3	Delay for 3-5 days
			2	Delay more than 6-8 days
			1	Delay more than 8 days
3	Low price	0,055	5	31-40% price is lower than owner estimates
			4	21-30% price is lower than owner estimates
			3	11-20% price is lower than owner estimates
			2	5-10% price is lower than owner estimates
			1	1-4% price is lower than owner estimates

No	Criteria	Weighting	Assessment scale	Description
4	Communication	0,136	5	Proactive in communication
			4	it is easy to be contacted via phone, email or fax and good respond
			3	Easy to contact but takes a long time to respond
			2	Difficult to contact by phone, fax, email, etc.
			1	Cannot be contacted
5	Documents and administration	0,036	5	All documents have been completed before delivery date and document contents are clear and detailed
			4	All documents have been completed before delivery but document contents are not clear and detailed
			3	Documents have been completed after delivery date and document contents is clear and detail
			2	Documents have been completed after delivery date but document contents is not clear and detailed
			1	A warning letter is released to supplier because document incomplete
6	Flexibility of payment and delivery time	0,034	5	All changes requests are full filled in accordance with company policy
			4	All changes requests are full filled in accordance with both party policy
			3	All changes requests are full filled in accordance with supplier policy
			2	Accept change requests for payment system but dont accept for delivery time
			1	Dont accept change requests of payment system and delivery time
7	Responsibility	0,158	5	Responsible to resolve the complaint on the same day and complaints can be resolved
			4	Responsible to resolve the complaint within 1-3 days and complaint can be resolved
			3	Responsible resolve the complaint within more than 3 days and complaints can be resolved
			2	Responsible resolve the complaint within more than 3 days but difficult to resolve complaints
			1	Not responsible for resolving complaints
8	Vendor location	0,024	5	Located in an area and less or equal to 20km
			4	Located in an area more than 20km
			3	Supplier are in different area/city
			2	Supplier are in different country but easy to communicate
			1	Supplier are in different country and difficult to communicate
9	Vendor reputation	0,140	5	History of performance and service is excellent
			4	History of performance and service is good
			3	Bad performance history but good service
			2	Good performance history but bad service
			1	History of performance and service is not good
10	Honest and maintenance of confidentiality of business	0,213	5	Very honest and transparent in giving information and Always have good commitment
			4	Honest and transparent in giving information and have good Commitment
			3	Sometimes honest and trasparent and sometimes have commitment
			2	Not transparent in giving information and do not have commitment
			1	Very not transparent in giving information and do not have commitment

4.6. Model Implementation

The last stage is to implement the model on 2 IT vendors for software hardware vendors and system development. each vendor is assessed based on the proposed model. It is found that the model can assist in evaluating vendor's performance and facilitate objective assessment. Previously they do not have a scoring system and vendor assessment has been done subjectively.

5. Discussion

In general, this study has the same flow as the research conducted by [7] that generates an evaluation model that can be used to select IT vendors. The difference is that this study focuses on performance appraisal and vendor monitoring. the other difference is in the determination of different rating scales as well as the different weights and types of criteria tailored to the needs of the SME IT vendors in Indonesia. Another difference is that the research is broader because the model in this study is divided into two categories, namely the vendor category of software & hardware providers and the vendor category of the system development service providers, while the model in the previous research only focus on IT outsourcing providers or system development. There are several common criteria, ie : System development tools and document completeness, ease of communication, vendor ability to solve problems, vendor's ability to manage projects, maintain confidentiality, flexibility of contract terms and time of system development and reputation. However, two criteria that are considered important by SMEs in Taiwan but considered not important by SME Indonesia, ie financial stability and no legal problems.

The four criteria of price, quality, delivery and service are the most widely used criteria in selecting or valuing vendors in general [6] and [10].These four criteria are also the criteria used in this study.This indicates that price, quality, delivery and service are the criteria that can be used in assessing the various vendors of both vendors in general, vendors in the automotive industry and also IT vendors. Similarly, [12] describes some important criteria for selecting vendors in the food industry. Some of the criteria used to select vendors in the food industry are also important criteria for SME IT vendors, including price, quality, delivery time and flexibility. However, there is one important criterion for selecting vendors in the food industry but it is not important for the SME vendor, that is, the stable financial condition of vendors. This criterion according to [9]is also an important criterion in choosing a vendor for education (university). In this study, honest and confidential criteria are the most important criteria in assessing the performance of IT vendors for SMEs. This is in line with other research features such as [13], [11] and [7] the location of the vendor is one of several criteria considered important in the selection and assessment of the vendor. This is in line with the results of other researchers that close location will facilitate monitoring vendor performance [14], [13],[15]. Reputation criteria are important criteria in this research. It also agrees with [16] which in his research use reputation as one of the six criteria used in selecting vendors for IT and [17] who argue that reputation is one of seven important criteria related to outsourcing Information Systems. reputation criteria are also important in assessing or selecting common vendors such as those found in [18] and [19]. The criteria of ease of communication and responsibility in resolving complaints are the two important criteria found in this study. In previous studies, both criteria were also widely used in selecting and assessing vendor performance both in general and IT vendors. [8], [16] and [9].

6. Conclusion

There are several conclusions from this study.The most important criteria for performance assessment of software & hardware vendors and system development is honest and can keep company secrets. while the least important criterion for vendor software & hardware assessment is vendor location and for system development category is system development cost. Criteria in vendor performance assessment for various industries are relatively similar, but with different importance level. The resulting model can be used by the SMEs in assessing the performance of vendor software & hardware providers and system development.The opportunity for further research is to examine the relationship between criteria to determine which criteria most influence other criteria. Interpretive structural modeling model and Equation Structural Modeling could be used for further research.

References

[1] Nair, A., Jayaram, J., & Das, A. *Strategic purchasing participation, supplier selection, supplier evaluation and purchasing performance.* International Journal of Production Research, **53(20) 6263-6278** (2015).

[2] Quinn, F,*"The Power of Procurement." Supply Chain Management Review* 9 (9): **6–8** (2005)

[3] Thorelli, H. B., *"Networks: Between Markets and Hierarchies."* Strategic Management Journal 7, **37–51** (1998)

[4] Earnest D and Bamford J, *Your Alliances are Too Stable.* Harvard Business Review June (2005)

[5] Schubert, P., & Leimstoll, U. *Importance and Use of Information Technology in Small and Medium-Sized Companies.* Electronic Markets, 17 (1), **38-55** (2007).

[6] Amil, M. A, *Supplier performance assessment tool in automotive industry using multivariate analysis.* (2009).

[7] Chang, S. I., Yen, D. C., Ng, C. S. P., & Chang, W. T, *An analysis of IT / IS outsourcing provider selection for small-and medium-sized enterprises in Taiwan. Information & Management*, 49 (5), **199-209** (2012).

[8] Shyur, H. J., & Shih, H. S, *A hybrid MCDM model for strategic vendor selection. Mathematical and Computer Modeling,* 44 (7), **749-761**(2006).

[9] Mirmousa, S., & Dehnavi, H. D, *Development of Criteria of Selecting the Supplier by Using the Fuzzy DEMATEL Method.* Procedia-Social and Behavioral Sciences, 230, **281-289** (2016).

[10] Dweiri, F., Kumar, S., Khan, S. A., & Jain, V. *Designing an integrated AHP based decision support system for supplier selection in automotive industry.* Expert Systems with Applications, 62, **273-283** (2016).

[11] Liu, F. H. F., & Hi, H. L. *The voting analytic hierarchy process method for supplier selecting. International journal of production economics*, 97 (3), **308-317** (2005).

[12] Żak, J. *Comparative analysis of multiple criteria evaluations of suppliers in different industries.* Transportation Research Procedia, 10, **809-819** (2015).

[13] Bush, A. A., Tiwana, A., & Tsuji, H. *An empirical investigation of the drivers of software outsourcing decisions in Japanese organizations. Information and Software Technology*, 50 (6), **499-510** (2008).
[14] Weber, C. A., Current, J. R., & Benton, W. C. *Vendor selection criteria and methods*. European journal of operational research, 50 (1), **2-18** (1991).
[15] Dargi, A., Anjomshoae, A., Galankashi, M. R., Memari, A., & Tap, M. B. M. *Supplier selection: A fuzzy-ANP approach. Procedia Computer Science*, 31, **691-700** (2014)
[16] Büyüközkan, G., & şakir Ersoy, M*Applying fuzzy decision making approach to IT outsourcing supplier selection*. International Journal of Mechanical, Mechatronic and Manufacturing Engineering, 3 (7), **815-819** (2009).
[17] Chen, Y. H., Wang, T. C., & Wu, C. Y*Strategic decisions using the fuzzy PROMETHEE for IS outsourcing*. Expert Systems with Applications, 38 (10), **13216-13222** . (2011).
[18] Sodenkamp, M. A., Tavana, M., & Di Caprio, D. *Modeling synergies in multi-criteria supplier selection and order allocation: An application to commodity trading*. European Journal of Operational Research, 254 (3), **859-874** (2016).
[19] Wan, S. P., Xu, G. L., & Dong, J. Y. *Supplier selection using ANP and ELECTRE II in interval 2-tuple linguistic environment*. Information Sciences, 385, **19-38** (2017).
[20] Nielsen, I. E., Banaeian, N., Golińska, P., Mobli, H., & Omid, M. *Green supplier selection criteria: from a literature review to a flexible framework for the determination of suitable criteria. In Logistics operations, supply chain management and sustainability* (pp. **79-99**). Springer International Publishing (2014).
[21] Wang, J. J., & Yang, D. L. *Using a hybrid multi-criteria decision aid method for information systems outsourcing*. Computers & Operations Research, 34 (12), **3691-3700** (2007).
[22] Udo, G. G. *Using analytic hierarchy process to analyze the information technology outsourcing decision. Industrial Management & Data Systems*, 100 (9), **421-429** (2000).

Workload analysis for Sanding Small Grand Piano Department in Manufacturing company

Amarria Dila Sari[1,a], Ahmad Azhar Basyir[1,b],M.Ragil Suryoputro[1,c]
[1]*Industrial Engineering Deparment, Faculty of Industrial Technology, Islamic University of Indonesia, Jl. Kaliurang km 14.5, Yogyakarta Indonesia*

Abstract. The production process is one of the main activities of manufacturing companies that need their continuous improvement and continuously. There are several factors that affect productivity, such as job stress and workload. The workload is shared with uneven to make the work environment uncomfortable. The production data in April 2017 explaining that the sanding small grand piano average produces for 20 units piano/day, with a production target of 21 units piano/day production target has not been achieved. The lack of achievement of these targets due to the difference between the expected level of ability with the capacity level. In this study focused to determine the operator workload on sanding small grand piano is included in the painting section. The workload is calculated using the method of Full-Time Equivalent (FTE). The results showed that the operators have overloaded with FTE scores of both 1.33, 1.35 and 1.34 While the operators have an underload workload with a value of FTE 0.94. 0.64, 0.72. Therefore, Kaizen activity such as balancing jobdesk and modification stopper were applied to reduce the waste. Workload after kaizen plan has decreased each 0.02 FTE with change of work element, each operator has a uniform workload with an average FTE value of 1.13 (underload).

1 Introduction

Manpower planning is one of the problems for the company, among others related to the workload analysis is not balanced between the number of workload and the amount of labour [1]. Workload is a management technique systematically performed to obtain information about the effectiveness and efficiency of the organization based on the work volume [2]. Full-time Equivalent (FTE) is one of the tools used to measure the workload, by comparing the time spent in completing various jobs on the effective working time, as well as simplifying the work measurement by changing hours of workload to the number of people required to complete a specific job [3].

Several preliminary studies related workload with various methods including, the study of mental workload using NASA-Task Load Index (TLX) for industrial employees in manufacturing company [4], [5]. Study about physical workload for laundy worker using cardio vascular load method have also been analyzed [6]. Research related workload using workload indicators of staffing need (WISN) in Namibia [7]. Study related workload analysis using Full-time Equivalent has been carried out at foundry SME [8]. Moreover, research that concerned to analyse workload measurement or staffing need has been conducted on mining industry [9]. Research about workload measurement using worksampling and task per job has been carried out at service company [10]

This research was conducted on manufacturing company engaged in the assembly of non-traditional instruments with piano, electone, pianica, organ, and others. The resulting products include Upright Piano (UP) and Grand Piano (GP). This study is focused on knowing the workload of operators on the sanding small grand piano which is included in the painting division. April 2017 production data explains that sanding small grand pianos produce an average of 20 pianos / day, with a production target of 21 units of piano / day, the production target cannot be achieved yet. The lack of achievement of the target is due to the difference between the level of ability expected with the level of capacity owned.

This study aims to find out the operational workload of sanding small grand piano operators with full-time equivalent method based on *kaizen* improvement so that it can be used as input for the company to improve the performance of existing human resources. *Kaizen* approach is an ongoing process to reduce waste, increase productivity, improve quality, and utilize existing resource utilization [11]

2 Theory

2.1 Human Resources Management

[a] Corresponding author: amarria@uii.ac.id

Human resource management is a strategic area of an organization [12]. Human resource management is the process of handling problems on various aspects ranging from employees, employees, labourers, managers, to other workforce to support organizational activities in order to achieve company goals that have been set [13]

2.2 Workload Analysis

The workload is the calculation and comparison between the needs of human resources or capacity required and the real capability of the human [14]. Workload is importance factor in determining human resource management policy in a system, such as personnel need planning [15].

2.3 Kaizen

Kaizen goal is a better product (improving quality), cheaper (lowering costs), safer (increased safety), faster (fix lead time), and more easily (increasing productivity) [16] . The most important element in *kaizen* is aware to *Muda*. *Muda* is some phenomena and effects that do not increase the value added.

3 Methods

3.1. Research Subject

This study was conducted on a sanding small grand piano division. Criteria subject in this research include male / female worker in sanding small grand piano, productive age workers range from 18 years to 45 years, workers with a minimum of one year's work experience, workers willing to be respondents.

3.2 Research Procedure

Research procedures were divided into the following phases:

a. Observation
   Observations are made through observations in the field, as well as interviews with experts in the company.
b. Determine the work unit and employees who will carry out workload analysis.
c. Determine the effective working time
d. Effective working time is divided into effective working hours and effective working days. The effective working hours are hours in a day reduced allowance. While, effective working days are determined by:
   1. Working days
   2. Annual leave
   3. Education and training
   4. National libraries
   5. Absence of work
   6. Working hours
e. Allowance
   Set up a standard allowance with the aim to determine the factors of worker / employee allowance which includes the type of activity and time requirements in completing a task / work that is not related to the main activity, such as allowance for personal needs, work environment.
f. Calculate the standard time
   The stage of calculating the standard time starts with processing the cycle time that has been obtained and calculating the rating factor size for each working element. Then, the cycle time will be multiplied by the rating factor to calculate normal time. The next stage of normal time will be added with allowance for each operator, according to the elements of the job.
g. Calculate the workload
   At this stage, the workload is calculated using the Full-Time Equivalent (FTE) method. The FTE value is useful for knowing the workload that is obtained by the operator. FTE value calculation itself use a standard time that takes into account daily production, total working time, and effective working time. In this study, the calculation was carried out in March 2017 - May 2017 while June 2017 was used as a recommendation from this study. The implication of FTE value is divided into three types : overload, normal, and underload. Based on the workload analysis guidance, the total FTE index value above 1.28 is considered overload, between 1 to 1.28 is considered normal whereas if the FTE index value is between 0 to 0.99 considered underload or workload is still less than the maximum [17].

4 Results and Discussion

Respondents in this study were 6 operators, 3 male and 3 female operators, aged between 21-35 years. Company hours are from 07.00 to 16.00 everyday except for Saturdays and Sundays which are holidays. Operator 1 and operator 2 work on a belt sander machine that can only be done by male operators. Operator 3, operator 4, and operator 5 work on the hand sanding section. Operator 6 also worked on the hand sanding section but was sent to the Top Stick Long and Top Stick Short cabinets

Effective working hours / month is a multiplication of effective working days (no days off) multiplied by working hours / days. While the total effective working hours is the multiplication of working hours / month multiplied by the average effectiveness factor. The effectiveness factor is a factor involving the aspect of workers' allowance. The average allowance of all operators is 21% so that the average effectiveness factor is 79% (1-21%).

Allowance is the adjustment done to the normal time to obtain the standard time for the purpose to recover the lost time due to personal needs, fatigue, and unavoidable delays.

The effective working time of March 2017 to June 2017 is shown by Table 1 below.

Table 1. Effective Working Time

Description	March	April	May	June	Unit
Work hour/month	22	19	21	15	Day
Work hour/day	8	8	8	8	Hour
Work hour/month	176	152	168	120	Hour
Effectiveness Factor	79%	79%	79%	79%	%
Total effective working hour	139	120	132	94	Hr/Month

4.1 Workload Analysis

The workload on sanding small grand piano for March 2017 - May 2017 is uneven due to the presence of operators with underload and overload workload. The second operator has an overload workload in April 2017 and May 2017 with an FTE of 1.33. For the sixth operator also has an overload workload in April 2017 and May 2017 with FTE values of 1.35 and 1.34 respectively. While the fourth operator has an underload workload in March 2017 with a value of FTE 0.94. And the fifth operator has an underload workload for March 2017 of 0.64 in April 2017 of 0.72 and May 2017 of 0.72, can be seen in the Figure 1-3.

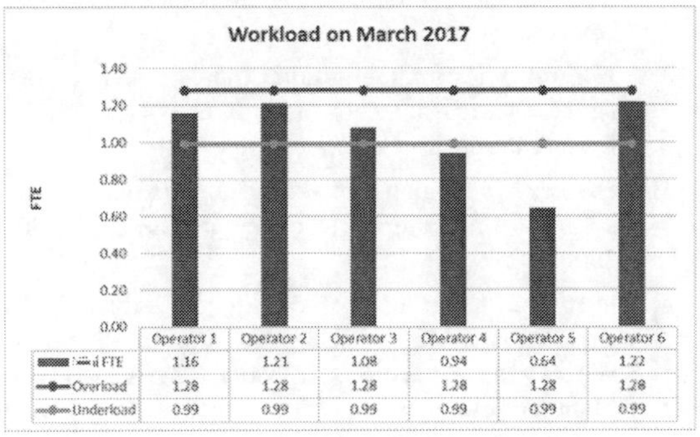

Figure 1. Workload on March 2017

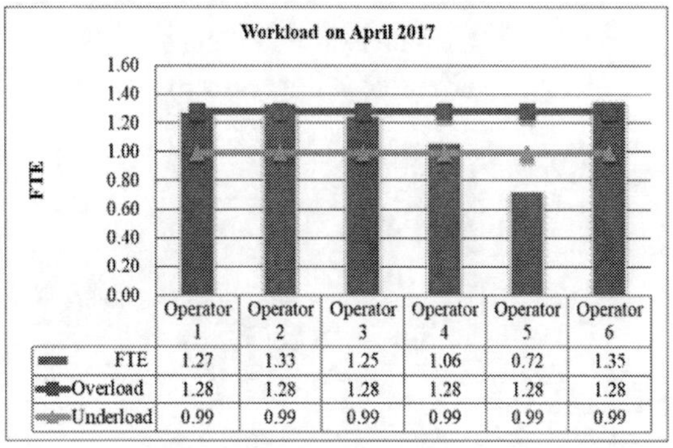

Figure 2. Workload on April 2017

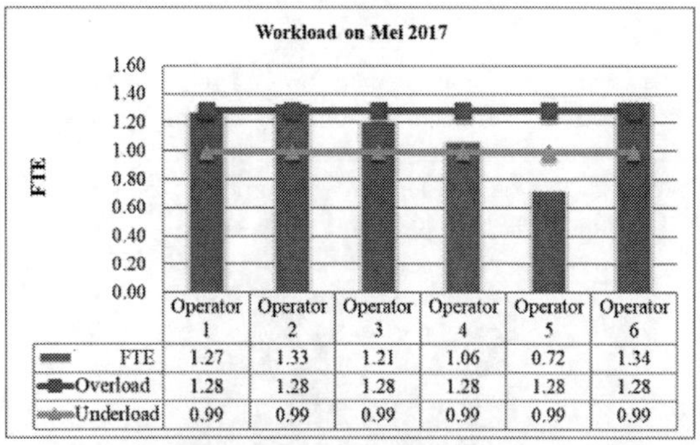

Figure 3. Workload on Mei 2017

4.2 Workload Calculations After *Kaizen* Plan

The *kaizen* plan on sanding small grand piano division by eliminating or combining work processes is useful to lower the operator's workload. Therefore, the workload for the operator becomes reduced. *Kaizen* is an ongoing process that can reduce waste, increase productivity, improve quality, and utilize existing resource utilization [10]. *Kaizen* has several concepts that can be used by companies to make improvements, one of them 3M or called *Muda*, *Mura*, and *Mura*. *Muda* is an activity that does not add value to the product, *mura* is an uneven process, and *muri* is an excessive worker burden [18]. In the present study there are several *kaizen* plans that are expected to decrease the workload of sanding small grand piano operators. The following comparison operator workload before and after *kaizen* plan in Figure 4.

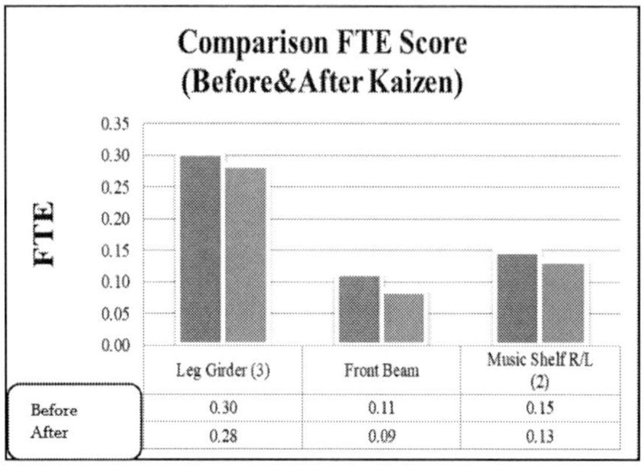

Figure 4. Comparison FTE Score (Before&After Kaizen)

The *kaizen* plan is undertaken for improvement of uneven workload measurement results. [1] activity carried out includes the incorporation of work elements on the belt sander # 500 cabinet leg girder ebony polish model with FTE value from 0.30 to 0.28 (Figure 4).

Figure 5. *Kaizen* Leg Girder

Second *kaizen* activity with stopper addition for belt sander element # 500 front beam cabinet polish ebony (PE) model with FTE value down from 0.11 to 0.09 (Figure 5).

Figure 6. *Kaizen* Front Beam

While the value of FTE on the element belt sander # 500 cabinet music shelf polish model ebony (PE) fell from 0.15 to 0.13 with stopper modifications (Figure 6).

Figure 7.*Kaizen* Music Shelf

5 Conclusion

Based on the research, could be concluded that all operator was categorized on FTE were not balanced. The operators have overloaded with FTE scores of both 1.33, 1.35 and 1.34 While the operators have an underload workload with a value of FTE 0.94. 0.64, 0.72. Therefore, *Kaizen* activity such as balancing jobdesk and modification stopper were applied to reduce the waste. Workload after *kaizen* plan has decreased each 0.02 FTE with change of work element, each operator has a uniform workload with an average FTE value of 1.13 (underload). Further research should be done by increasing the number of samples to further strengthen the research.

References

[1] H. Purnomo , "Workload Analysis for Determining the Number of Employees at Banking Companies," Gema, p. 50, 2015.

[2] J. Hurd, FTE for Definition.Calculations,Examples And System Usage Information For HRMS and GL Budget Purpuses, Colorado, 2004.

[3] Depkeu, Pedoman Pelaksanaan Analisis Beban kerja di Lingkungan Departemen Keuangan. RI, 2006.

[4] Puspawardhani, "Mental Workload Analysis Using NASA-TLX Method Between Various Level of Work in Plastic Injection Division of Manufacturing Company," in In Advances in Safety Management and Human Factors, Orlando, 2016

[5] T. Sealetsa, "Ergonomics issues among sewing machine operators in the textile manufacturing industry in Botswana," Work , vol. 38, no. 3, pp. 279-289, 2011.

[6] Sari, A.D., Suryoputro, M.R., Pramaningtyas, M.D., Putra, P.S. and Maulidyawati, S.B.,," in IOP Conference Series: Materials Science and Engineering , Yogyakarta, 2016.

[7] P. Mc QUide, R. Kolehmainen-Aitken, R.-L. and N. Foster, "McQuide, P. A., Kolehmainen-Aitken, R.-L., & Forster, N. (2013). Applying the workload indicators of staffing need (WISN) method in Namibia: challenges and implications for human resources for health policy," 2013.

[8] A. Sari, F. Hardiansal and M. Suryoputro,

Workload assessment on foundry SME to enhance productivity using full-time equivalent," in MATEC Web of Conferences, Yogyakarta, Indonesia, 2017.

[9] Hutagalung, R., and Gustomo, A., Workload Analysis for Planning Needs of Personnels in the Corporate Administration Unit PT Timah (Persero) TBK. The Indonesian Journal of Business Administration vol. 2, no. 19, 2290-2297, 2012

[10] A. Syamsul and Jasril, The Measurement Of Workload And Optimal Number Of Education Personnel Using Work Sampling And Task Per Job Methods (A Case Study At Pt X.) in IOP Web Conferences, 2017

[11] Bwemelo and Gordian, "*Kaizen* as a strategy for improving SSME's peformance: assessing its acceptability and feasibility in tanzania.," European journal of business and management, 2014.

[12] Ulrich, D. Human resource champion: the next agen for adding value and delivering result. harvard business school press. 1991

[13] Moekijat. Manajemen personalia dan sumber daya manusia. Yogyakarta: BFFE. 2008

[14] OrientPoint. FTE Analysis and Models, Retrived from : http://www.orientpoint.com/FTE.htm. 2014

[15] Niebel, B.W and Freivalds, A. Methods, Standards, and Work Design. John-Wiley & Sons, New York.2002

[16] Liker, J. K.,. The Toyota Way. (Gina Gania & Bob Sabran, Penerjemah). Jakarta: Erlangga.2004

[17] Dewi, U., & Satya, A. Analisis kebutuhan tenaga kerja berdasarkan beban kerja karyawanpada PT PLN (Persero) distribusi Jakrta raya dan Tangerang bidang sumber daya manusia dan organisasi. 2012

[18] Imai, Masaaki. Gemba Kaizen: A commonsense approach to a continuous improvement strategy. New York: McGraw Hill, 2012.

Power Translations for Deterrence or Stimulation of Wheelie/Wheel-stand for a Straight Moving Vehicle

Gursagar Singh

Mechanical Engineering Department, Thapar Institute of Engineering and Technology, India

Abstract- The objective of this paper is to identify and study various factors which induce the possibility of wheelie/wheel-stand of a rear wheel drive vehicle. The purpose is to assist race and stunt drivers with a mathematical indicator (M) which could predict if a vehicle will undergo a wheel-stand or not, and if it will, what are the safe limits to avoid rollover and loss of control. The principle idea behind this research is that during a wheelie, the total power [P] supplied by an engine is utilized in uplifting [P1] and manoeuvring [P2] of the vehicle. After determining the coordinates of centre of gravity of the vehicle, P1 and P2 are calculated and on the basis of these values, factor of power translation [M] is calculated.

1. INTRODUCTION

We all have seen vehicles saluting up at some point of our life, may it be in a drag racing event or a stunt show on internet or TV, wheelies are always a pleasure to watch. For a normal person, who has nothing to do with automobile tech and who has no interest in physics or mechanics will only see a wheelie as front tire off the ground with roar of the high-power engine. But in actual a wheelie is not as simple as it seems. Various force equilibriums are broken during a wheel stand to make the vehicle ride on its rear rotor.

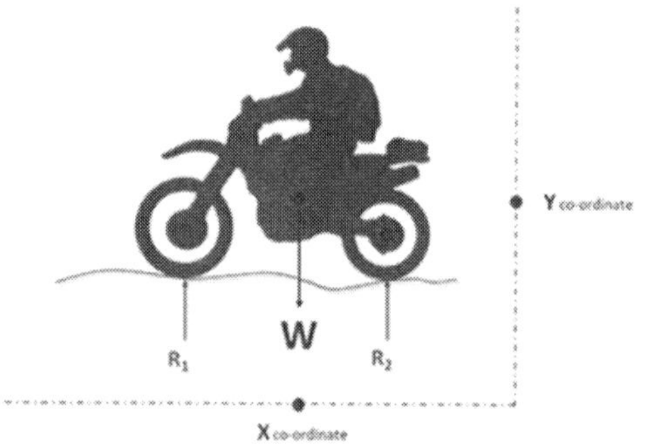

Figure 1(a). Weight and reaction forces.

W= Weight of vehicle in newtons
R_1= Reaction force at front wheel(s) in newtons
R_2= Reaction force on rear wheel(s) in newtons
Let us consider a motorbike.
The whole weight of this two-wheeler will act on its point of Center of Gravity (CG), which lies somewhere between the X-co-ordinates and Y- co-ordinates if viewed from side of the motorbike. In this case the co-ordinates of CG in Z direction have been ignored because it will not be needed in further calculations. The front and the rear wheels experience reaction forces, combinely equivalent to weight. *In order to undergo a wheelie or wheel stand the torque of the rear wheel must be equivalent to moment generated by weight vector on the center of rear wheel.*

Figure 1(b). A two-wheeler undergoing a wheel-stand

2. TO FIND CG OF THE VEHICLE

As mentioned previously, for calculations on moments and torques, it is required that CG of the undertaken four-wheeler or two-wheeler vehicle must be known. For cars and other four wheelers it is has been assumed that their CG point is known. Moreover, the reactions at various tires of a four-wheeler must also be known to apply this theory. It is quite interesting to note that the point of CG of any two-wheeler can be found out with the help of some known dimensions. For this, the following assumptions and considerations must be made:

1. The CG co-ordinate in Y direction lies on the axis AB which adjoins rotation centres for both wheels.

2. The total weight W includes the weight of rider as well.

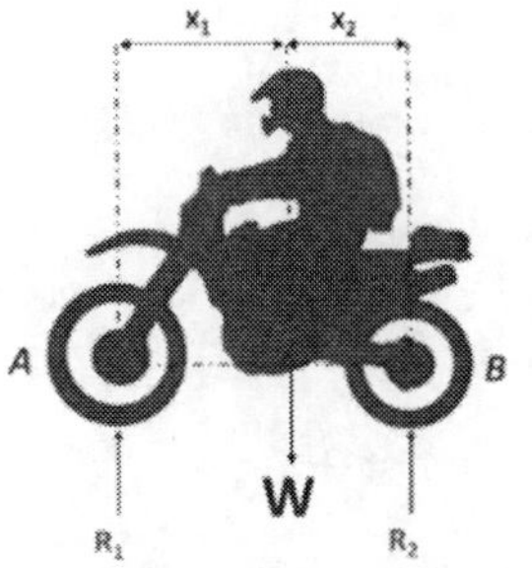

Figure 1(c). Axis AB adjoining rotation centres of both wheels.

2.1 CG co-ordinates along X-axis:

Let x_1 and x_2 be distances of point of CG from A and B respectively.

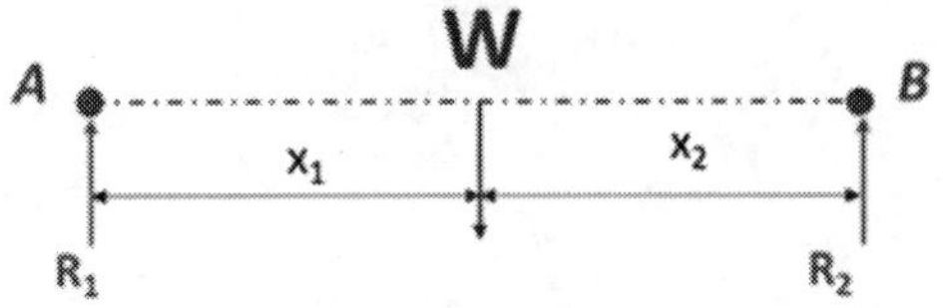

Figure 2. Axis AB adjoining rotation centres of both wheels.

Taking Moments about B,

$$R_1 (x_1+x_2) = Wx_2$$

$$R_1 = Wx_2/(x_1+x_2)$$ [1][2][3]

As we know,

$$R_1+R_2 = W$$

$$R_2 = W- R_1$$ [1][2][3]

So R_1 and R_2 can be found if W, x_1 and x_2 are known.

From fig.2 it can be noted that,

$$R_1+R_2 = W \quad \text{- (Eq.1)}$$

$$x_1+ x_2 = AB \quad \text{- (Eq.2)}$$

$$R_1\, x_1 = R_2x_2 \quad \text{- (Eq.3)}$$

From eq. (1), (2) and (3),

$$x_1= AB-x_2$$

$$R_1(AB-x_2)-R_2X_2 = 0$$

$$R_1AB-R_1x_2-R_2x_2 = 0$$

$$R_1AB = R_1x_2-R_2x_2$$

$$R_1AB = x_2\,(R_1+R_2)$$

$$R_1AB/(R_1+R_2) = x_2$$

So $\mathbf{x_2}$, the CG co-ordinate along X-axis is =

$$\mathbf{R_1AB/(R_1+R_2)} \quad \text{- (Eq.4)}$$

2.2 CG co-ordinates along Y-axis:

Let us now consider the motorbike is tilted at an angle θ, such that the front wheel is lifted 'P' distance from ground. In order to find the coordinates along Y-axis we will have to repudiate the assumption[1] made in 1.2.1 (The CG co-ordinate in Y direction lies on the axis AB which adjoins rotation centers for both wheels), because in actual scenario the point of CG will always lie above the AB axis.

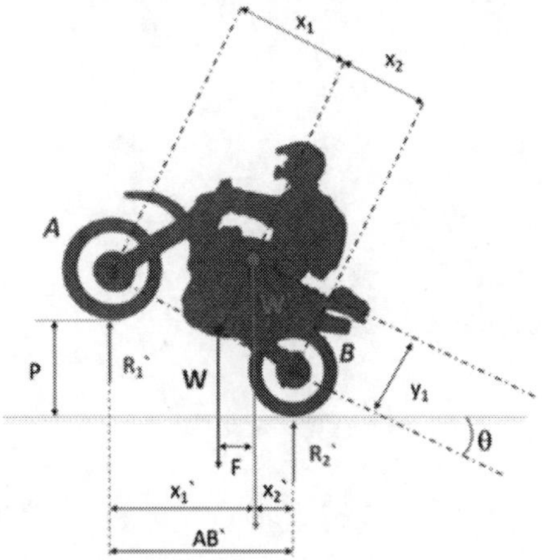

Figure 3. Various forces when the motorbike is tilted at an angle θ.

Referring to fig. 3

W` = W = Weight of motorbike

R_1`= Reaction force due to weight at front wheel while the vehicle is tilted.

R_2`= Reaction force due to weight at rear wheel while the vehicle is tilted.

x_1, x_2 = Co-ordinates of CG along X-axis (when the bike was not tilted).

x_1`, $x_2$` = Apparent co-ordinates of CG along X-axis (when the bike was tilted).

y_1 = Co-ordinates of CG along Y-axis.

Applying equation (4) on fig. 3,
We will get,

$$x_2` = R_1`AB`/ (R_1`+ R_2`)$$

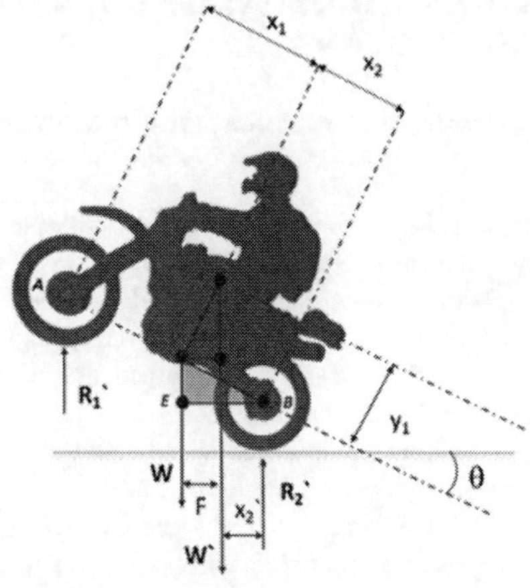

Figure 4(a). △DEB and △DCF on bike

Referring to fig. 4(b),
In *ΔDEB,*

$$DB = x_2$$

$$EB = x_2\, Cos\theta$$

$$x_2\, Cos\theta - x_2` = F$$

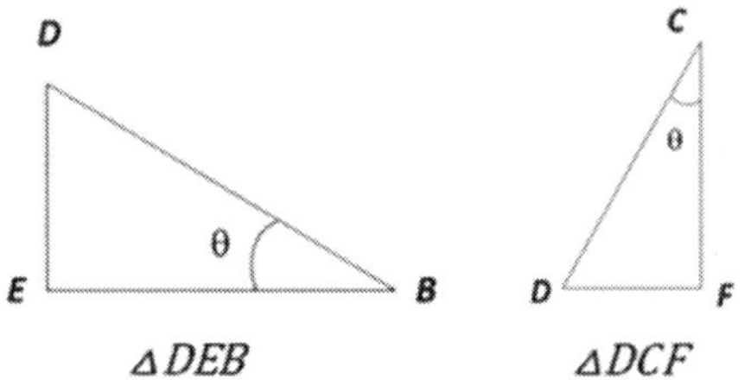

Figure 4(b). △DEB and △DCF

In *ΔDCF*,

$$DF = F$$
$$CD = y_1$$
$$y_1 = F/\mathrm{Sin}\theta \quad \text{- (Eq.5)}$$

So y_1, the CG co-ordinate along Y-axis is = **F/Sin*θ***

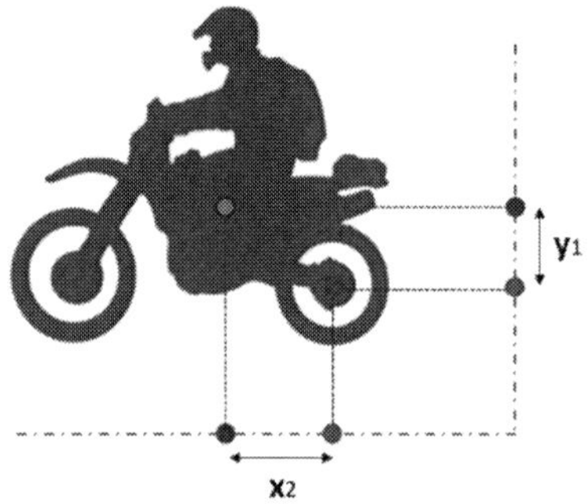

Figure 5. Position of Centre of gravity with reference to rear wheel axle.

3. POWER TRANSLATIONS IN A WHEEL-STAND

3.1 Understanding the physics behind the wheel stand:

When a vehicle is in the state of rest, it isn't possible for such vehicle to reach some positive value of final velocity in an instant. If this happens then the acceleration of the vehicle will be infinite. [4] The inertia of rest opposes quick movement of vehicle, thus in actual practice it is not possible to attain infinite acceleration.

In case of obnoxious or undesired wheel-stand/wheelie, a highly powerful vehicle engine tends to quickly accelerate the vehicle from rest. The vehicle simply fails to completely overcome its inertia of rest and the torque supplied to wheels results in uplift from the front. But as the vehicle maneuver and start gaining velocity the power supplied by engine redirects itself from the state of wheel stand to desired state of motion.

So, it can be said that the overall scenario is about translation of power. The total power **(P)** generated by engine is divided into two work streams:

a) Power utilized in uplifting of vehicle, $\mathbf{P_1}$
b) Power utilized in forward motion of the vehicle, $\mathbf{P_2}$

Neglecting the transmission and frictional losses,
$\mathbf{P = P_1 + P_2}$

3.2 Determination of P1 and P2:

In order to understand the complex power translations which happens during the wheel-stand, the values of P_1 and P_2 must be known.

3.2(a) Determination of P_1:

Let us again consider a motorbike,

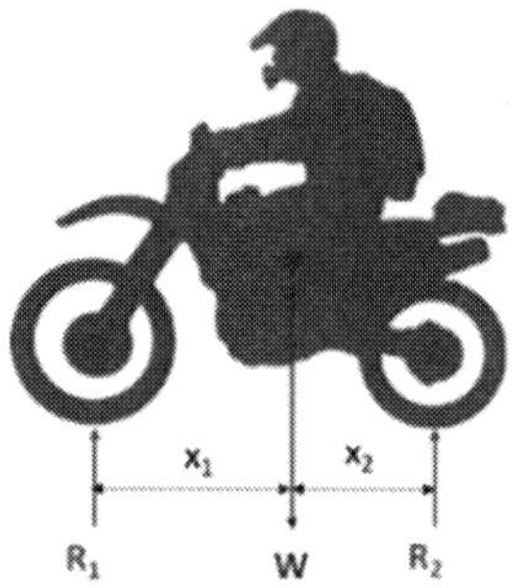

Figure 6. Weight and reaction forces.

Torque required to uplift the bike from front (T):

$$T = Wx_2 \text{ [5]} \quad \text{- (Eq.6)}$$

Power required to uplift the bike from front (P):

$$P_1 = T\omega \quad (\omega = \text{angular velocity})$$

Substituting Eq. (6) in this,

$$P_1 = Wx_2\,\omega \quad \text{- (Eq.7)}$$

Since, $\pi \text{rad} = 180°$

$(\pi/180)$ rad = 1° [4][5]

$(\pi/180)\,\theta$ rad = $\theta°$

and, ω = rad/sec

$$\omega = (\pi/180)\,\theta\ °/\text{sec} \quad \text{- (Eq.8)}$$

From Eq. (7) and Eq. (8)

$$\mathbf{P_1 = Wx_2\,(\pi/180)\,\theta} \quad \text{- (Eq.9)}$$

The value of x_2 is a function of θ, as the θ increases the value of x_2 changes as its cosine function.

Moreover,

$$W = mg$$

Thus, Eq. 9 will become

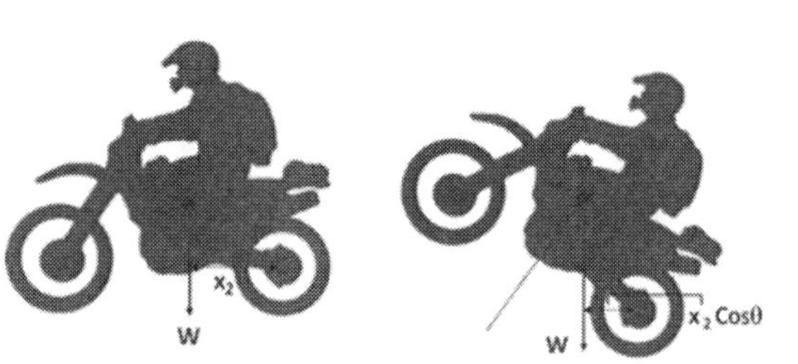

Figure7. Change in distance between weight vector and rear axle with change in angle θ

$$P_1 = mgx_2\cos\theta\ (\pi/180)\ \theta \qquad \text{- (Eq.10)}$$

This equation tells us about the power utilization during a wheel-stand at a specific value of θ. But since the value of θ varies from 0 to maximum tilt angle, the actual value of power utilization could be found out using integration. Integrating P_1 with respect to small angle $d\theta$,

$$\int_0^\theta \frac{P1}{d\theta} = \int_0^\theta [mgx2\cos\theta \left(\frac{\pi}{180}\right)\theta]\, d\theta$$

$$\int_0^\theta \frac{P1}{d\theta} = mgx2\left(\frac{\pi}{180}\right)\int_0^\theta [\theta.\cos\theta]\, d\theta$$

$$P1 = mgx2\left(\frac{\pi}{180}\right)|\theta.\sin\theta + \cos\theta|_0^\theta$$

$$P1 = mgx2\left(\frac{\pi}{180}\right)[\theta.\sin\theta + \cos\theta - 0 - 1]$$

$$P1 = mgx2\left(\frac{\pi}{180}\right)[\theta.\sin\theta + \cos\theta - 1] \qquad \text{-(Eq.11)}$$

So, it can be stated that "*A rear wheel drive vehicle of mass 'm' kg and CG co-ordinate along X-axis being 'x_2' meters will need P_1 watts of power at rear wheel(s) to undergo a wheelie of θ degrees*"

3.2(b) Determination of P_2:

The power required for forward motion of bike (P_2) can be calculated using following formula;

$$P_2 = mav^{[5][6]} \qquad \text{-(Eq.12)}$$

Where,
m = Mass of vehicle in kg
a = Acceleration of vehicle in m/sec^2
v = Velocity of vehicle in m/sec
P_2 = Power required to move the vehicle for the distance of 1 meter for 1 second

3.3 Factor of power translations (M):

The complete event of a vehicle undergoing a wheelie can be divided in three time stages:

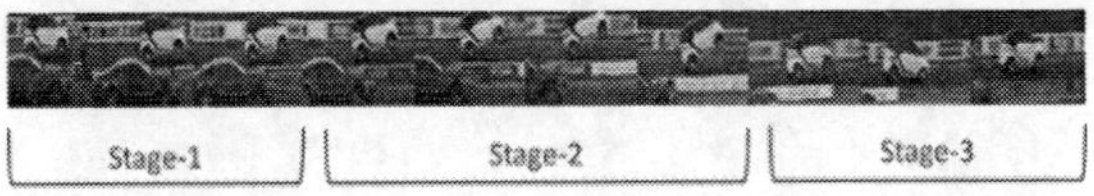

Figure 8. Different time stages of a wheel-stand

Stage-1:
As soon as the vehicle tends to start off from rest, inertia comes into play and it opposes quick forward motion or acceleration of vehicle. But since very high torque is supplied to rear wheel(s) and in order to form a force equilibrium, the front of the vehicle gets uplifted. In other words, it can be said that while moving from rest, the power supplied by engine gets utilized to undergo a wheelie instead of giving it a straight motion. ***(P_1>P_2)***

Stage-2:
When the vehicle reaches the maximum angle of tilt (θ_{max}) while undergoing a wheel-stand, a moment equilibrium is created for a very small time, if a vehicle maintains this equilibrium theoretically it can then wheelie for infinite time but in actual constant acceleration is not possible. So, at θ_{max} the power utilized for front lift (P1) and power utilized to increase the momentum (P2) equally contributes to consumption of total power. ***(P_1=P_2)***

Stage-3:
After reaching θ_{max} the vehicle undergoing the wheelie fails to maintain the moments equilibrium due to insufficient acceleration and as a result of this value of θ start decreasing and ultimately it reaches 0, which means the vehicle has be grounded. The decrement of θ also shows that the distribution of power is uneven. The power utilized to maneuver the vehicle exceeds the power required for a wheelie. ***(P_1<P_2).***

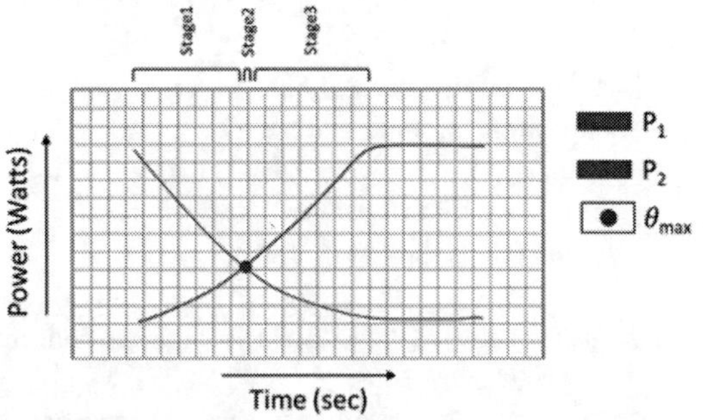

Figure 9. Variation of P_1 and P_2 with respect to power and time

As we know now, the total power generated by the engine (P) is distributed in P_1 and P_2. At this stage a question arises, how can we find out how much power is being directed to P_1 and P_2 at any particular point of time. Because only at θ_{max} P_1 equates with P_2, at all other angles $P_1 \neq P_2$.
So, it can be said that at any instant of time,

$$P_1 = (M)P_2$$

Where,
M = Factor of power translation.

Hence $$M = \frac{P1}{P2}$$

From Eq. 11 and 12

$$M = \frac{mgx2\left(\frac{\pi}{180}\right)[\theta.\sin\theta + \cos\theta - 1]}{mav}$$

$$M = \frac{gx2\left(\frac{\pi}{180}\right)[\theta.\sin\theta + \cos\theta - 1]}{av}$$

The value of M can give us the idea of if the vehicle under consideration will undergo wheelie or not. For those who want to desirably stimulate a wheel-stand, like professional stunt drivers, should make sure that value of M remains close to 1. On the other hand, those who want to avoid a wheelie for the sake of handling, like drag racers, should make sure that center of mass of the vehicle is away from the rear wheel to prevent wheel-stand under maximum acceleration.[7]

4. CONCLUSIONS

4.1 After the above discussions and calculations, it has been found that:

a) In order to undergo a wheelie or wheel stand the torque of the rear wheel must be equivalent to moment generated by weight vector on the center of rear wheel.

b) In case of a two-wheeler, if distance AB between axis of rotation of both wheels is known and the reaction forces (R_1 and R_2) due to weight of bike and rider are also known the C.G co-ordinates along X-axis will be equal to:
$R_1AB/(R_1+R_2)$ (Units same as of distance AB).

c) Similarly, if distance AB between axis of rotation of both wheels is known and the reaction forces R1 and R2 are also known while the bike is tilted at angle θ. The C.G co-ordinates along Y-axis will be:
$F/\mathrm{Sin}\theta$ (Units same as of distance AB).

d) The total power (P) generated by engine is divided into two work streams:
   1. Power (P_1) utilized in uplifting of vehicle and is equal to:
      $[mgx_2\mathrm{Cos}\theta\ (\pi/180)\ \theta]$ watts.
   2. Power (P_2) utilized in forward motion of the vehicle and is equal to:
      $[mav]$[1][2] watts.

e) A rear wheel drive vehicle of mass 'm' kg and CG co-ordinate along X-axis being 'x2' meters will need $\{mgx2\left(\frac{\pi}{180}\right)[\theta.Sin\theta + Cos\theta - 1]\}$ watts of power at rear wheel(s) to undergo a wheelie of θ degrees.

f) Percentage of the total engine power consumed in wheelie could be found by $\left(\frac{P1}{P1+P2}\times 100\right)$.

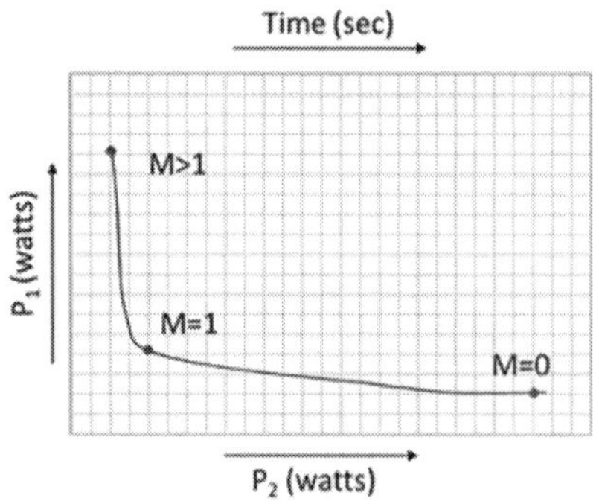

Figure 10. Variation of M with respect to P_1 P_2 and time

g) The ratio of P_1 and P_2 is known as **Factor of power translations (M).** The value of M can give us the idea of if the vehicle under consideration will undergo wheelie or not;
   1. If **M>1:** Vehicle will undergo an aggressive wheelie.
   2. If **M<1:** Vehicle will not undergo a wheelie.
   3. If **M=1:** Vehicle will undergo a wheelie of θ if accelerated with '**a**' m/sec^2 to reach the final velocity of '**v**' m/sec.

5. WAY FORWARD

The objective of this paper is to study about the various forces which stimulate a wheel-stand or a wheelie in a rear wheel drive vehicle. A mathematical indicator (M) has been introduced to predict the rollover of a vehicle during a wheelie. This research could not only contribute to safety for drag racers and stunt but the derived mathematical relations, as an outcome of this research, can also contribute to the field of design and stress analysis of mechanical parts like wheelie bars[8], reducer housing and transmission systems.

Figure 11. A wheelie bar mounted to chassis of a drag racer.

REFERENCES

[1] S.S. Rattan (July 1, 2017). *Theory of Machines* (4th ed.) McGraw-Hill Education.
[2] Birinder Singh (2007) *Applied Mechanics* (1st ed.) SK Kat Publication.
[3] N Khurmi & R S Khurmi (2014) *Introduction to Applied Mechanics*. S. Chand Publishing.
[4] Willard W. Pulkrabek (2014) *Engineering Fundamentals Of The Internal Combustion Engine* (2nd ed.) Pearson Publication.
[5] John Pearley Huffman (June 2010) "*The physics of the wheel stand*". Car and driver Retrieved.
[6] C. Ashish (April 21, 2016) "*What's the physics behind the wheelie*". Scienceabc.com.
[7] UNSW "*Wheelstands or wheelies*". Physclips .
[8] Gary Grant (July 10, 2012) "*This is why drag cars have wheelie bars*" Wheels.ca.

Kano Model for Identification Car Modification Features of Mobile Application

Muchamad Sugarindra[1] and Rian Notia Pratama [2]

[1&2] *Industrial Engineering, Universitas Islam Indonesia, Indonesia*

Abstract: Modification was an activity to replace the object more attractive especially car modification. Lately, car modification become the one of the promising business since that were providing high profits. The owner of car look for the accessories for their car frequently. Then, the owner of car modification should increasing their facility to make the consumer satisfied. Furthermore, if the consumer satisfied with the service than it provided high profit. For increasing high profit than the owner of car modification should made the innovation. One of the innovation was car modification with mobile application (apps.) by Kano model. The aims of this study was to identify the features in car modification using Kano model. Kano model used to identify consumer needs that divided into 3 category. There are must be, one dimensional and attractive. The result from open questionnaire would be proceed by Kano model. This study revealed that feature of selection car body colour, feature of selection lamp model and feature engine cap model of into one dimensional category, feature of selection car velg, and matching the tire into must be category and feature of selection cutting sticker, and selection of bumper model into attractive category, and priority to develop are feature of selection car body colour, cutting sticker and selection of bumper model. This category should be the reference to be the resource for car modification apps.

1 Introduction

In this era, business competitions in Indonesia were getting strict. From that reason above business people should have some strategic that the product would be accepted in the market. One of the strategic which used by the business people were providing the best services and expanding the market in unreached area [1]. As well as business competition in car market. Lately, based on Gabungan industri kendaraan bermotor Indonesia (Gakindo) the sales of car on 2016 reached 1, 06 million. It increased about 4,5 % from year 2015 to 2016. This finding shows that the consumers of car were increasing. Department of design and sales have the high contribution to increase the car sales. The consequence of high sales was swelling of the machine shop. It was for not only machine maintenances but also car modification. Generally, Car Company releasing original car. However, some of consumer needs to modify their car for better performance although small or huge modification. Modification was the activity, which changed the object to be more interesting. The car modification such as changing the colour of car or alloy wheels. Lately, business modification very promising. Some of machine shop got turnover around 40-50 million per month and profit about 30% from that [2].

Because the growths of workshop then the owner must to attract the consumer with good services and low price. Nowadays, even though the consumer got good services and price, they could not satisfices the consumer. It happened because of the first target was they satisfied about the result even though they got high price. If they were not satisfied yet, they would remain to modify their car until pleased. If that happened consequently the consumer would spending a lot of money. For keep away from repeated modification and good services accordingly the workshop need applications (apps.) for car modification on android based. This app. could help the consumer with simulation modification on android. With that, simulation would be helpful for consumer. These apps. Will help the consumer to choice the modification and for workshop workshop would give the best services for consumer.

To create the apps., needs identification for user. There are model to apps. identification. One of them is Kano model. Kano model used to identify the consumer needs frequently. This model could give the priority for the needs based on three category. [3], [4], [5], [6]. [7]. Kano model is methods of overcoming bias of result from survey to find consumer needs traditionally [8]. In the Kano model there are 3 main requirements that affect the needs or customer satisfaction is must be requirement, one dimensional requirement, and attractive requirements [9]. Must be requirement was the basis requirement or minimal criteria that should be full filled by the produce. However if the requirement weren't that will not affect the product; one dimensional requirement was if the fulfilment more completed then the costumer more pleasure and attractive requirements was if they weren't then the product or services couldn't influences but if they were could be increasing the consumers satisfaction. Those requirements would determine the apps. Which is necessary or not.

From the reason above, the purposes for this study was to identify the app., of car modification which required by the workshop owner and consumer. These apps., be expected to improve the profit for the workshop owner, improving the consumer satisfaction, save cost for consumer and reduce the wasted work.

2 Material and Methods

2.1 Research Design

Design of this study was survey by using primary and secondary data.

2.2 Data Collection

These data collected by interview with open questionnaire for primary data and study literature for secondary data. The questionnaire content was about the features, which needed in these apps. and desire analysis about the desire for car modification apps. Study literature would support the information about apps. These studies have four questionnaires that distributed to the consumers. First questionnaires were the open questionnaires which used to know the customer needs on information systems which would be created; second questionnaires was the questionnaires for the decision level, are they valid enough to be attributed for design; third questionnaires was Kano model questionnaires which ask about functional and dysfunctional questions, the purposes for this Kano model questionnaires were to identification the features which needed in this apps. such as positive for functional and negative for dysfunctional question; fourth questionnaires ask about the features specification which proposed to car modification apps.

2.3 Subject

Subject on this study were the consumer who has ever had modification their car and the owner of workshop, which served car modification with inclusion criteria, was the people, which could make the decision by age range between 17-35 years old. The number of samples based on the formula 1 and the minimum sample based in formula 2 [10]:

$$n = \frac{Z^2.P(1-p)}{E^2} \quad (1)$$

The formula of number sampling (n: number of sample; Z: level of confidence; P (1-p): variation of population; E: sampling error) [10].

$$n = \frac{1{,}645^2.0{,}5(1-0{,}5)}{0{,}1^2} \quad (2)$$

n = 67, 65~68 sample

2.4 Data processing and analyses

After data were collected, then the data processing and analyses were done. The data that obtained would be analyses by validity and reliability using Pearson Product Moment. The data was valid if p value > 0.05 and reliable if r count > r table the criteria of reliability were 0,00 – 1,19: reliability was very low; 0,20 – 0,39: low reliability; 0,40 – 0,59: moderate reliability; 0,60 – 0,70: high reliability; 0,80 – 1,00: very high reliability.

2.4.1. Kano model category

The valid and reliable data would be process with Kano model for evaluation attribute. These methods to determine the attribute for the customer satisfaction then categorization were needed. The result of categorization would be used for features of apps. The evaluation of Kano model by input the data based on evaluation Kano model table (Table 1).

Table 1. Kano model evaluation [11]

Consumer needs		Dysfunctional				
		L	E	N	T	D
Functional	L	Q	A	A	A	O
	E	R	I	I	I	M
	N	R	I	I	I	M
	T	R	I	I	I	M
	D	R	R	R	R	Q

A : Attractive
O : One dimensional
M : Must be
I : Indifference
R : Reverse
Q : Questionable

L : Like
E : Expect
N : Neutral
T : Tolerance
D : Dislike

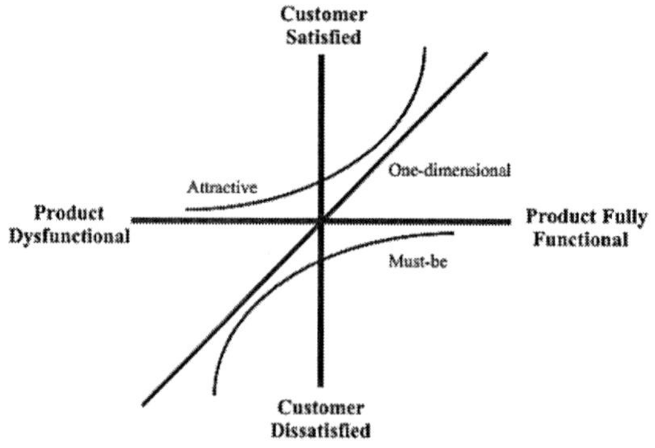

Figure 1. Kano Diagram [12]

In the Kano model, there are 5 categories of quality factors: must-be, one-dimensional, attractive, indifference and reverse which each have different characteristics and levels of influence on consumers (M, O, A) [3], Indifference (I): this feature will not have an impact on consumers and; Reverse (R): a feature that if held will cause dissatisfaction on the consumer.

Kano model can know the influence of each attribute by identifying relative values to meet user satisfaction or not. In the same conditions, Kano model can be separated into two conditions, namely customer satisfaction and customer dissatisfaction following formula 3 & 4 [11]:

Customer Satisfactions (CS)=A+O/(A+O+M+I) (3)

Customer Dissatisfactions (CD)=O+M/(A+O+M+I) (4)

If CS value is closed to 1, then attribute increasingly influence consumer satisfaction, If CS value is closed to 0, that means that attribute does not affect consumer satisfaction. If the CD value is closed to -1 then the attribute if not met will greatly affect customer satisfaction and if close to 0 means that the attribute has no effect on customer disappointment [13]

2.4.2. Conclusion and recommendation

After the step have done, furthermore the features for apps. Were decided.

3 Result and discussion

3.1. Feature attributes determination

From the open questionnaire about feature attribute, the subject explained that these feature were needed. They are features of selection car Body Colour (**BC**), selection of Car Velg model (**CV**), selection of Tire model (**T**), selection of colour Car Seat (**CS**), selection of Cutting Sticker (**CSt**), selection of design Interior for Audio (**IA**), selection of Bumper Model (**BM**), selection of Windshield Glass (**WG**), selection of Lamp Model (**LM**), selection of Cap Engine model (**CE**), and selection of Steering Wheel model (**SW**). These results would be needed as attributes for car modification app.

3.2 Validity

Table 2. Validity and reliability test of feature attributes for car modification app.

Feature attributes	R-table	R-Count	Explanation	Reliability	Explanation
BC	0.235	0.473	valid	0.830	reliable
CV	0.235	0.460	valid	0.830	reliable
T	0.235	0.544	valid	0.830	reliable
CS	0.235	0.613	valid	0.830	reliable
CSt	0.235	0.602	valid	0.830	reliable
IA	0.235	0.689	valid	0.830	reliable
BM	0.235	0.672	valid	0.830	reliable
WG	0.235	0.606	valid	0.830	reliable
LM	0.235	0.742	valid	0.830	reliable
CE	0.235	0.663	valid	0.830	reliable
SW	0.235	0.635	valid	0.830	reliable

The results of open questionnaire about the feature attributes have been done validity and reliability test. The result shows that all of the features attributes were valid and reliable (Table 1).

3.3 Kano model category

Table 3. Features attributes category for car modification app. By Kano model

Attributes	A	O	M	I	R	Q	Total (n)	Category
BC	9	46	8	6	1	0	70	O
CV	10	15	37	8	0	0	70	M
T	8	16	34	5	6	1	70	M
CS	4	17	8	38	3	0	70	I
CSt	39	12	8	10	1	0	70	A
IA	15	11	8	32	3	1	70	I
BM	32	18	6	12	1	1	70	A
WG	8	13	6	37	6	0	70	I
LM	10	32	9	14	5	0	70	O
CE	15	28	10	14	3	0	70	O
SW	14	20	6	25	5	0	70	I

Based on table 3, selection of car velg model (**CV**), and selection of tire model including categories must-be. Feature of selection car body colour (**BC**), selection of lamp model and selection of cap engine model (**CE**) including categories one-dimensional. Selection of cutting sticker (**CSt**), selection of bumper model (**BM**) including categories attractive. Selection of colour car seat (**CS**), selection of windshield glass (**WG**), selection of steering wheel model (**SW**) including categories indifference. Indifference category has low influence for customer satisfaction.

After obtained the categorization of each attribute, continued processing data to determine the level of consumer satisfaction of product features based on the formula 3 and 4. Tabulation data can show in table 4:

Table 4. Satisfaction level of respondents.

Attributes	Category	CS	CD
BC	O	0,797	-0,783
CSt	A	0,739	-0,290
BM	A	0,735	-0,353
LM	O	0,646	-0,631
CE	O	0,642	-0,567
SW	I	0,523	-0,400
AI	I	0,394	-0,288
T	M	0,381	-0,794
CV	M	0,357	-0,743
WG	I	0,328	-0,297
CS	I	0,313	-0,373

Base on table 4 found attributes include to the priority category to develop, because these category can influence customer satisfaction. Feature selection car body colour became the first priority, feature of selection cutting sticker became the second priority and feature of selection bumper model became third priority.

Feature which has been identified from this research continued with observation and identify special feature for workshop owner. The result of observation was found that the selected features need to add the spare part price. Then, the consumer can calculate directly how much it will cost to modify his car. This apps in the design connected with the database owner of the workshop to know the inventory spare part, invoices to customers and monthly reports to the owner of the workshop.

4 Conclusion

Kano model effective to divide attributes into 3 categories, making it easier for programmers to prioritize the attributes that should be developed. The features application that recommended for car modification could be divided are:

1. Must be category: feature of selection car velg, feature of selection tire.
2. One dimensional category: feature of selection car body colour, feature of selection lamp model
3. Attractive category: feature of selection cutting sticker, and selection of bumper model.
4. Priority to develop are feature of selection car body colour, feature of selection cutting sticker and selection of bumper model.
5. The addition of a price list to selected features for the consumer user interface, inventory features, invoices and monthly reports to the user interface of the workshop owner.
6. Further research can be done in making the application and testing usability of application design that has been made

References

1. Ulus, A. A. 2013. Bauran Pemasaran Pengaruhnya terhadap Keputusan Pembelian Mobil Daihatsu pada PT. Astra International Manado. Universitas Sam Ratulangi, Manado. Jurnal EMBA ISSN 2302-1174 Vol. 1 No. 4, pp.1134-1144
2. Meraup untung bisnis modifikasi mobil, http://bandung.bisnis.com/read/20110611/3/57387/meraup-untung-bisnis-modifikasi-mobil, accessed on a date 6th April, 2018
3. Hashim, A.Md.; Dawal, S.Z.Md. (2012). "Kano model and QFD integration approach for Ergonomic Design Improvement". International (summer) Conference on Business Innovation and Technology Management, Vol.57, pp.: 22 – 33.
4. Zong, L.; Yu, S.; Li, B. (2013). "The model building research of product innovation design quality based on QFD and KANO". Trans Tech Publications, Switzerland, Vol. 271-272, pp.: 1467 – 1472.
5. Choirumah, Moses L. Singgih, 2014, Pengembangan Model Integrasi QFD-Kano-Axiomatic Design-Fuzzy Goal Programming Serta Aplikasi Triz Dalam Memaksimalkan Kepuasan Dan Meminimalkan Biaya Pengembangan Produk, http://digilib.its.ac.id/public/ITS-paper-40145-2510100044-paper.pdf accessed on a date 15th May 2018
6. Shinta Oktriviani,Sari Wulandari, Ika Arum Puspita, 2017, The Need Analysis Of I-Gracias Application Using The Integration Of Web Usability And Kano's Model (Study Case On By Telkom University Engineering Faculty College Student), e-Proceeding of Engineering : Vol.4, No.3 December 2017
7. Aprilia P.D, Amani H, Wulandari S., 2017 Needs Analysis Of E-Commerce Website Services On Dkantin.Com Using Integration E-Service Quality And Kano, e-Proceeding of Engineering : Vol.4, No.3 December 2017
8. Huang, J. 2017. Application of Kano Model in Requirements Analysis of Y Company's Consulting Project. American Journal of Industrial and Business Management, pp. 910-918.
9. Tan, K. C., and Pawitra, T. A. 2001. Integrating SERVQUAL and Kano's Model into QFD for Service Excellence Development. Managing Service Quality Vol. 11 No. 6, pp. 418–430
10. Eriyanto. 2007. Teknik Sampling Analisis Opini Publik. Yogyakarta: PT. LkiS Pelangi Aksara.
11. Berger, C.; Blauth, T.; Boger, C.; Burchill, G.; Du Mouchel, W.; Pouliot, F.; Richter, R.; Rubinoff, A.; Shen, D.; Timko, M.; Walden, D. (1993). "Kano's methods for understanding customer-defined quality". Center for Quality Management Journal, Vol. 2 (4), pp.: 3 – 36
12. Walden, D. 1993. A Special Issue on Kano's Methods for Understanding Customer Defined Quality. The Center for Quality of Management Journal, pp. 3-35.
13. Hasan M., 2016 Analisis Kebutuhan Robot Rehabilitasi Pasien Pasca Stroke dengan Menggunakan Metode Kano, JITI, Vol.15 (2), December 2016, 151 – 15

https://doi.org/10.1051/matecconf/201822103004

Automated programming robot component transfer system utilizing machine vision as detection interface

H J Vermaak and L Rogers

RGEMS Research Unit, Department of Electronic Engineering, Central University of Technology, Free State, Bloemfontein, South Africa

Abstract. Modern day automation systems rely on fixed programming routines to carry out their operations. If an automated flexible system is introduced onto such a production line, the complete reprogramming process required for new products needs could be automated with limited loss in production time. Therefore, instead of reprogramming each new position for the robot system the system takes over real-time control of the robot and carries out the required steps autonomously. The benefit with such a system would be that the robot would not need to be reprogrammed for every new routines but is controlled in a real-time environment to carry out new procedures based on external vision sensors. Using a real-time system could remove the need for a fixed programming environment and replace it with an automated changing programming setup. This could result in a system automatically adapting to a new product introduction through real-time machine vision processing techniques.

1 Introduction

The Assembly industry is a demanding, yet expanding market. The influence of the 4th Industrial Revolution is creating new opportunities within this sector. Flexibility within the product assembly industry is one of the most important features of the modern day assembly system. The assembly industry is driven by the need for new consumer products and the ability to respond to product changes rapidly[1]. Flexible assembly systems (FAS) must be able to convert quickly to the production of new product families and thus produced an increased variety of products in unpredictable quantities [2]. A FAS system can be a single product line or it can be reconfigurable to be flexible as needed by product demand [3] [4]. Flexibility can be expressed as the ability of a system to reconfigure itself for different product lines in response to changing requirements with limiting the down time [5].

The challenge for a flexible system that includes a robot arm is the reprogramming of the routines to adapt to the new product that need to be assembled [6]. Maintaining accuracy, speed and safety is very importance to ensure flawless operation of the system. This project will investigate an automated self-learning environment to allow for the control and programming of an assembly system.

2 Automated programming system

The aim of this project is to investigate the use of a vision system as detection sensor in developing a system that can learn new product designs in real-time, allowing the robot system to adjust accordingly without the need for manual reconfiguration of the system. This in result can lead to an increased output in production in newly introduced products, while also having alternative implementation options. The vision system would scan the design pallet detecting the different components and calculating optimal movement paths for the robot system.

The designed system should be able to build a new product design completely autonomously without any user intervention or manual robot reprogramming. Once a new product is introduced onto the existing assembly line, the system should automatically adapt and build the new product considering that all preconditions are met such as the components needed to build the design is available within the component catalogue on the component pallets on the conveyor.

2.1 Test environment

The test setup needs to be developed to accommodate the vision system, the robot arm as well as the conveyor system that acts as the component transfer system. The test setup need to use the vision system to learn new product designs in real-time, allowing the robot arm to adjust accordingly without the need for manual reprogramming of the robot system. The test setup is shown below in figure 1.

2.2 Automated learning robot system

A KUKA robot was used in this case study with the RobotSensorInterface (RSI) software installed on it. The RSI software allows an external sensor system to take real-time control of an industrial robot. This ability of the software allows an external sensor system to take real-time control of the robot and this creates the opportunity that the fixed programming of the robot can be replaced with an automated changing programming setup. A flexible component transfer system can be created using this self-learning robot system.

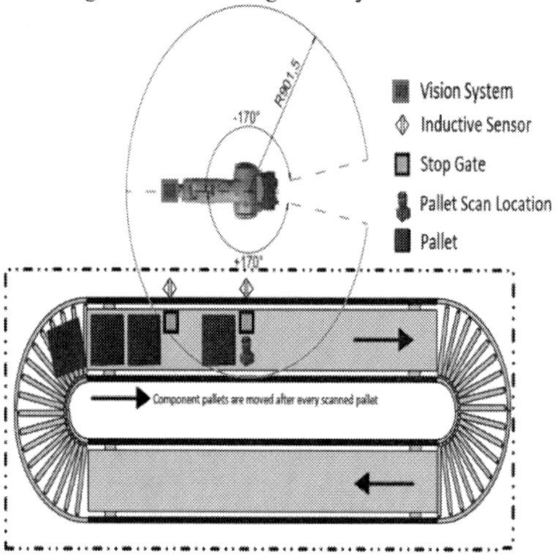

Figure 1. Test system layout including a KUKA Robot Arm.

RSI allows for continual influence over the robot motion by means of sensor data. In figure 2 the fundamental principles regarding Ethernet exchange are depicted. An object relates to a certain property or variable that will be updated through the RSI system. A certain object will be configured as either an inbound object or alternatively an outbound object. Different modes of correction can be configured namely [6]:
Motion suppressed sensor correction:

- Axis angle correction (absolute or relative),
- Cartesian correction (absolute or relative),

Sensor-guided motion

- Axis angle correction, absolute or relative
- Cartesian correction, absolute or relative

2.3 RSI system setup

This principle relating to the RSI setup utilizing a visual aid with its inbound and outbound objects will now be explained. The sensor cycle rate is the rate at which the signal processing is calculated. In other words, it's the cycle time defined that all work should be completed to and from the sensor system. The RSI system supports two sensor processing rates namely Input Processing Output (IPO) mode which sets the processing rate to 12ms. A faster processing rate is available using IPO_FAST mode which fixes the processing rate to 4ms. All data needs to be processed within this cycle rate in order for it to be a valid operation. There is a constant packet transfer between the robot and the sensor system even if no correction is currently in progress [7].

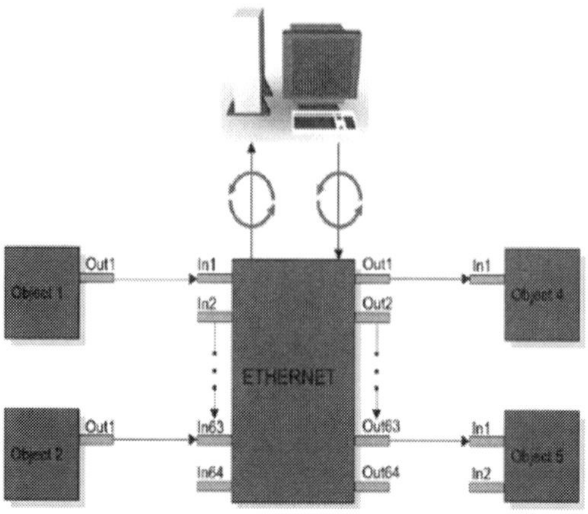

Figure 2. Data exchange configuration setup via Ethernet [6].

3 Vison system

The vision system is used as a remote detection and image acquiring sensor to the RSI software. It acquired images of components on the pallet and calculated each items relative position to the KUKA gripping device [6]. The centre of the pallet is taken as default position. This process was done by calculating the centre of gravity for each component. The centre of gravity was needed as it was used as the area to pick the components to ensure stability during moving operations [6]. The Accord.net framework software was implemented and used to calculate the orientation data for each component. These raw and central moments to the blob in question are implemented by the framework in order for a resultant angle to be calculated [8]. This calculation is very important when the need arise to rotate a components to fit in the new product template. The calculated centre of gravity for different shape components can be seen in figure 3 below.

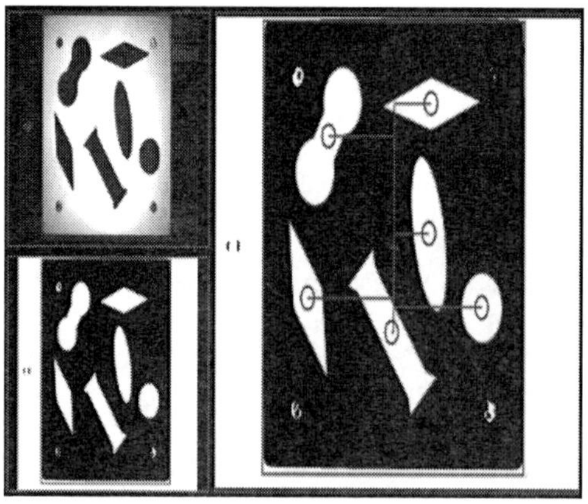

Figure 3. The calculated center of gravity for each component (small circle in each component).

3.1. Vision and robot calibration

First the robot calibration (positioning) was done taking into account that the camera need to be directly above the pallet when taking an image. The region of interest was then calibrated for the camera taking in account that the region need to be within the borders of the pallet for the calculating purposes. Combining the calibration of the robot and the vision system produce a flexible vision-processing environment that can be used in different situations when needed.

Table 1. Overall Average accuracy results based on all builds

Build	X (mm)	Y(mm)	Angle(°)
1	1.7810555	1.391437	0.36042
2	2.6764228	1.134341	1.159303
3	1.3676975	1.421224	0.799562
4	1.5639524	1.54429	0.774123
5	1.7865532	1.797749	0.639775
6	1.289422	0.904378	0.965501
Overall Average	**1.7441839**	**1.365569**	**0.783114**

Table 2. Overall average time results based on all builds

Build	Time Taken(s)
1	342.54
2	340.94
3	342
4	334.6
5	342.51
6	341.01
Overall Average	**340.6**

4 System validation

The system compromise of multiple sub-systems (that can operate independently) but are required to work as one unit to achieve the objectives for this case study. In figure 4 below the block diagram for the system is shown. It is clear from the figure that all systems communicate is over Ethernet as previously explained. Each sub-system has the ability to operate independently, but the operating instructions are produced by the software controller as indicated in figure 4.

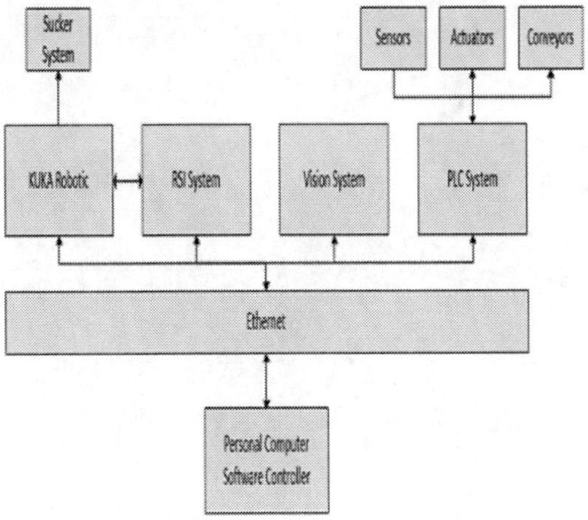

Figure 4. Block diagram of complete system indication the different sub-systems.

The system was challenged through various experiments that would require it for example to build a "Design" in the quickest most efficient manner possible. The speed and accuracy were recorded under different circumstances such as unique designs, similar components within the designs. Comparisons were conducted between the manual procedure of configuring the robot system and the automated process that this project presents when introducing a new design on an existing assembly system. Factors such as ease of setup, time consumption, reliability were all taken into account while maintaining an innovative thought process to allow leeway for the introduced system to thrive.

5 Results discussion and conclusion

It should be noted that the speed when picking and placing components was reduced due to the real-time control of the KUKA robotic system, movements between fixed points are still capable of running at 100% speed as these are fixed points within the coordinate system on the KUKA Robot.

The tables below summarize the builds data in order to get a more overall average of the accuracy and speed of the system. In terms of average the system achieved an average of within 1.74mm of the X-axis and within 1.37mm on the Y-axis. The angle of placement of the components achieved a more desirable average of within 0.78°. The system was calibrated to the best possible level, but theoretically if an even more accurate calibration is achieved, the system could certainly improve on these averages. Many factors are taken into account that could affect calibration, such as movement of the system, camera accuracy (focus, lens) etc. with all these factors possibly affecting the calibration of the system. The speed of the system also yielded excellent results, with the system taking an average of 5.67 minutes to learn a design and re-build it into the correct configuration. This is most definitely beneficial in environments where constant unique pick and placing tasks are required, as time is saved reprogramming the KUKA robot system and the system can immediately rebuild the seen design in a short period of time with no changes in software needed. This time included the learning of the new design as well as pick and placing six components from the component conveyor over to the design conveyor based on the required configuration.

Overall the system performed optimally and completed all given tasks without any issues occurring. The systems accuracy and speed was impressive with "everyday" calibration in a changing environment. As stated before if the system could be calibrated to perfection, the results could be even more impressive. The accuracy of the system depends on the accurate calibration and the system could respond even more accurately when picking and placing components if the calibration is done more accurately. Future work may include providing the system with multiple components present and then need to search for correct components to assemble new product design.

References

1. Raj T, Shankar R and Suhaib M, 2007 "*A review of some issues and identification of some barriers in the implementation of FMS*", *Int. J. Flex. Manuf. Syst.*, vol. 19, no. 1, pp. 1–40.
2. Vermaak H and Niemann J, 2017 "*Virtual Commissioning: A Tool to Ensure Cost-Effective System Integration" Int. Works. of Electr., Control, Measurement, Signals and their application to Mechatronics (San Sebastion)(Spain: IEEE), pp. 247-255.*
3. Koren Y, 2006 "*General RMS Characteristics. Comparison with Dedicated and Flexible Systems*," *Reconfigurable Manufacturing Systems and Transformable Factories*, ed Berlin, Heidelberg: Springer, pp. 27-45.
4. Koren Y, 2010, "*The Global Manufacturing Revolution: Product-Process-Business Integration and Reconfigurable Systems* vol. 80: John Wiley & Sons.
5. Gyulai D, Vén Z, Pfeiffer A, Váncza J and L. Monostori, 2012 "*Matching Demand and System Structure in Reconfigurable Assembly Systems*," *Procedia CIRP*, vol. 3, pp. 579-584.
6. Rogers L and Vermaak H, 2017 "*Automated Adapting component transfer system using real-time robot control within a KUKA ROBOTSENSORINTERFACE environment*," *Int. Conf. on Elec. Elect. and IT research (Cape Town)(Cape Town: IEEE)*, pp. 1474-1479.
7. Martin Edeberg PN, 2010 "*Implementing Multi-Touch Screen for Real-time Control of a Robotic-Cell*," Chalmers University of Technology, Sweden.
8. Luwes NJ, *2010 "Artificial Intelligence Machine Vision Grading System, DTech Thesis,* Bloemfontein: Central University of Technology.

A novel geometric error identification methodology for the tilting head of five-axis machine tools based on double ball bar

Jinwei Fan, Changjun Wu, Zhongsheng Li

College of Mechanical Engineering & Applied Electronics Technology, Beijing University of Technology, Beijing100124, China

Abstract. The paper proposes a novel geometric error identification methodology for the tilting head of five-axis machine tools using double ball bar(DBB).Firstly, based on the motion condition of the tilting head of five-axis machine tools, three measurement patterns in Y direction, X direction and Z direction are proposed respectively. Then, the relative displacement equations of two balls of DBB in three measurement patterns are established respectively on the basis of homogeneous transform matrix (HTM) and multi-body system (MBS) theory. Finally, the geometric error parameters of the tilting head are identified totally. The presented method is universal and provides a reference for the error identification for the tilting head of five-axis machine tools.

1 Introduction

With a rapid development of precision machining for the complex parts, five-axis machine tools (FAMTs) are preferred with the advantages of higher material removal rate, better the ability of positioning and orienting the tool with respect to the workpiece and lower production cost. However, the existing FAMTs cannot provide the same machining accuracy as their three-axis counterparts due to the additional rotary axes leading more geometric error sources, thus hindering the development and practical implementation of five-axis precision machining.

Over the past few decades, many researchers have investigated some efficient calibration approaches to identify the geometric error components of FAMTs based on double ball bar (DBB), which has been proved to be competent for rotary axes calibration. Zargarbashi et al. [1] presented a method to measure the position-dependent geometric errors of a trunnion-type A-axis using a DBB. Lee et al. [2] measured position-independent geometric errors of a rotary table axis(C-axis) and a trunnion axis (A-axis) of a five-axis machine tool using a DBB. Zhu et al. [3] proposed a new identification approach to recognize geometric errors of an A-tilting table by using a DBB. Zhang et al. [4] developed a novel measuring method to identify five geometric errors caused by the rotary table (C-axis) with a DBB. Lasemi et al. [5] provided a theoretical method to identify geometric errors of a rotary axis and C rotary axis utilizing a DBB. Chen et al. [6] proposed a comprehensive geometric error identification method for a tilt table using a DBB.

As can be observed in the above-mentioned researches, the geometric error identification methods for rotary axes of FAMTs are rarely aimed at the tilting head (B-axis). Therefore, an adequate geometric errors identification methodology for the tilting head (B-axis) of FAMTs should be developed.

2 The measuring principle based on the DBB

As shown in Figure 1, Every DBB measurement system has two sockets. The one clamped by the tool holder on the spindle is denoted as Socket 1, and the other set on the worktable is denoted as Socket 2. Two ball of the DBB, which is denoted as Ball 1 and Ball 2 respectively, are installed on the ball bowls of the sockets. The measurement of the rotary axes is usually realized by detecting position changes of two balls of the DBB.

Fig. 1 The measuring diagram for the tilting head based on DBB

According to Ref. [3], kinematic chain and homogeneous transform matrix (HTM) have been widely used as a general modelling method. In addition, Fan et al. [7] proposed a generalized kinematics modelling method, regarding a machine tool as multi-body system (MBS)

including many rigid bodies. These theories can describe the motion relationship between adjacent bodies simply and conveniently (see Yang et al. [8] and Wu et al. [9]). Meanwhile, the position coordinates of two balls of the DBB can be transformed to the same coordinate system through kinematic chain respectively. Therefore, the mapping relationship between the theoretical model and actual measurement values can be established by calculating the distance between Ball 1 and Ball 2.

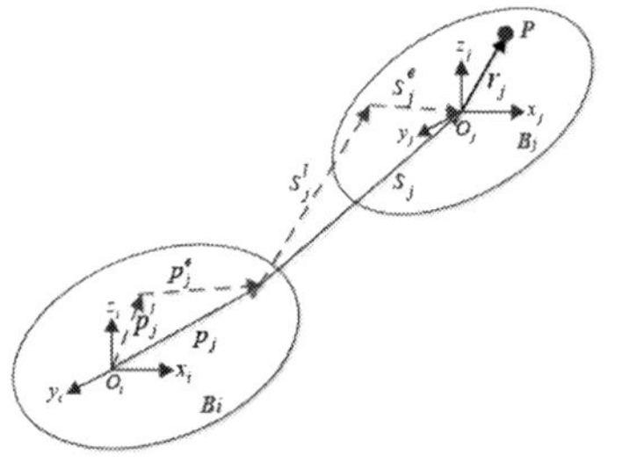

Fig.2 Schematic diagram of motion relation between the adjacent bodies

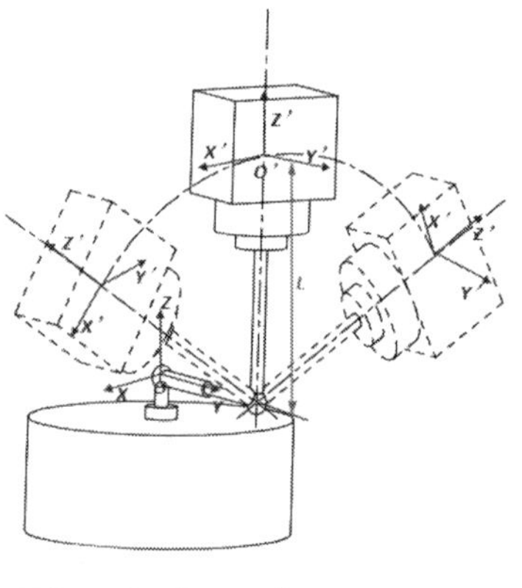

Fig. 3 The motion diagram of measurement pattern of Y direction for B axis

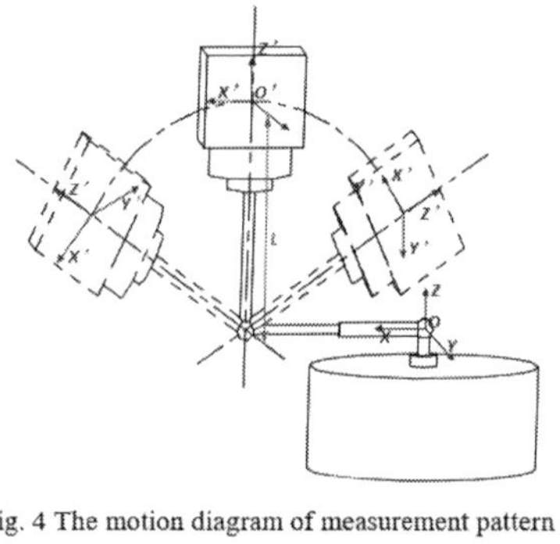

Fig. 4 The motion diagram of measurement pattern of X direction for B axis

As shown in Figure 2, the position array of P in the Bj body coordinate system is represented as $\{r_j\}=\{r_x \; r_y \; r_z \; 1\}^T$,whereas the position array of P in the Bi body coordinate system is represented as $\{r_{ji}\}=\{x_{ji} \; y_{ji} \; z_{ji} \; 1\}^T$. Hence, $\{r_{ji}\}$ is described as,

$$\{r_{ji}\}=\{p_j^l\}\{p_j^e\}\{s_j^l\}\{s_j^e\}\{r_j\} \tag{1}$$

Where $\{p_j^l\},\{p_j^e\},\{s_j^l\},\{s_j^e\}$ describe the relative position transformation matrix, the relative position error transformation matrix, the relative motion transformation matrix, the relative motion error transformation matrix between the Bi and the adjacent lower body Bj respectively.

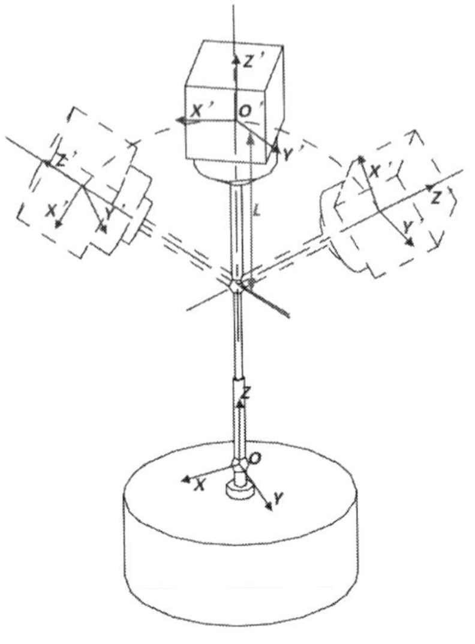

Fig. 5 The motion diagram of measurement pattern of Z direction for B axis

3 Results and Discussion

3.1 The Identification Principle of Geometric Error Parameters

Based on MBS theory, the tilting head (B-axis) has eight geometric errors, composed of six position-dependent geometric errors ($\delta_x(B),\delta_y(B),\delta_z(B)$, $\varepsilon_x(B)$,$\varepsilon_y(B)$, $\varepsilon_z(B)$), and two position-independent geometric errors ($\varepsilon_{xB},\varepsilon_{Bz}$).

In the measuring process of the DBB, B axis is moving by multi-axes synchronous linkage to drive the spindle rotating around the centre point of Ball 1, whose position is required to keep unchanged. Figure 3-5 show three measurement patterns in Y direction, X direction and Z direction respectively. For the convenience of the research, the coordinate system of the centre point of Ball 2 is represented as $O-XYZ$, whereas, the coordinate system of the gyration center of B axis is represented as $O'-X'Y'Z'$. Note that $O-XYZ$ keeps stationary, and $O'-X'Y'Z'$ moves with the B axis. Fig.6 shows the position relation diagram of three measurement patterns. Due to the existence of geometric errors of B axis, the centre point of Ball 1 deviates from the deal position, denoted as A, C, and D respectively, to the actual position, denoted as A', C', and D' respectively. Moreover, in $O'-X'Y'Z'$, the coordinate of A, C, and D is represented as, $\{A\}_{o'}=[0 \;\; 0 \;\; -L \;\; 1]^T$, $\{C\}_{o'}=[0 \;\; 0 \;\; -L \;\; 1]^T$,

$\{D\}_{o'}=[0\ \ 0\ \ -L\ \ 1]^T$,where L indicates the tool length between the gyration center of B axis and the centre point of Ball 1.

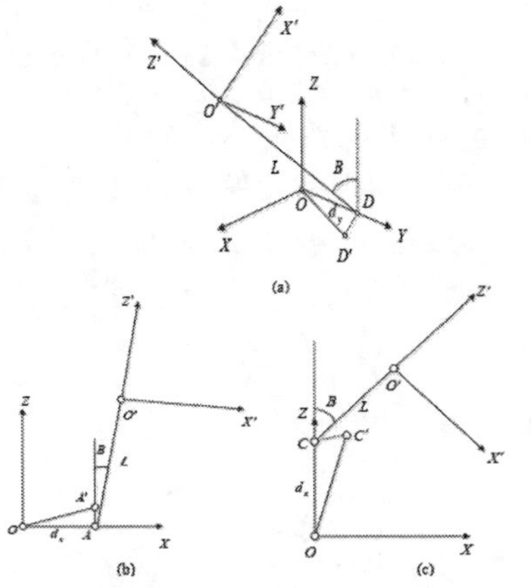

(a) Y-direction (b) X-direction (c) Z-direction
Fig.6 The position relation diagram of three measurement patterns for B-axis.

3.2 Measurement Pattern in Y Direction

As shown in Figure 2 and 6(a), by utilizing (1), D' is described to $O-XYZ$, then equation (2) is obtained.

$$\{D'\}_o=\{p_j^l\}\{p_j^e\}\{s_j^l\}\{s_j^e\}\{D\}_{o'}$$
$$=\begin{bmatrix}1 & -\varepsilon_{xB} & 0 & 0\\ \varepsilon_{xB} & 1 & -\varepsilon_{Bz} & 0\\ 0 & \varepsilon_{Bz} & 1 & 0\\ 0 & 0 & 0 & 1\end{bmatrix}\begin{bmatrix}\cos B_j & 0 & \sin B_j & L\sin B_j\\ 0 & 1 & 0 & d_y\\ -\sin B_j & 0 & \cos B_j & L\cos B_j\\ 0 & 0 & 0 & 1\end{bmatrix}$$
$$\begin{bmatrix}1 & -\varepsilon_z(B_j) & \varepsilon_y(B_j) & \delta_x(B_j)\\ \varepsilon_z(B_j) & 1 & -\varepsilon_x(B_j) & \delta_y(B_j)\\ -\varepsilon_y(B_j) & \varepsilon_x(B_j) & 1 & \delta_z(B_j)\\ 0 & 0 & 0 & 1\end{bmatrix}\begin{bmatrix}0\\0\\-L\\1\end{bmatrix} \tag{2}$$

Where Bj indicates the rotating angle of B axis when the spindle moves from the initial position to the j-th position. d_y indicates the original length of the ball bar in the measurement pattern of Y direction.

By simplifying Eq. (2) and ignoring the higher order term, the distance between D' and O in coordinate system $O-XYZ$ is able to be represented as, $L_{D'O}=\sqrt{-2L\sin B_j\varepsilon_{xB}+2Ld_y\cos B_j\varepsilon_{Bz}+2Ld_y\varepsilon_x(B_j)+2d_y\delta_y(B_j)+d_y^2}$,

Since $L_{D'O}$ is equal to the actual length of the ball bar in the j-th position, the following Equation can be obtained.

$$\sqrt{-2L\sin B_j\varepsilon_{xB}+2Ld_y\cos B_j\varepsilon_{Bz}+2Ld_y\varepsilon_x(B_j)+2d_y\delta_y(B_j)+d_y^2}$$
$$=d_y+\Delta d_y \tag{3}$$

Where Δd_y indicates the length variation of the ball bar from the initial position to the j-th position in the measurement pattern of Y direction.

Then, Equation (4) is obtained by simplifying (3) and ignoring the higher order term.

$$L(\varepsilon_{Bz}\cos B_j-\varepsilon_{xB}\sin B_j+\varepsilon_x(B_j))+\delta_y(B_j)=\Delta d_y \tag{4}$$

Assuming that $Q_j=\varepsilon_{Bz}\cos B_j-\varepsilon_{xB}\sin B_j+\varepsilon_x(B_j)$,then equation (5) is obtained as follows,

$$LQ_j+\delta_y(B_j)=\Delta d_y \tag{5}$$

With two different values of L, the following equation is obtained.

$$\begin{cases}L_1Q_j+\delta_y(B_j)=\Delta d_{y1}\\ L_2Q_j+\delta_y(B_j)=\Delta d_{y2}\end{cases} \tag{6}$$

According to Eq. (6), $\delta_y(B_j)$ and Q_j are represented respectively as follows,

$$\delta_y(B_j)=\frac{L_2\Delta d_{y1}-L_1\Delta d_{y2}}{L_2-L_1},\quad Q_j=\frac{\Delta d_{y2}-\Delta d_{y1}}{L_2-L_1} \tag{7}$$

When $B_j=0°$, i.e. B axis is at the initial position, $\varepsilon_x(B_j)=0$ .Hence, $\varepsilon_{Bz}=Q_0$ .When $B_j=90°$ or $270°$, equation (8) is obtained.

$$\begin{cases}Q_j\big|_{B_j=90°}=-\varepsilon_{xB}+\varepsilon_x(B_j)\big|_{B_j=90°}\\ Q_j\big|_{B_j=270°}=\varepsilon_{xB}+\varepsilon_x(B_j)\big|_{B_j=270°}\end{cases} \tag{8}$$

Due to the periodic distribution of $\varepsilon_x(B_j)$, ε_{xB} and $\varepsilon_x(B_j)$ are represented respectively as,

$$\begin{cases}\varepsilon_{xB}=\left(Q_j\big|_{B_j=270°}-Q_j\big|_{B_j=90°}\right)/2\\ \varepsilon_x(B_j)=Q_j-\varepsilon_{Bz}\cos B_j+\varepsilon_{xB}\sin B_j\end{cases} \tag{9}$$

3.3 Measurement Pattern In X Direction

As shown in Figure 3 and 6(b), by utilizing (1), A' is described to $O-XYZ$ as,

$$\{A'\}_o=\{p_j^l\}\{p_j^e\}\{s_j^l\}\{s_j^e\}\{A\}_{o'}$$
$$=\begin{bmatrix}1 & -\varepsilon_{xB} & 0 & 0\\ \varepsilon_{xB} & 1 & -\varepsilon_{Bz} & 0\\ 0 & \varepsilon_{Bz} & 1 & 0\\ 0 & 0 & 0 & 1\end{bmatrix}\begin{bmatrix}\cos B_j & 0 & \sin B_j & d_x+L\sin B_j\\ 0 & 1 & 0 & 0\\ -\sin B_j & 0 & \cos B_j & L\cos B_j\\ 0 & 0 & 0 & 1\end{bmatrix}$$
$$\begin{bmatrix}1 & -\varepsilon_z(B_j) & \varepsilon_y(B_j) & \delta_x(B_j)\\ \varepsilon_z(B_j) & 1 & -\varepsilon_x(B_j) & \delta_y(B_j)\\ -\varepsilon_y(B_j) & \varepsilon_x(B_j) & 1 & \delta_z(B_j)\\ 0 & 0 & 0 & 1\end{bmatrix}\begin{bmatrix}0\\0\\-L\\1\end{bmatrix} \tag{10}$$

Where d_x indicates the original length of the ball bar in the measurement pattern of X direction.

By simplifying Eq. (10) and ignoring the higher order term, the distance between A' and O in coordinate system $O-XYZ$ is represented as,

$$L_{A'O}=\sqrt{-2L\cos B_jd_x\varepsilon_y(B_j)+2\cos B_jd_x\delta_x(B_j)+d_x^2+2\sin B_jd_x\delta_z(B_j)} \tag{11}$$

Since $L_{A'O}$ is equal to the actual length of the ball bar in the j-th position, Eq. (12) can be obtained.

$$\sqrt{-2L\cos B_jd_x\varepsilon_y(B_j)+2\cos B_jd_x\delta_x(B_j)+d_x^2+2\sin B_jd_x\delta_z(B_j)}$$
$$=d_x+\Delta d_x \tag{12}$$

Where Δd_x indicates the length variation of the ball bar from the initial position to the *j*-th position in the measurement pattern of X direction.

Then, Equation (13) is obtained by simplifying (12) and ignoring the higher order term.

$$-L\cos B_j\varepsilon_y(B_j)+\cos B_j\delta_x(B_j)+\sin B_j\delta_z(B_j)=\Delta d_x \quad (13)$$

Assuming that $P_j=\cos B_j\delta_x(B_j)+\sin B_j\delta_z(B_j)$, then Equation (14) is represented as follows,

$$-L\cos B_j\varepsilon_y(B_j)+P_j=\Delta d_x \quad (14)$$

With two different values of L, the following equation is obtained.

$$\begin{cases}-L_1\cos B_j\varepsilon_y(B_j)+P_j=\Delta d_{x1}\\ -L_2\cos B_j\varepsilon_y(B_j)+P_j=\Delta d_{x2}\end{cases} \quad (15)$$

According to Eq. (15), $\delta_y(B_j)$ and Q_j are represented as respectively,

$$\varepsilon_y(B_j)=\frac{\Delta d_{x2}-\Delta d_{x1}}{(L_2-L_1)\cos B_j},\ P_j=\frac{L_2\Delta d_{x1}-L_1\Delta d_{x2}}{L_2-L_1} \quad (16)$$

3.4 Measurement Pattern in Z Direction

As shown in Figure 4 and 6(c), by utilizing (1), C' is described to $O-XYZ$ as,

$$\{C'\}_o=\{p_j^l\}\{p_j^e\}\{s_j^l\}\{s_j^e\}\{C\}_o$$

$$=\begin{bmatrix}1&-\varepsilon_{xB}&0&0\\ \varepsilon_{xB}&1&-\varepsilon_{Bz}&0\\ 0&\varepsilon_{Bz}&1&0\\ 0&0&0&1\end{bmatrix}\begin{bmatrix}\cos B_j&0&\sin B_j&L\sin B_j\\ 0&1&0&0\\ -\sin B_j&0&\cos B_j&d_z+L\cos B_j\\ 0&0&0&1\end{bmatrix}$$

$$\begin{bmatrix}1&-\varepsilon_z(B_j)&\varepsilon_y(B_j)&\delta_x(B_j)\\ \varepsilon_z(B_j)&1&-\varepsilon_x(B_j)&\delta_y(B_j)\\ -\varepsilon_y(B_j)&\varepsilon_x(B_j)&1&\delta_z(B_j)\\ 0&0&0&1\end{bmatrix}\begin{bmatrix}0\\ 0\\ -L\\ 1\end{bmatrix} \quad (17)$$

Where d_z indicates the original length of the ball bar in the measurement pattern of Z direction. By simplifying Eq. (17) and ignoring the higher order term, the distance between C' and O in $O-XYZ$ is represented as,

$$L_{C'O}=\sqrt{2L\sin B_jd_z\varepsilon_y(B_j)+2\cos B_jd_z\delta_z(B_j)-2\delta_x(B_j)d_z\sin B_j+d_z^2} \quad (18)$$

Since $L_{C'O}$ is equal to the actual length of the ball bar in the j-th position, Eq. (19) can be obtained.

$$\sqrt{2L\sin B_jd_z\varepsilon_y(B_j)+2\cos B_jd_z\delta_z(B_j)-2\delta_x(B_j)d_z\sin B_j+d_z^2} \quad (19)$$

Where Δd_z indicates the length variation of the ball bar from the initial position to the *j*-th position in the measurement pattern of Z direction.

Then, Equation (20) is obtained by simplifying (19) and ignoring the higher order term.

$$L\sin B_j\varepsilon_y(B_j)+\cos B_j\delta_z(B_j)-\sin B_j\delta_x(B_j)=\Delta d_z \quad (20)$$

If assuming, $R_j=\cos B_j\delta_z(B_j)-\sin B_j\delta_x(B_j)$, Then, there is $R_j=\Delta d_z-L\sin B_j\varepsilon_y(B_j)$. Hence, $\delta_z(B_j)$ and $\delta_x(B_j)$ are represented respectively as,

$$\begin{cases}\delta_z(B_j)=\cos B_jR_j+\sin B_jP_j\\ \delta_x(B_j)=\cos B_jP_j-\sin B_jR_j\end{cases} \quad (21)$$

So far, according to the above measured results in three different patterns in the X, Y, and Z direction, the five position-dependent geometric error parameters ($\delta_x(B)$, $\delta_y(B)$, $\delta_z(B)$, $\varepsilon_x(B)$, $\varepsilon_y(B)$), and two position-independent geometric error parameters (ε_{xB}, ε_{Bz}) for the tilting head of five-axis machine tools can be solved totally.

4 Conclusions

Based on the motion condition of the tilting head of five-axis machine tools, three measurement patterns in Y direction, X direction and Z direction are proposed respectively using DBB . Then, the relative displacement equations between two balls of DBB in measurement patterns in Y direction, X direction and Z direction are established respectively on the basis of HTM and MBS theory. Finally, the geometric error parameters of the tilting head (B-axis) are identified totally. The presented method in this paper is universal and provides a reference for the error identification for the tilting head of five-axis machine tools.

Acknowledgments

This work is financially supported by the National Natural Science Foundation of China (No. 51275014), Science and Technology Major Projects of High-end CNC Machine Tools and Basic Manufacturing Equipment of China (No. 2016ZX04003001).

References

1. S H H Zargarbashi, and J.R.R Mayer 2006 Assessment of Machine Tool Trunnion Axis Motion Error using Magnetic Double Ball Bar *Int J Mach Tools Manuf* **46** 1823–1834
2. Kwang-Il Lee, and Seung-Han Yang 2013 Measurement and verification of position-independent geometric errors of a five-axis machine tool using a double ball-bar *Int J Mach Tools Manuf* **70** 45–52
3. Shaowei Zhu, Guofu Ding, Shengfeng Qin, Jiang Lei, Li Zhuang, and Kaiyin Yan 2012 Integrated geometric error modeling, identification and compensation of CNC machine tools *Int J Mach Tools Manuf* **52** 24-29
4. Yi Zhang , Jianguo Yang , and Kun Zhang 2013 Geometric error measurement and compensation for the rotary table of five-axis machine tool with double ball bar *Int J Adv Manuf Technol* **65** 275–281
5. Ali Lasemi, Deyi Xue, and Peihua Gu 2016 Accurate identification and compensation of geometric errors

of 5-axis CNC machine tools using double ball bar. Meas *Sci. Technol.* **27** 055004
6. Jianxiong Chen, Shuwen Lin , Xiaolong Zhou, and Tianqi Gu 2016 A ballbar test for measurement and identification the comprehensive error of tilt table *Int J Mach Tools Manuf* **103** 1–12
7. J W Fan, J L Guan, W C Wang, Q. Luo, X L Zhang, and L Y Wang 2002 A universal modeling method for enhancement the volumetric accuracy of CNC machine tools. *Journal of Materials Processing Technology* **129** 624-628
8. Jixiang Yang, J R R. Mayer, and Yusuf Altintas 2015 A position independent geometric errors identification and correction method for five–axis serial machines based on screw theory *Int J Mach Tools Manuf* **95** 52-66
9. Changjun Wu , Jinwei Fan, Qiaohua Wang and Dongju Chen 2018 Machining accuracy improvement of non-orthogonal five-axis machine tools by a newiterative compensation methodology based on the relative motionconstraint equation *Int J Mach Tools Manuf* **124** 80–98

Motion Error Analysis and Modeling Technology of CNC Camshaft Grinder

Jinwei Fan, Haohao Tao, Changjun Wu, Yuhang Tang

College of Mechanical Engineering & Applied Electronics Technology, Beijing University of Technology, Beijing100124, China

Abstract. Taking certain CNC camshaft grinder as the subject, its motion error analysis and modeling were studied. The movement forms and types of errors between the moving parts were analysed, the movement between coordinate systems of the adjacent bodies was used to express the movement between the two adjacent bodies, and the ideal motion equation and actual motion equation in case of errors between the adjacent bodies were set up. Actual motion equation between adjacent bodies was made and further extended to analyse arbitrary low-order body arrays, which provides a theoretical basis for studying the multi-branch CNC camshaft grinding machine error modeling. Complex multi-branch chain CNC camshaft grinder simplified was made as a simple multibody system, and corresponding body coordinate systems and motion reference coordinate systems for moving parts were established to calculate the corresponding transformation matrix between adjacent bodies. Moving parts of the machine were divided into "workpieces-bed" and "wheel-bed" two kinematic chains. The precision constraint equation of the machine is $P_t = P_w$ in case of using errors influence, and the constraint equation was solved. It provides necessary conditions for study of CNC camshaft grinder error compensation. Results show that the machine precision is significantly improved after error compensation.

1 Introduction

High machining precision is critically important and has increased demand in recent years in aerospace, military, ships, cars, and other industries [1,9]. CNC camshaft grinder as an efficiency, high-precision machining equipment of the camshaft which precision retaining ability is essential. However, Due to the existence of different kinds of error source, all machine tools will be faced with the problem of lower accuracy. It is necessary to analyze the errors of machine tools and take appropriate measures to realize error compensation. Therefore, the method of reducing the effect of error on the machining precision of machine tool can be divided into error prevention and error compensation. By establishing the error model of machine tools, identifying all errors of machine tools and reducing the influence of machine errors on machining accuracy by means of correcting NC instructions, which is called error compensation to reduce the effect of the error of machine tool on the machining precision of the workpiece. The method can overcome the shortcomings of the traditional error prevention method, such as high cost and poor applicability, so it has been developed rapidly.

The kinematic error analysis and modeling of machine tools are the basis of error compensation, the analysis of kinematic error is the key of error modeling. In order to improve the machining accuracy and precision retaining ability of the CNC camshaft grinder, the kinematic error analysis and modeling of the CNC camshaft grinder was proposed. At present, considerable research work for error compensation of multi-axis machine tools is devoted. Su[1] adopted the multi body system (MBS) theory for the precision modeling and error compensation of multi axis CNC machine tools. Fan et al.[2] studied the geometric error compensation technique for improving the accurcy of precision cam grinding. According to ref.[3], the software design of the error analysis and compensation technology for the camshaft NC grinding are carried out. Wang[4]adopted polynomial fitting and linear fitting method to the geometric error and thermal error modeling of CNC machine tools and the on-line error compensation. According to ref.[5], the geometric error compensation and error traceability of CNC machine tools were analyzed. Lechniak et al.[6] proposed the offline software error compensation method. Chnan et al.[7] presented a error compensation method for geometric error and force error of three-axis grinding machine. Zhang et al.[8] used the double ball bar to detect and compensate the error of the rotary table of five axis CNC machine tool. Machine tools are regarded as complex mechanical systems which made up of multiple moving bodies[10]. In recent years, along with the application of follow-up type camshaft grinder is more extensive, in addition, the error analysis and modeling method is too complex and lack of general model. Therefore, the theory of multi-body system can be used to reduce the difficulty of the error modeling of machine tool. However, the multi body system theory is seldom applied to the error modeling of machine tool. So this paper uses the theory of multi-body system to establish model of follow-up type camshaft grinder geometry error.

2. Kinematic error analysis of CNC camshaft grinder

In this study, a CNC camshaft grinder is chosen as the research object, as shown in Fig. 1. The machine tool consists of machine bed (body 1), Z-axis slide carriage (body 2), grinding wheel heads (body 3), spindle (body 4), workpiece (body 5), X-axis slide carriage (body 6), grinding wheel frame (body 7), grinding wheel (body 8) and tailstock (body 9). The grinding wheel moves along the X-axis, worktable moves along the Z-axis, and the spindle rotation around the C-axis. The cam profile is accomplished by simultaneously controlling the X-axis and the C-axis. Geometric errors are usually composed of position-independent geometric errors (PIGEs), and position-dependent geometric errors (PDGEs). PIGEs don't vary with the moving position of the machine tool, but PDGEs are the opposite ones. Fig.2 is the schematic diagram of PIGEs between linear axes. Based on the theory of the rigid body motion, each moving part of a machine tool has six DOFs in the Cartesian coordinate system. Hence, each moving axis of the CNC camshaft grinder has six PDGEs. Fig.3 shows six PDGEs of X-axis, where $\delta_i(x)$ and $\varepsilon_i(x)$ $(i = x, y, z)$ represent the linear errors and the angular errors of X-axis along the i direction respectively. Therefore, all the geometric errors of the CNC camshaft grinder are listed, as shown in the Table 1.

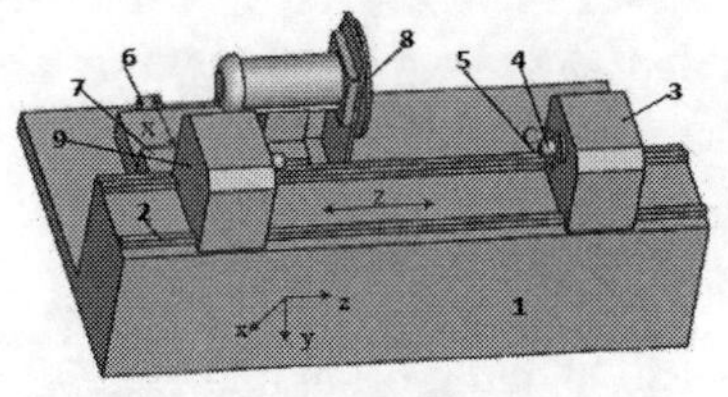

Fig. 1 Structure map of CNC camshaft grinder

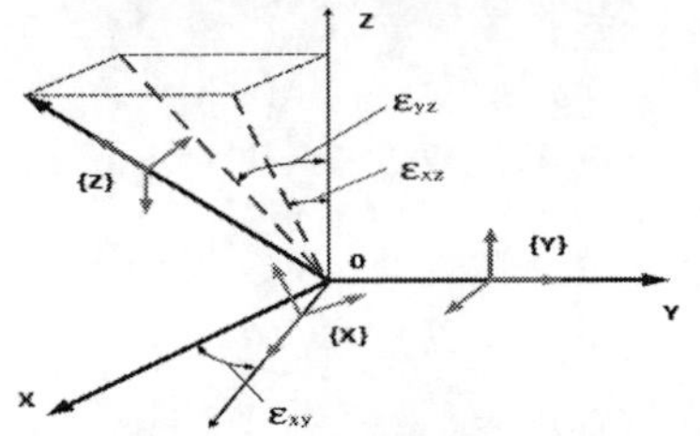

Fig.2 the schematic diagram of PIGEs between linear axes

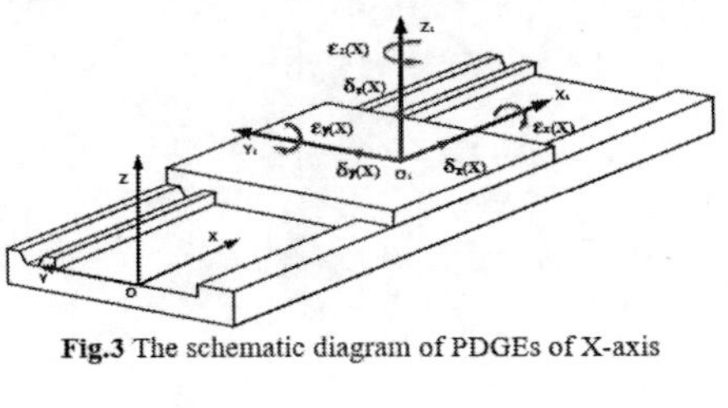

Fig.3 The schematic diagram of PDGEs of X-axis

3. Topological structure and lower body array

Based on MBS theory, a machine tool can be assumed to be composed of various rigid bodies. Fig.4 shows the topology structure map of the CNC camshaft grinder, where "B_i" means the i-th body. As shown in Fig. 4, the kinematic chain of the CNC camshaft grinder can be divided into two branches. One is workpiece branch (B_1-B_2-B_3-B_4-B_5), and the other is wheel branch (B_1-B_6-B_7-B_8).

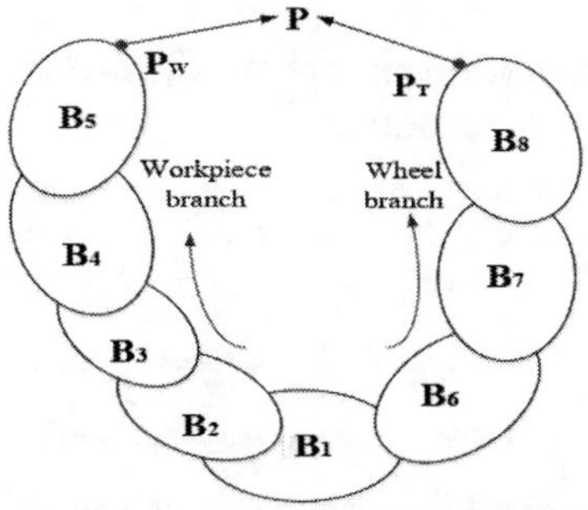

Fig.4 the topology structure of the CNC camshaft grinder

4. Establishment of the relationship between adjacent body motion

Suppose that the workpiece and tool are clamped without errors. Point O is origin of B_1 body coordinate and also base origin of the machine. When $j = 1,2,3,4,5,6$, Q_j^1 and Q_j are ideal and actual motion reference points of B_j, respectively. O_j^1 and O_j are ideal and actual position origin of B_j body coordinate in Q_j coordinate system. Thus, q_j^1 and q_j are ideal and actual position vectors of B_j relative to B_{j-1} and q_j^e is position error vector. S_j^1 and S_j are ideal and actual motion vectors of B_j relative to B_{j-1} and S_j^e is displacement error. q_w is position matrix of workpiece relative to the worktable and q_t is position matrix of tool relative to B-axis. r_w and r_t are position vectors of tool center in the workpiece coordinate system and tool coordinate system.

Table. 1 Geometric errors of the CNC camshaft grinder

	Axis	Linear error x direction	y direction	z direction	Angular error Around x	Around y	Around z
PDGES	X	$\delta_x(x)$	$\delta_y(x)$	$\delta_z(x)$	$\varepsilon_x(x)$	$\varepsilon_y(x)$	$\varepsilon_z(x)$
	Z	$\delta_x(z)$	$\delta_y(z)$	$\delta_z(z)$	$\varepsilon_x(z)$	$\varepsilon_y(z)$	$\varepsilon_z(z)$
	C	$\delta_x(c)$	$\delta_y(c)$	$\delta_z(z)$	$\varepsilon_x(c)$	$\varepsilon_y(c)$	$\varepsilon_z(c)$
PIGES						ε_{xz} , ε_{xc}	ε_{zc}

5. Kinematic error modeling of CNC camshaft grinder

The CNC camshaft grinder is a special multi body system with only two branches, and the components are connected in a certain form. Based on the theory of multibody system, the error modeling of the machine tool is established. the actual and theoretical tool center in the machine coordinate system can be obtained,

$$P_w = P_{wo} = S_{21}S_{32}S_{43}S_{54}S_{5r} = S_{12p}S_{12pe}S_{12s}S_{12se}S_{23p}S_{23pe}S_{23s}S_{23se}S_{34p}S_{34pe}S_{34s}S_{34se}S_{45p}S_{45pe}S_{45s}S_{45se}S_{5r} \quad (3)$$

$$P_t = P_{to} = S_{61}S_{76}S_{87}S_{8r} = S_{16p}S_{16pe}S_{16s}S_{16se}S_{67p}S_{67pe}S_{67s}S_{67se}S_{78p}S_{78pe}S_{78s}S_{78se}S_{8r} \quad (4)$$

where P_w is actual grinding position in workpiece branch and P_t is actual Grinding position in tool branch. S_{ijp} is ideal position matrices, S_{ijs} is ideal motion matrices. S_{ijpe} is actual position matrices, S_{ijse} is actual motion matrices. S_{5r} represent position matrix of actual grinding point in workpiece branch, S_{8r} represent position matrix of actual grinding point in tool branch. An important ideal condition is that the actual grinding position in workpiece branch is equal to the actual grinding position in tool branch, as shown in Eq. 5,

$$P_w = P_t \quad (5)$$

Suppose that the position matrix of the coordinate system of the workpiece relative to the coordinate system of the spindle faces is $\begin{pmatrix} q_{5x} & q_{5y} & q_{5z} & 1 \end{pmatrix}^T$, the position matrix of the coordinate system of the tool relative to the coordinate system of the spindle faces is $\begin{pmatrix} q_{8x} & q_{8y} & q_{8z} & 1 \end{pmatrix}^T$. Let E represents a unit matrix. Then, D-H matrices can be built, which are shown in Table 2.

Table 2. The D-H matrices of the CNC Camshaft Grinder (parts)

Ideal position and motion matrices	Actual position and motion matrices
$S_{12p} = E$	$S_{12pe} = E$
$S_{12s} = \begin{bmatrix} 1 & 0 & 0 & 0 \\ 0 & 1 & 0 & 0 \\ 0 & 0 & 1 & 0 \\ 0 & 0 & 0 & 1 \end{bmatrix}$	$S_{12se} = \begin{bmatrix} 1 & -\varepsilon_z(z) & \varepsilon_y(z) & \delta_x(z) \\ \varepsilon_z(z) & 1 & -\varepsilon_x(z) & \delta_y(z) \\ -\varepsilon_y(z) & \varepsilon_x(z) & 1 & \delta_z(z) \\ 0 & 0 & 0 & 1 \end{bmatrix}$
$S_{23p} = E \quad S_{23s} = E$	$S_{23pe} = E \quad S_{23se} = E$
$S_{16p} = E \quad S_{16s} = E$	$S_{16pe} = E \quad S_{16se} = E$
$S_{67p} = E$	$S_{67pe} = \begin{bmatrix} 1 & 0 & \varepsilon_{xz} & 0 \\ 0 & 1 & 0 & 0 \\ -\varepsilon_{xz} & 0 & 1 & 0 \\ 0 & 0 & 0 & 1 \end{bmatrix}$
$S_{67s} = \begin{bmatrix} 1 & 0 & 0 & x \\ 0 & 1 & 0 & 0 \\ 0 & 0 & 1 & 0 \\ 0 & 0 & 0 & 1 \end{bmatrix}$	$S_{67se} = \begin{bmatrix} 1 & -\varepsilon_z(x) & \varepsilon_y(x) & \delta_x(x) \\ \varepsilon_z(x) & 1 & -\varepsilon_x(x) & \delta_y(x) \\ -\varepsilon_y(x) & \varepsilon_x(x) & 1 & \delta_z(x) \\ 0 & 0 & 0 & 1 \end{bmatrix}$

Transform D-H matrices into the Eqs. ((3)-(5)), then the precision machining constraint equation of CNC camshaft grinder can be obtained,

$$\begin{aligned}&\delta_x(z)+q_{5z}(\varepsilon_{xc}+\varepsilon_y(z)+\varepsilon_x(c)\sin c-\varepsilon_y(c)\cos c)-q_{5y}(\sin c+\varepsilon_z(x)\cos c+\cos c)-q_{5x}(\varepsilon_z(x)\sin c-\cos c-\varepsilon_z(c)\sin c)\\&-x_w(\varepsilon_z(x)\sin c-\cos c+\varepsilon_z(c)\sin c)-y_w(\sin c+\varepsilon_z(x)\cos c+\varepsilon_z(c)\cos c)+z_w(\varepsilon_{xc}+\varepsilon_y(z)-\varepsilon_x(c)\sin c+\varepsilon_y(c)\cos c)\\&+\delta_x(c)\cos c-\delta_y(c)\sin c=x+q_{8x}+x_t+\delta_x(x)+z\varepsilon_{xz}-q_{8y}\delta_z(x)-y_t\varepsilon_z(x)+q_{8z}(\varepsilon_{xz}+\varepsilon_y(x))+z_t(\varepsilon_{xz}+\varepsilon_y(x))\end{aligned} \tag{6}$$

$$\begin{aligned}&\delta_y(z)-q_{5z}(\varepsilon_{yc}+\varepsilon_x(z)-\varepsilon_x(c)\cos c-\varepsilon_y(c)\sin c)-q_{5y}(\varepsilon_z(z)\sin c+\varepsilon_z(c)\sin c-\cos c)-q_{5x}(\varepsilon_z(x)\cos c\\&+\sin c+\varepsilon_z(c)\cos c)+x_w(\varepsilon_z(z)\cos c+\sin c+\varepsilon_z(c)\cos c)-y_w(\varepsilon_z(z)\sin c-\cos c+\varepsilon_z(c)\sin c)-z_w(\varepsilon_{yc}+\\&\varepsilon_x(z)-\varepsilon_x(c)\cos c+\varepsilon_y(c)\sin c)+\delta_x(c)\sin c+\delta_y(c)\cos c=y_t+q_{8y}+\delta_y(x)-(q_{8z}+z_t)\varepsilon_x(x)+(q_{8x}+x_t)\varepsilon_z(x)\end{aligned} \tag{7}$$

$$\begin{aligned}&z+\delta_z(z)+\delta_z(c)-q_{5z}\varepsilon_y(c)\cos c+q_{5y}((\varepsilon_{xc}+\varepsilon_y(z))\sin c+(\varepsilon_{yc}+\varepsilon_x(x))\cos c+\varepsilon_x(c))+\\&q_{5x}((\varepsilon_{yc}+\varepsilon_x(z))\sin c-((\varepsilon_{xc}+\varepsilon_y(z))\cos c-\varepsilon_y(c))+x_w((\varepsilon_{yc}+\varepsilon_x(z))\sin c-\\&(\varepsilon_{xc}+\varepsilon_y(z))\cos c-\varepsilon_y(c))y_w((\varepsilon_{xc}+\varepsilon_y(z))\sin c+(\varepsilon_{yc}+\varepsilon_x(z))\cos c+\varepsilon_x(c))+z_w=\\&z+q_{8z}+z_t+\delta_z(x)-x\varepsilon_{xz}+(q_{8y}+y_t)\varepsilon_x(x)-q_{8x}(\varepsilon_{xz}+\varepsilon_y(x))-x_t(\varepsilon_{xz}+\varepsilon_y(x))\end{aligned} \tag{8}$$

In the model of the linkage between the X axis and the C axis of machine tools, z'、z'' represent the camshaft outline curve, grinding wheel radius is R, the rotation angle of the grinding wheel is θ. According to the ref. [1], the position matrix of actual grinding point in workpiece branch and in tool branch can be obtained, $S_{8r}=\left[-\left|R\cos(\arcsin(\rho\sin(\theta-\varphi)/R))\right|,\rho\sin(\varphi-\theta),z'',1\right]^T$

$S_{5r}=\left[\left|\rho\cos\varphi\right|,-\rho\sin\varphi,z',1\right]^T$,Where φ represent the cam rotation angle, z'、z'' represent the Z axis coordinate values of the grinding point P in the workpiece coordinate system and in the tool coordinate system, respectively.

The 21 error parameters in the precision machining constraint equation can be measured and identified by the dual ball bar instrument and multi-body system theory. By mapping the relationship between tool route and NC instruction, and between NC instruction and practical tool path, the NC instructions under the condition of error can be corrected. The machine tool after error compensation was chosen as the research object to process the automobile engine camshaft, the maximum error has been dropped from 22μm to 9μm, thus, the machining accuracy of the machine tool after error compensation have been significantly improved.

6 Conclusion

Based on the motion condition of the tilting head of five-axis machine tools, three measurement patterns in Y direction, X direction and Z direction are proposed respectively using DBB . Then, the relative displacement equations between two balls of DBB in measurement patterns in Y direction, X direction and Z direction are established respectively on the basis of HTM and MBS theory. Finally, the geometric error parameters of the tilting head (B-axis) are identified totally. The presented method in this paper is universal and provides a reference for the error identification for the tilting head of five-axis machine tools.

Acknowledgments

This work is financially supported by the National Natural Science Foundation of China (No. 51275014), Science and Technology Major Projects of High-end CNC Machine Tools and Basic Manufacturing Equipment of China (No.2014ZX04011031).

References

1. LIU Y J, FAN J W, LI Y, et al. Manufacturing precision reliability maintenance technology of the CNC camshaft grinder[J]. *Journal of Beijing University of Technology*, 2013,**39**(10):1446-1451.
2. HU Y. The research on geometric error software compensation method of the SK-21 CNC grinding tool[D]. *Beijing: Beijing University of Technology*, 2006.
3. SU S P. Study on the methods of precision modeling and error compensation for multi-axis CNC machine tools [D]. *Changsha: National University of Defense Technology*, 2002.
4. FAN J W, GUAN J L, YAN S Z. Technology for enhancing the grinding accuracy of precision cam by geometric error compensation method [J]. *China Mechanical Engineering*,2004, **15**(14) :1223-1226.
5. LI J. Software design and technical study of error analysis and compensation of camshaft NC grinding [D]. *Changsha: Hunan University*, 2009.
6. WANG W, YANG J G, YAO X D, et al. Synthesis modeling and real-time compensation of geometric error and thermal error for CNC machine tools [J]. *Journal of Mechanical Engineering*, 2012, **48**(7):165-170, 179.
7. WANG X F. Geometric error compensation and error influence tracing analysis for complex machine tool [D]. *Beijing: Beijing University of Technology*, 2014.
8. LECHNIAK Z, WERNER A, SKALSKI K, et al. Methodology of off-line software compensation for errors in the machining process on the CNC machine tool [J]. *Journal of Materials Processing Technology*, 1998, **76**(1 /2 /3) : 42-48.

9. Wu C, Fan J, Wang Q, Chen D (2018) Machining accuracy improvement of non-orthogonal five-axis machine tools by a new iterative compensation methodology based on the relative motion constraint equation. *Int J Mach Tools Manuf.* **124**(1):80-98
10. Wu CJ, Fan JW, Wang QH, Pan R, Tang YH, Li ZS (2018) Prediction and compensation of geometric error for translational axes in multi-axis machine tools. *Int J Adv Manuf Technol.* **95**:3413–3435.

Effects of tooth profile and belt speed on cogged V-belt vibration and stress

Zechen Zeng, Xianglong Zhang, Jiangming Kan and Wenbin Li

School of Technology, Beijing Forestry University, Beijing 100083, China

Abstract. With the aim of reducing the vibration and maximum stress of a cogged V-belt and determining the appropriate cogged V-belt for the vertical disc stump grinder (RG25HD), this paper analyses the relationship between, on the one hand, the horizontal displacement of the vibration and node stress and, on the other, the tooth profile and belt speed. In addition, the pattern of the vibration and node stress of the cogged V-belt is investigated using various tooth profiles and belt speeds. Our approach uses a rigid/flexible coupling simulation of belt power transmission based on a finite element flexible body method and employs Recurdyn multi-body dynamics software. After the simulation, we obtain the stress distribution of the belt, the lateral displacement curve and the node stress curve. Our results indicate that cog shape affects belt stress and vibration. When selecting or designing a cogged V-belt, a belt with trapezoidal cogs is preferable. Regarding the effect of belt speed, the trapezoidal cogged V-belt's lateral vibrating displacement and maximal stress increase with an increase in belt speed. Therefore, when the stump grinder is in operation, its revolving speed should be maintained at a low rate to reduce vibration and prolong the belt's life.

1 Introduction

Most Chinese fast-grown, high-yield forests have entered a mature or over-mature period. Forests that have many cutting roots must be maintained after rotation, and tree stumps must be cleared during landscaping. The vertical disc stump grinder, a light-duty machine used to remove stumps and build caves, has been extensively employed. The machine can not only cut stumps to increase soil fertility but also creates a deep hole to facilitate reforestation and subsequent production. The RG25HD-type vertical disc stump grinder (RAYCO; Wooster, OH 44691, USA) transmits power to the cutting element by a cogged V-belt. However, the cogged V-belt vibrates considerably when the machine is at work, which has a substantial impact on the belt fender and leaves a black mark (figure 1). A cogged V-belt has better flexibility and heat dissipation performance [1] than an ordinary V-belt. Tooth form and belt speed affect the vibration and stress of the belt, thus affecting the belt's transmission stability and service life. However, the GB does not specify the tooth form of the cogged V-belt. In the literature, most studies have been performed on synchronous, flat or multi-wedge belts. Several scholars have studied the dynamic performance of synchronous belts with different tooth forms and the transmission optimization of a synchronous belt [2-6] and obtained the vibration and stress distribution of such a belt for various tooth forms. A number of researchers have performed transmission simulations of the ordinary V-belt, flat belt and multi-wedge belt [7-9]. Although the various experts use different research methods [10-12], the most commonly used approach is rigid flexible coupling simulation based on a modal flexible body [13, 14]. In the study of the influence of the tooth form on belt transmission characteristics, the belt undergoes greater displacement and deformation in the transmission process because of the large centre distance, which is more obvious under high-speed operation. Therefore, rigid flexible coupling simulation based on a modal flexible body confronts substantial limitations when describing the nonlinear behaviour of the belt, such as large deformations and contact [15].

With the development of the dynamics of rigid flexible coupled multibody systems and the gradual improvement in computer hardware, rigid flexible coupling simulation based on a finite element flexible body has become a useful tool for analysing belt transmission. This paper focuses on cogged V-belt power transmission and provides a basis for the selection and design optimization of a cogged V-belt by rigid flexible coupling simulation with different tooth forms (i.e., trapezoidal, circular and triangular) at various speeds.

2 Principle

2.1 Principle of finite element flexible body modeling

Different from the method of using a modal flexible body to solve large degree of freedom systems, the finite element flexible body uses node displacement and node slope as nodal coordinates instead of the dynamic substructure modal synthesis method. In the modal synthesis method, the flexible body is typically viewed as a set of finite element model nodes, and the elasticity and deformation of the object are described by the model. The flexible body established by this method is more effective for systems with small elasticity and small deformation. Different from the modal flexible body, the deformation of a flexible body is described by the relative displacement and rotation between flexible body nodes. The absolute nodal coordinate equation is based on the geometric nonlinear theory of continuous structure. Because of the high-order term of the longitudinal strain and the elastic force, the absolute nodal coordinate equation can be used to analyse the large deformation of the structure. Considering the system dynamics, this method can accurately predict contact problems between flexible bodies and rigid bodies. Additionally, effective stress can be directly obtained by this method.

Figure 1. Cogged V-belt and shelter of stump grinder

2.2 Influence of vibration and stress on belt transmission

Practice shows that the failure modes of belt power transmission include skid, fatigue fracture and wear. Among them, skidding and fatigue fracture are the most common failure modes. Therefore, the design principle of belt transmission is to guarantee sufficient fatigue strength and service life with no skidding.

Since the belt is a large flexible component, it causes unexpected lateral vibration when it operates. Lateral vibration not only produces noise and accelerates belt wear but also causes belt removal and affects transmission accuracy. Therefore, it is reasonable to adopt vibration at the start of and during transmission as an evaluation criteria for the tooth form of the cogged V-belt.

According to the design principle of belt transmission, the maximum stress of belt $\sigma_{max} \leq \sigma_{rN}$, σ_{rN} is the finite life fatigue limit, where the subscript mark r represents the stress ratio of the differential force and n represents the corresponding number of stress cycles. It can be experimentally observed that the maximum stress and the number of stress cycles have the following relations:

$$\sigma_{rN}^{m} N = C \qquad (1)$$

In this equation: σ_{rN} is finite life fatigue limit, MPa; m is material constant, whose value is determined experimentally; N is stress cycle times; C is a constant.
By equation (1), it is known that for a belt of definite material and structure, when C is the fixed value, the larger that the fatigue limit of finite life is, the smaller the stress cycle is, that is, the shorter the life of the belt. Therefore, the smaller that the limit of fatigue life of the designed tooth form of a cogged V-belt is (that is, the smaller the maximum stress), the better the tooth form is.

Comprehensive analysis shows smaller lateral vibration and maximum stress are beneficial to improving belt transmission stability and belt service life.

3 Establishment of rigid flexible coupling model of cogged V-belt

3.1 Determination of belt and pulley parameters

The "Mechanical design manual" and "GB/T 6931.3-2008" only describe the cogged V-belt in terms of structural features and applications and do not specify its teeth. Based on the RG25HD type vertical disc stump grinder, this paper designs three tooth forms for a cogged V-belt with reference to the synchronous toothed belt: trapezoidal, circular and triangular. The tooth form parameters of the cogged V-belt refer to the L type synchronous belt with trapezoidal teeth (table 1). The cross-sectional parameters of the cogged V-belt refer to the A type plain V-belt (table 2). The tape line is 1066.8 mm, with a total of 112 teeth. The longitudinal section of each type cogged V-belt is shown in figure 2.

Table 1. The parameters of trapezoid tooth.

parameter	value
Pitch	9.525 mm
Tooth profile angle	40°
Tooth thickness	4.65 mm
Tooth height	1.91 mm
Fillet radius	0.51 mm
Tip radius	0.51 mm

Table 2. The parameters of cogged V-belt's cross section.

parameter	value
Pitch width	11 mm
Tip width	13 mm
height	8 mm
Wedge angle	40°
Applicable groove reference width	11 mm

According to the type of cogged V-belt, the belt pulley also chooses the A-type V-belt pulley. The pulley's groove parameters are shown in table 3. The datum width system is adopted in this paper. The reference diameter of the driving pulley d_d is 112 mm, and the reference diameter of the driven pulley is 150 mm.

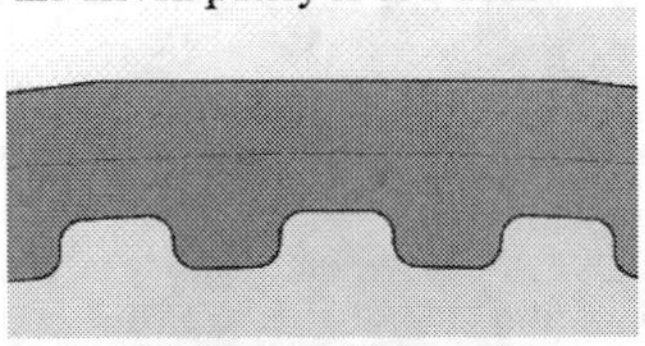

(a) Trapezoidal tooth

(b) Circular tooth

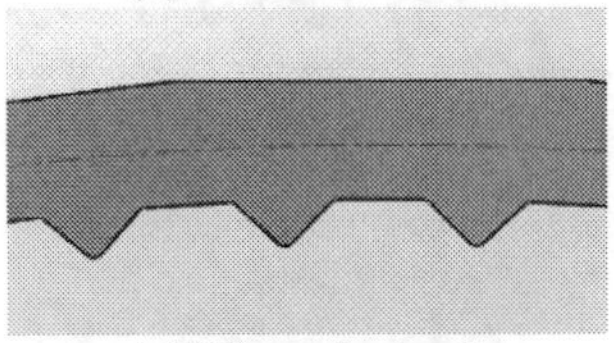

(c) Triangular tooth

Figure 2. Cross sections of three cogged V-belts.

Table 3. Parameters of the belt pulley's groove shape.

parameter	value
b_d	11.0 mm
h_{amin}	2.75 mm
h_{fmin}	8.7 mm
e	15±0.3 mm
f_{min}	9 mm
δ_{min}	6 mm
r_2	0.5~1 mm
Wedge angle	34°

3.2 Model import and mesh generation

First, we import the pulleys and belts with the stp format into the Recurdyn software program. To speed up the process and make the flexible body easy to edit, we use the Mesher function to mesh the belt into a flexible body. The three cogged V-belts are selected with Solid4 units in the meshing process. Each Solid4 unit has 4 nodes, and each node has 3 degrees of freedom, which can be used for 3D solid body model. The modulus of elasticity is 2000 MPa, the density is 1.65×10^3 kg/m^3, and the Poisson's ratio is 0.47. The pulley material is steel. The final meshing units and the number of nodes are shown in table 4.

Table 4. Number of nodes and elements for the three cog files.

Type	trapezoidal	circular	triangular
Nodes	21958	31673	25323
Elements	80794	103304	92082

3.3 Drive and touch settings

An initial tension is applied to the cogged V-belt with the cylinder as the transition part. We add a rotating pair between the cylinder and the driven pulley and a moving pair between the cylinder and the ground. Then, we add a revolute pair between the driving pulley and the reference coordinate system (i.e., the ground). Next, a start-up function for the driving pulley is executed via the step function. Finally, we establish the centre distance between the driving pulley and the driven pulley as 330 mm.

In actual operation, the cogged V-belt transmits power by the friction between the side of the belt and the side of the pulley groove. Therefore, a contact constraint is required. The contact between the belt and the pulley is the contact between the flexible surface and the rigid surface. Therefore, the command Ssuf.ToFsuf is selected in the contact option under Flexible, i.e., the contact between the flexible surface and the rigid surface. First, we select two groove surfaces of the driving pulley as Surface and then select two flexible transmission surfaces as Patchset (i.e., chip set). Thus, the contact between the flexible surface and the rigid surface is generated. Then, we set the coefficient of elasticity between the cogged V-belt and the driving pulley as 3000 N/mm and the friction coefficient as 0.5. In addition, we add a contact between the flexible surface and the rigid surface between the cogged V-belt and the driven pulley. The contact parameters are the same as between the driving pulley and the cogged V-belt. The complete simulation model of the drive and contact is shown in figure 3. The simulation model in figure 3 is a trapezoidal cogged V-belt, and the steps of the simulation model of the triangular and circular cogged V-belts are the same as those of the trapezoidal cogged V-belt (the details are omitted for brevity) [16-18].

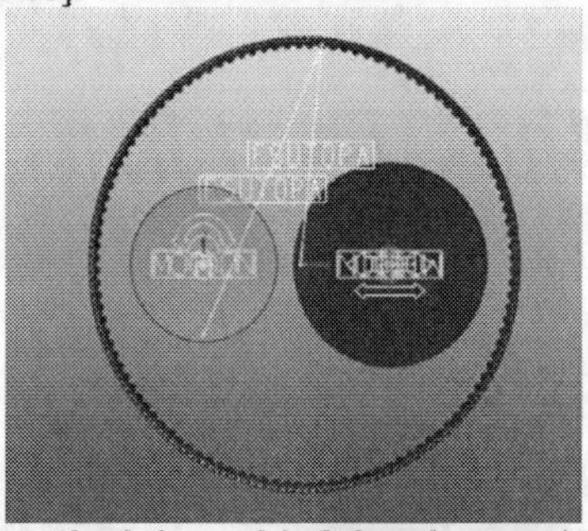

Figure 3. Simulation model of triangular cogged V-belt.

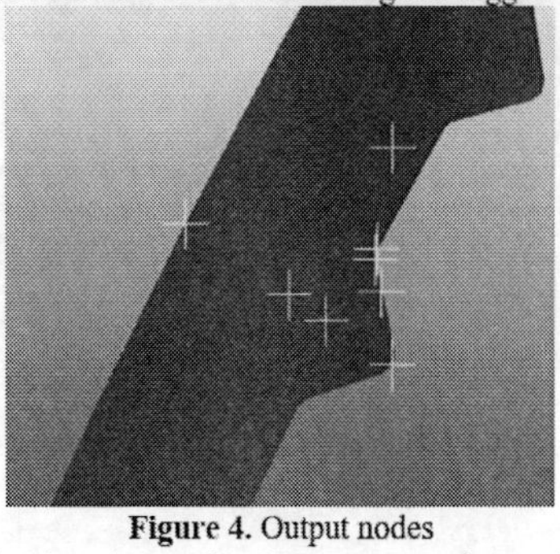

Figure 4. Output nodes

To study the change in the stress on the nodes with different positions, we select the output nodes at the root,

 https://doi.org/10.1051/matecconf/201822104003

back and back of the belt (figure 4). We set the simulation time as 0.6 s and the simulation step as 1000.

4 Results analysis

4.1 Simulation results for rigid flexible coupling of trapezoidal, circular and triangular cogged V-belts

After the simulation, the equivalent stress nephogram of the cogged V-belt can be viewed by the Contour option. Under the condition that the driving pulley speed is 700 r/min, figure 5 shows the stress distribution of the trapezoidal cogged V-belt and the belt teeth. The AB segment is the tight side of the cogged V-belt, and the belt body generally withstands tight side pulling stress. The BC segment is a driven pulley segment whose bent follower is the driven pulley, and the belt body generally withstands bending stress and the frictional stress between the driven pulley and itself. The CD segment is the loose side of the cogged V-belt, and the belt body generally withstands loose side pulling stress. The stress of the DA segment is consistent with that of the BC segment. However, the small pulley is smaller in size, and the bending is more serious. Therefore, the bending stress is greater. As shown in figure 5, the maximum stress occurs at 0.0048 s. At this moment, the driven pulley and the belt begin to come into contact with one another, the stress increases due to the collision, and the maximum stress value is 252 N/mm^2. Regarding the stress distribution of the circular cogged V-belt and the triangular cogged V-belt with the same method, the results indicate that the maximum stress of the circular cogged V-belt also occurs at 0.0048 s. That is, when the driven pulley and the belt are in contact with one another and when the driven wheel starts, the maximum stress value at 0.0048 s is 315 N/mm^2. The maximum stress of the triangular cogged V-belt occurs at 0.169 s. The animation shows that serious belt slipping occurs at 0.169 s, with serious extrusion from the driven pulley part, which causes the bending stress to dramatically increase to a level far higher than that of the trapezoidal and circular cogged V-belts, whereby the maximum stress is 954 N/mm^2.

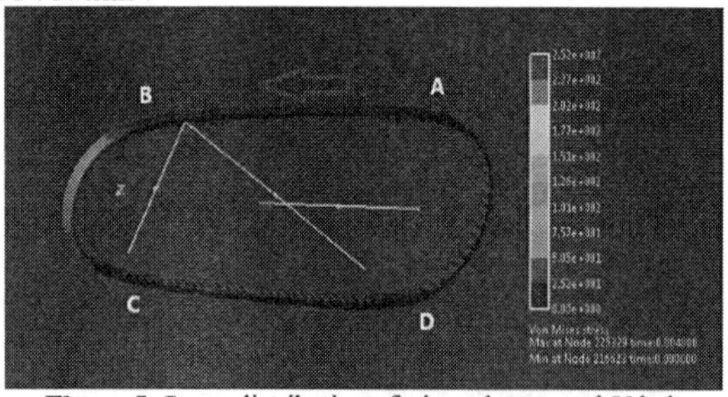

Figure 5. Stress distribution of triangular cogged-V belt

To analyse the influence of tooth form on the transverse vibration of the belt power transmission, the displacement curve and stress curve along the Y direction of the selected output node under different belt speed conditions is plotted. Under the condition that the driving pulley speed is 700 r/min, figure 6 shows the displacement curve along the Y direction of the trapezoidal cogged V-belt back node, and figure 7 shows the stress curve of the trapezoidal cogged V-belt node. In figure 6, the maximum stress occurs at the moment that the node moves to the driving pulley. Because the radius of the driving pulley is smaller, the belt withstands a greater bending stress. When the node moves to the tight side and the loose side, the tensile stress is smaller, and the stress on the tight side is approximately the same as on the loose side. When the driven pulley is large, the stress value is close to the tight side and the loose side, and both are less than the stress at the moment that the node moves to the small pulley. In figure 7, the 235740 value is for the back node of the tooth, 235735 is for the root node, and 229096 is for the back node. Combining the form feature of the trapezoidal cogged V-belt and figure 7, it is found that the stress at the root node is the largest, followed by the stress at the back node of the tooth, and then the stress at the back node. This outcome shows that the stress concentration occurs at the root of the tooth, and the choice of proper fillet radius can reduce the influence of stress concentration.

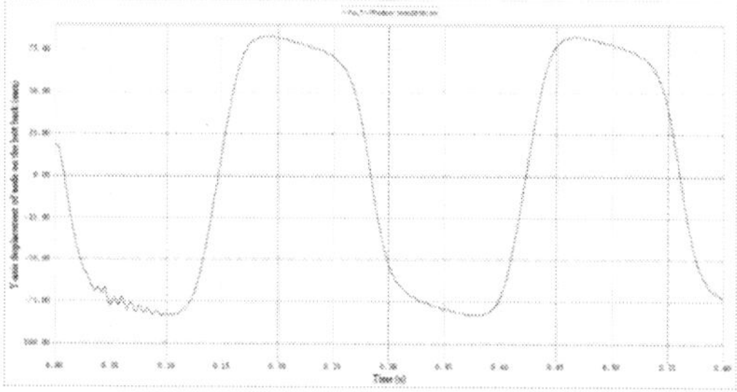
Figure 6. Y-axis displacement curve of node on the belt back at 700 r/min.

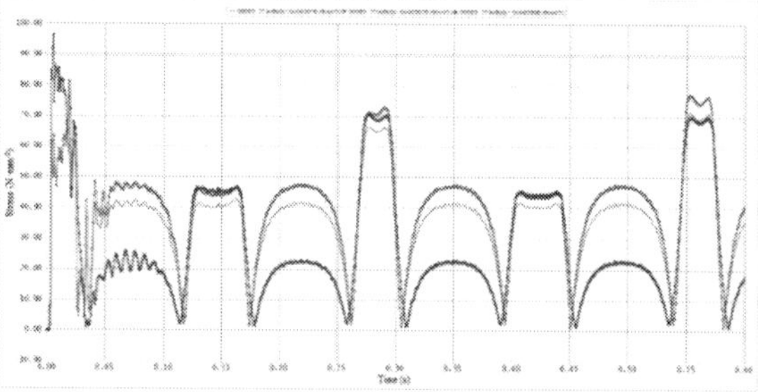
Figure 7. Stress curve of nodes at 700 r/min.

Under other belt speed conditions, the simulation results for the trapezoidal, circular and triangular cogged V-belts are similar to the previously described results. Thus, they are omitted for brevity. The data are shown in table 5, 6 and 7. The transverse vibration displacement of three cogged V-belts varies with the speed (figure 8). The maximum bending stress of the root node varies with the speed (figure 9).

4.2 Tooth form contrast conclusion

When three types of cogged V-belt are compared according to 700 r/min stress distribution nephogram results, the maximum stress of the trapezoidal cogged V-belt is 252 N/mm^2, that of the circular cogged V-belt is

315 N/mm^2, and that of the triangular cogged V-belt is 954 N/mm^2. These results indicate that the maximum stress of the trapezoidal cogged V-belt is the lowest. Compared with the vibration at 700 r/min, the vibration of the trapezoidal cogged V-belt and the circular cogged V-belt are mostly consistent in the transmission process, and the amplitude is smaller. In contrast, the triangular cogged V-belt vibration is violent, and the slip is serious. The maximum bending stress of the trapezoidal cogged V-belt is 78 N/mm^2, that of the circular cogged V-belt is 98 N/mm^2, and that of the triangular cogged V-belt is 425 N/mm^2. These outcomes indicate that the trapezoidal cogged V-belt has the least bending stress. Similarly, the vibration and stress of the different tooth forms at low speed and high speed can be compared. A comprehensive comparison reveals that at various speeds the maximum transverse vibration of the trapezoidal cogged V-belt is smaller, the vibration is relatively stable with a smaller increase, and the maximum bending stress is smaller and increases slightly with increasing speed. The vibration of the circular cogged V-belt is consistent with the trapezoidal cogged V-belt at low speed. However, the maximum transverse vibration first increases and then decreases with increasing speed, the vibration is unstable, the maximum bending stress first increases and then decreases with increasing speed, and the increase and decrease are higher. The maximum transverse vibration and maximum bending stress of the triangular cogged V-belt decrease with the increase in the speed, which shows that the triangular cogged V-belt is more suitable for high-speed belt power transmission. The comprehensive analysis shows that the overall transmission performance of the trapezoidal cogged V-belt is better than that of the circular and triangular cogged V-belts. When choosing or designing a cogged V-belt, the trapezoidal cogged V-belt format is preferable.

Table 5. Simulation results for trapezoidal cogged V-belt at various speeds.

	700 r/min	1225 r/min	1750 r/min	2275 r/min	2800 r/min
Maximum amplitude (mm)	5	28	37	40	50
Maximum bending stress at the back node of a tooth (N/mm^2)	72	66	60	50	74
Maximum bending stress at the root node (N/mm^2)	78	90	89	91	115
Maximum bending stress at the back node (N/mm^2)	72	74	74	80	83
Maximum transverse vibration displacement (mm)	83.68	90.25	96.17	96.2	108.61

Table 6. Simulation results for circular cogged V-belt at various speeds.

	700 r/min	1225 r/min	1750 r/min	2275 r/min	2800 r/min
Maximum amplitude (mm)	5	66.11	49.77	44.33	50
Maximum bending stress at the back node of a tooth (N/mm^2)	33	193.28	476.41	80.38	93.01
Maximum bending stress at the root node (N/mm^2)	98.05	206.92	539.66	121.71	147.21
Maximum bending stress at the back node (N/mm^2)	68	120.17	252.87	94.14	84.86
Maximum transverse vibration displacement (mm)	83.31	124.3	109.4	107.08	109.44

Table 7. Simulation results for triangular cogged V-belt at various speeds.

	700 r/min	1225 r/min	1750 r/min	2275 r/min	2800 r/min
Maximum amplitude (mm)	70	68.84	41.91	63.55	100
Maximum bending stress at the back node of a tooth (N/mm^2)	425	261.43	367.09	404.84	105.41
Maximum bending stress at the root node (N/mm^2)	426.74	283	543.17	566.67	135.57
Maximum bending stress at the back node (N/mm^2)	245	227.78	311.60	302.02	127
Maximum transverse vibration displacement (mm)	129.02	135.2	128	116.18	111.05

4.3 Belt speed contrast conclusion

When figure 8 and 9 are combined, it is found that the belt speed has different influences on the various tooth forms. For the trapezoidal cogged V-belt, the transverse vibration becomes more severe with the increase in the speed. However, the increase is not obvious, and the maximum bending stress of the root node slightly increases with the increase in the speed. For the circular cogged V-belt, the vibration first increases and then decreases with the increasing speed, which shows that this belt is not suitable for operation under the condition of 1225 r/min speed. For this belt, the maximum bending stress of the root node first increases and then decreases at the speed of 1225 r/min-2275 r/min and reaches a maximum value at 1750 r/min. For the triangular cogged V-belt, vibration decreases with the increase in speed, which shows that this belt is not suitable for operation at a low speed. For this belt, the maximum bending stress of the root node is larger at low speed and decreases significantly at a high speed to a value close to the values of the other two cogged V-belts. Therefore, when a trapezoidal cogged V-belt is selected as the transmission mechanism, to reduce vibration and prolong the belt's working life, the driving pulley should be at a low speed during the actual use of a stump grinder. In this way, the transverse vibration can be effectively lowered, and the working life can be prolonged. Simultaneously, the stability of the belt transmission can be improved.

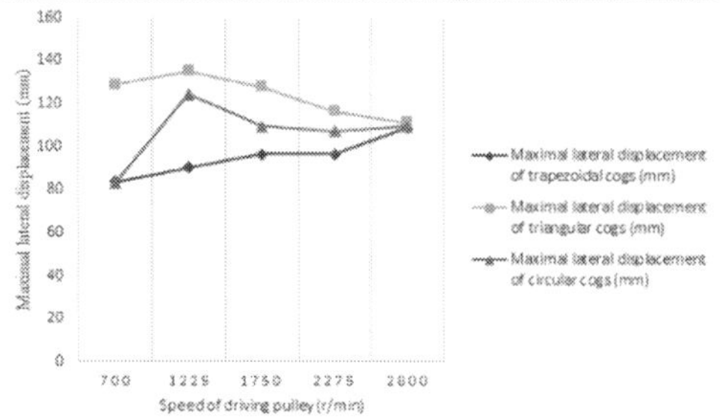

Figure 8. Maximal lateral displacement at various speeds.

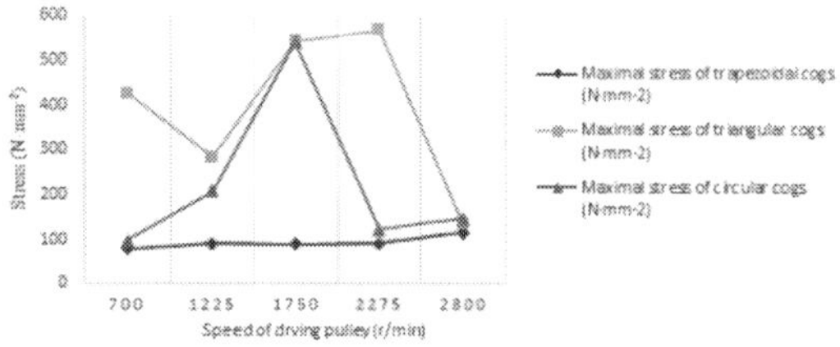

Figure 9. Maximal stress at different speeds.

5 Conclusion

In this paper, the virtual prototype simulation of cogged V-belt power transmission was performed by finite element flexible body and rigid flexible coupling simulation, and the influence of tooth form and belt speed on the vibration and stress distribution of the cogged V-belt was investigated.

First, 3D models of the belts, pulleys and cylinders were constructed using 3D software. Then, the models were imported into Recurdyn multi-body dynamics software to simulate rigid flexible coupling. After comparing and analysing the stress distributions, displacement curves and stress variation curves of trapezoidal, circular and triangular cogged V-belts under various belt speeds, it was found that tooth form affects the vibration and stress of the belt. The stability of the triangular cogged V-belt was the worst at the 700 r/min belt speed, and the maximum stress was 954 N/mm^2, which was much higher than that of the trapezoidal and circular cogged V-belts. This outcome means that the belt is not suitable for low-speed operation. For high-speed operation, the vibration of the three cogged V-belts was more consistent. However, the performance of the triangular cogged V-belt was poor, and the bending stress of the trapezoidal cogged V-belt was the lowest. When the simulation results for the three types of cogged V-belts were compared for different speeds, it was found that the vibration and maximum stress of the trapezoidal cogged V-belt increased with increasing speed; those of the circular cogged V-belt first increased and then decreased; and those of the triangular cogged V-belt decreased with increasing speed. However, all of the data for the circular and triangular cogged V-belts were larger than those of the trapezoidal cogged V-belt. In a comprehensive consideration, the trapezoidal cogged V-belt should be preferred in the design or selection of the tooth form because its comprehensive performance is better than that of the circular and the triangular cogged V-belts. Additionally, the speed of the stump grinder cutting transmission system should be maintained at a low rate during actual operation to reduce lateral vibration, improve transmission stability and prolong the belt's working life.

Acknowledgments

First and foremost, I appreciate my college which gives me a comfortable learning atmosphere. Second, I would like to show my deepest gratitude to my tutor, Mr. Kan, who has walked me through all the stages of the writing of this thesis. Without his illuminating instruction and patience, this thesis could not have reached its present form. I am also greatly indebted to all my teachers who have helped me to develop the fundamental and essential academic competence. My sincere appreciation also goes to all my classmates, who are my proud of my life. Last but not least, I want to thank all my friends for their encouragement and support.

References

1. Jinwen Liu and Haixia Liu 2007 *China Rubber/Plastics Technology and Equipment* **33(10)** 34-8
2. Zhanguo Li, Yaochen Shi, Shu'an Qin and et al 2012 *Journal of Mechanical Transmission* **36(09)** 25-7
3. Yan Li, Zhanguo Li and Yaochen Shi 2014 *Journal of Changchun University* **24(02)** 156-9
4. Zhanguo Li, Hao Li, Yaochen Shi and et al 2015 *Journal of Changchun University* **25(04)** 9-12
5. Jianhua Guo, Qingxin Meng, Xin Yu and et al 2015 *International Conference on Mechatronics, Electronic, Industrial and Control Engineering* (Atlantis Press)
6. Dong Jiao, Zhenghao Ge, Tongyang Zhang and et al 2014 *Journal of Mechanical Transmission* **38(06)** 71-3
7. G L Zhou and S S Hao 2009 *Materials Science Forum* 628-9; 61-6
8. Y Chen and H Xue 2010 *International Conference on Intelligent System Design and Engineering Application* (IEEE Computer Society) pp 949-51
9. Zhanguo Li, Yunhe Zhang and Yaochen Shi 2013 *Journal of Mechanical Transmission* **37(08)** 34-6
10. Kang Yuan, Qunfeng Fu and Guoxiang Huang 2015 *Journal of Mechanical Transmission* **39(07)** 94-7
11. Tao Wang and Xiangyi Guan 2009 *Science & Technology Information* **31** 115-48
12. Dengyun Ma, Biao Xiong, Fantao Fang and et al 2014 *Journal of Mechanical Transmission* **38(09)** 118-22
13. Zhenghao Ge, Kaikai Zhang, Jinsheng Liang and et al 2011 *Journal of Mechanical Transmission* **35(03)** 64-6
14. Yuandong Liu, Wenlin Wang and Mingjun Luo 2013 *Journal of Mechanical Transmission* **37(06)** 28-32

15. Yi Liu 2013 *Application and improvement of multi body dynamics simulation based on Recurdyn* (Beijing: Publishing House of Electronics Industry)
16. Xiaolong Wang 2014 *Optimization Design of Toothed Wrapping Narrow-V Belt and its Relevant Experimental Study* (Harbin: Harbin Institude of Technology)
17. Yongbin Du 2011 *Design and Reaearch of Banded Wedge and Synchronous V Belt for Beam Pumping* (Dongying: China University of Petroleum)
18. Zhanguo Li, Xiuguang Yang, Jinfang Wang and et al 2015 *Journal of Mechanical Transmission* **39(07)** 154-7

An Improvement in Welded Joint Using Vibration Assisted Arc Welding

Mohammad Ahsan Habib[1], Md. Anayet U Patwari[1], Nuruzzaman Rakib[1], Mahmudul H. Pavel[1], Ahmed Y. Sanin[1], Asif Salman[1] and Fahrial Alam[1]

[1]*Department of Mechanical and Chemical Engineering (MCE), Islamic University of Technology (IUT), Board Bazar, Gazipur 1704, Dhaka, Bangladesh*

Abstract. Welding is one of the important processes in mechanical engineering. In engineering sector, welding is a common phenomenon. To improve the quality of joining process, essential changes are necessary in regular welding process. According to the research in welding process, it has been found that vibration can improve mechanical properties of the joining section. However, welding joint quality varies with different ranges of parameters in different conditions. This study has performed to analyze those ranges of parameters in some selected conditions. Here are some input parameters like vibration amplitude, vibration frequency, welding speed and electrode angle are selected to analyze the variation in welded joint. For analyzing the results, output parameters – hardness and deposition rate are calculated. It can be seen that welding process assisted by vibration increases strength and quality of welding joint.

1 Introduction

A weld is made when distinct pieces of material to be fused and form one piece when heated to a temperature high enough to cause softening or melting. In arc welding, vibration can be applied either during or after the process. For sophisticated fabrication, researchers are now driving their effort to develop the process of arc welding during vibration, i.e. vibration assisted welding (VAW), which can cut most of the costs related to post weld vibrations or heat treatments. Production lead time can be considerably reduced due to the parallel processing of vibration and welding. Moreover, VAW leads to improved microstructure [1] and better mechanical properties [2]. There are different modes in which vibration is applied to workpiece such as longitudinal, transverse, vertical and random vibrations. In their oldest available paper, it is reported that the effect of longitudinal vibration on mechanical properties such as yield strength, ultimate tensile strength and hardness [3]. In another investigation, breaking strength, ultimate tensile strength (UTS) and yield strength of specimen made under longitudinal vibratory conditions registered an improvement of [4] and 2% respectively [5]. Electromagnetic stirring of molten pool was found to be favoring the grain refinement in gas tungsten arc (GTA) welds of thin sheet made of aluminum alloys [6]. A two-dimensional mathematical model was developed to study the effect of electromagnetically oscillated arc on the microstructure and temperature distribution of bead on plate welds on thin tantalum sheets made through GTA Welding [7]. The ultrasonically VAW electrode could produce more arc pressure in TIG welding [8].

The main aim of this study is to study the effect of vibration and other welding parameter on weld joint properties, as necessary study on this area has not done yet. Here, some input parameters like vibration amplitude, vibration frequency, welding speed and electrode angle are selected to analyze the variation in welded joint. For analyzing the results, output parameters – hardness and deposition rate are selected. The results suggest the way to select the vibration welding parameters to get better joint.

2 Experimental Setup and Methodology

2.1 Concept and Setup

Figure 1 and 2 show the experimental setup, where the wooden frame includes a large table over which all the accessories like base table, a large wooden plate for vertically up and down movement, nylon rack and pinion, pulley, and motors are attached. On the wooden plate, there is a pulley, which can move upward and downward and it is controlled by a dc 12V motor. On that plate, there is a pinion made of Nylon. A Nylon rack is set on an aluminum frame which is connected with the wooden frame. Rack and pinion are meshed together, that is why this connection can move back and forth. On that nylon rack, welding holder is joined. The vertical wooden plate has space in front of it, where the small wooden base table is kept. This wooden base table has a square hole where metallic rods are placed. At that place, job piece will be kept for the experiment. Sidewall of the small base table are connected with two springs with main table. The small base table has four wheels and those wheels can roll on two rail lines. One vibration device is established in front of that small base plate. This arrangement allows the base table and the job piece to vibrate in back and forth direction. The system is made fully automatic by using an Arduino micro-controller and relay circuits.

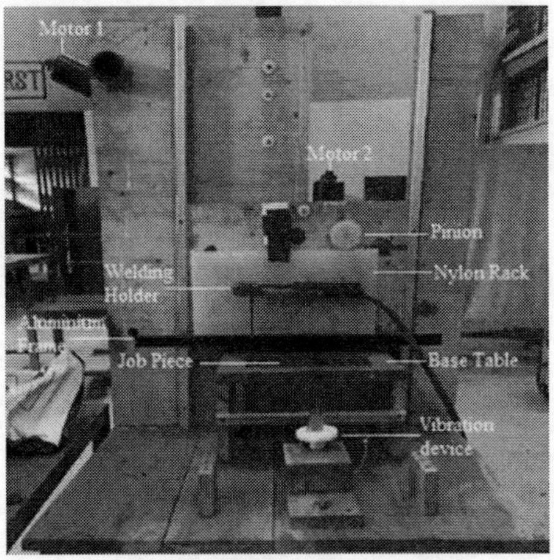

Figure 1. Experimental setup

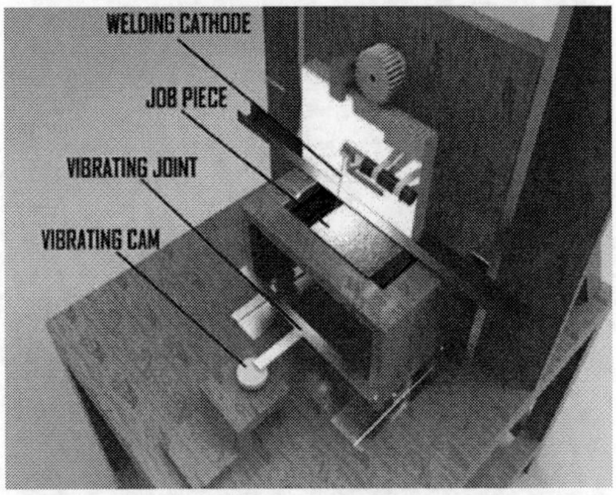

Figure 2. Vibration system designed in CAD Software

Table 1. Process parameters with their values at different levels

Parameters		Levels				
		$-\sqrt{2}$	-1	0	+1	$+\sqrt{2}$
Vibration amplitude (μm)	V_A	12.9	15.0	20.0	25.0	27.1
Vibration Frequency (Hz)	V_F	79.29	100.0	150.0	200.0	220.7
Welding speed (mm/sec)	W_S	1.4	2.0	3.5	5.0	5.6
Electrode Angle (°)	E_A	11.9	15.0	22.5	30.0	33.1

Table 2. Experimental results for without vibration

No	Input parameters				Output parameters (Experimental)	
	V_A	V_F	W_S	E_A	Hard. (HRC)	Depo. rate (gm/sec)
1	0	0	1.4	15.0	34.0	0.50
2	0	0	2	15.0	33.0	0.48
3	0	0	3.5	15.0	30.0	0.45
4	0	0	5	15.0	28.0	0.41
5	0	0	5.6	15.0	27.0	0.38
6	0	0	2	11.9	34.5	0.50
7	0	0	2	15.0	33.0	0.48
8	0	0	2	22.5	31.0	0.45
9	0	0	2	30.0	29.0	0.47
10	0	0	2	33.1	28.0	0.48

Table 3. Experimental results using vibration

No	Input parameters				Output parameters (Experimental)	
	V_A	V_F	W_S	E_A	Hard. (HRC)	Depo. rate (gm/sec)
1	15.0	100.0	2.0	15.0	34.54	0.68
2	25.0	100.0	2.0	15.0	41.07	0.70
3	15.0	200.0	2.0	15.0	41.14	0.72
4	25.0	200.0	2.0	15.0	48.00	0.75
5	15.0	100.0	5.0	15.0	29.19	0.48
6	25.0	100.0	5.0	15.0	34.67	0.50
7	15.0	200.0	5.0	15.0	34.40	0.53
8	25.0	200.0	5.0	15.0	41.20	0.55
9	15.0	100.0	2.0	30.0	27.21	0.59
10	25.0	100.0	2.0	30.0	35.33	0.61
11	15.0	200.0	2.0	30.0	34.27	0.63
12	25.0	200.0	2.0	30.0	41.07	0.66
13	15.0	100.0	5.0	30.0	22.13	0.42
14	25.0	100.0	5.0	30.0	29.19	0.44
15	15.0	200.0	5.0	30.0	28.60	0.47
16	25.0	200.0	5.0	30.0	34.40	0.50
17	12.9	150.0	3.5	22.5	28.40	0.57
18	27.1	150.0	3.5	22.5	34.40	0.60
19	20.0	79.29	3.5	22.5	29.19	0.62
20	20.0	220.7	3.5	22.5	34.67	0.65
21	20.0	150.0	1.4	22.5	34.40	0.33
22	20.0	150.0	5.6	22.5	28.79	0.49
23	20.0	150.0	3.5	11.9	34.27	0.57
24	20.0	150.0	3.5	33.1	29.19	0.58
25	20.0	150.0	3.5	22.5	34.80	0.40
26	20.0	150.0	3.5	22.5	34.90	0.39
27	20.0	150.0	3.5	22.5	34.70	0.41
28	20.0	150.0	3.5	22.5	34.60	0.39
29	20.0	150.0	3.5	22.5	35.00	0.40
30	20.0	150.0	3.5	22.5	34.80	0.41

2.2 Methodology

Experimental design is widely used for controlling the effects of parameters in many processes. For all input parameters vibration amplitude, vibration frequency, welding speed and electrode angle, five levels of data using CCD is used (Table 1). Total thirty experiments are conducted, where, 24 of them are non-center points and 6 of them are center points. Moreover, for comparing the vibration results with non-vibration welding, 10 separate experiments are conducted (Table 2). After completing the welding process, hardness of every specimen is measured with Rockwell Hardness tester. Moreover, the deposition rate is calculated by dividing weight of the deposited metal by the welding time. As a workpiece mild steel bar and 2.5 mm filler rod is used. All the experiments are repeated five times to get an average data. The average hardness and deposition rates for different conditions are shown in Table 2 and 3.

2.3 Mathematical Model

For developing the mathematical model, first the fitness function for the output parameters are calculated. To find out the fitness function the natural log transformation is selected and quadratic type equation is used for this case. By using these conditions, the ANOVA for response surface quadratic model shows significant behavior and for lack of fit is shows non-significant behavior. This proves that, the fitness equation is working with in the reasonable range. The model (Eq. 1 and 2) as suggested for the hardness and deposition rate by the fit and summary tests is shown below:

$$\ln(Hardness) = 3.34656 + 0.018552 \times V_A + 1.71317 \times 10^{-3} \times V_F - 0.055885 \times W_D - 0.011791 \times E_A \quad (1)$$

$$\ln(Depsition\,Rate) = 3.62962 - 0.19605 \times V_A - 0.019767 \times V_F + 0.10054 \times W_D - 0.10213 \times E_A + 9.14818 \times 10^{-6} \times V_A V_F + 3.04939 \times 10^{-4} \times V_A W_D + 6.76318 \times 10^{-5} \times V_A E_A + 1.36819 \times 10^{-4} \times V_F W_D + 1.06895 \times 10^{-5} \times V_F E_A + 3.56317 \times 10^{-4} \times W_D E_A + 4.90434 \times 10^{-3} \times V_A^2 + 6.54561 \times 10^{-5} \times V_F^2 - 0.028737 \times W_D^2 + 2.02903 \times 10^{-3} \times E_A^2 \quad (2)$$

3 Results and Discussion:

3.1 Effect of Vibration Amplitude

In figure 3 (a), it can be seen that with the increase of vibration amplitude, hardness increases. The reason is, amplitude gives the job piece perfect movement range so that the job piece can oscillate smoothly which makes remaining bubble in the welded pole out of the joint, cause the porosity to decreases at the joint. Figure 4 (a) indicates the welded joint is irregular and porous when vibration amplitude has been kept low. Figure 4 (b) shows when the amplitude has been increased then welded joint is in regular shape and less porous. Figure 3 (b) shows that, deposition rate is decreasing with the increment of amplitude but after a certain period, the curve is going up, as the deposition rate starts to grow after a certain period.

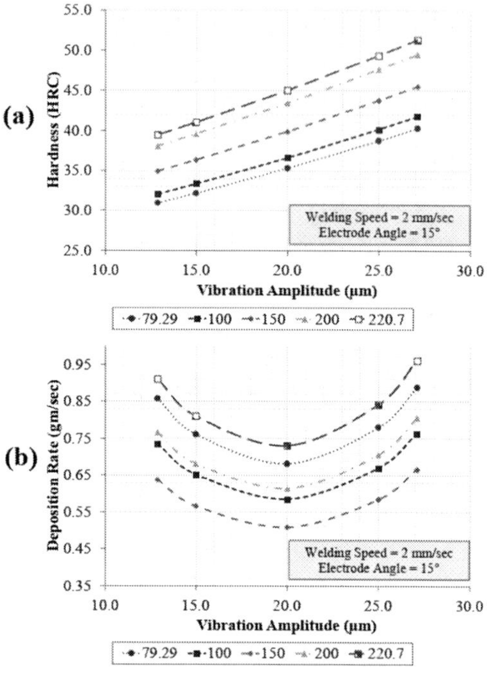

Figure 3. Effect of vibration amplitude for different vibration frequency on (a) Hardness (b) Deposition Rate

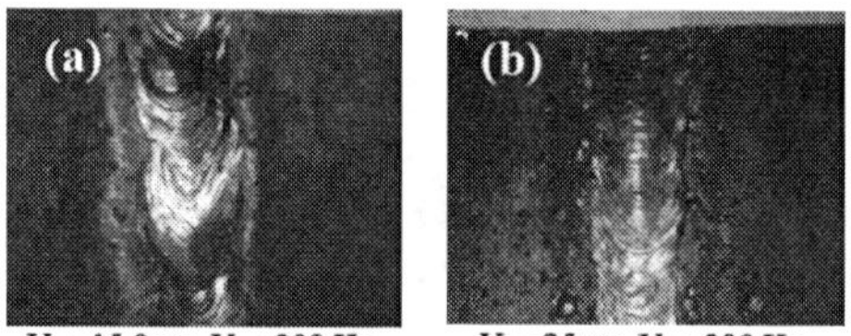

V_A=15.0µm, V_F=200 Hz, W_S = 2.0 mm/sec, E_A=15^0 V_A=25µm, V_F=200 Hz, W_S = 2.0mm/sec, E_A=15^0

Figure 4. Porosity in welded joint with the change of vibration amplitude (a) high porosity (b) less porosity

3.2 Effect of Vibration Frequency

Figure 5 (a) shows hardness of the joining point is improving with the increase of frequency. Frequency helps the joining point to release its bubble from the molten metal. In figure 6 (a), if low frequency is provided with low amplitude then high amount of porous welded joint is seen with irregular shape but in figure 6 (b), high frequency with high amplitude is given then it shows better result because now those bubbles are not there for vibration. Figure 5 (b) represents deposition rate is decreasing first but after a certain point it is increasing with the increase of frequency and amplitude.

 https://doi.org/10.1051/matecconf/201822104004

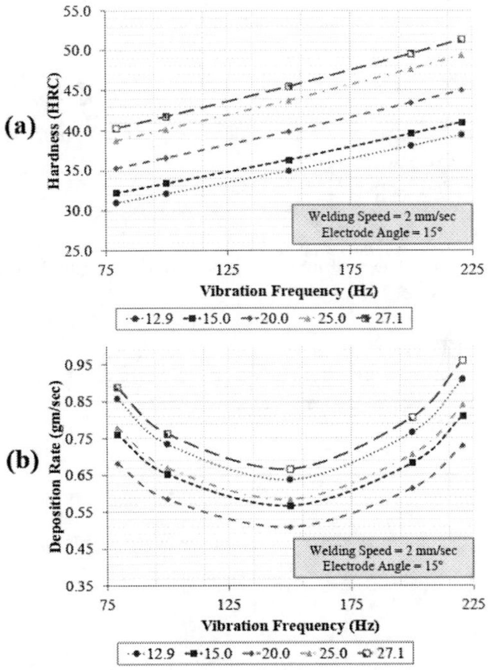

Figure 5. Effect of vibration frequency for different vibration amplitude on (a) Hardness (b) Deposition Rate

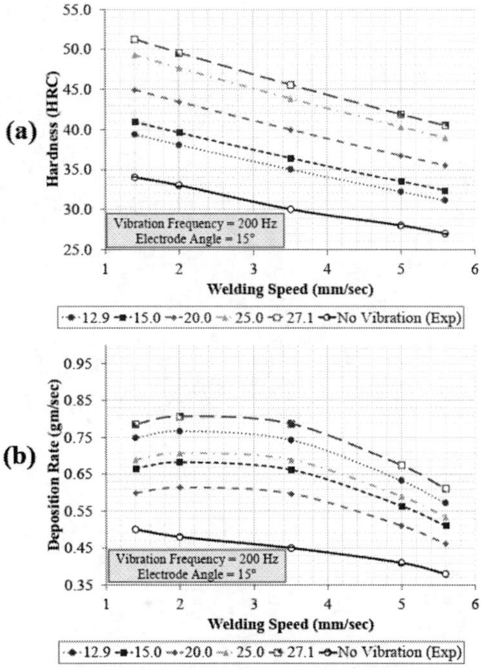

Figure 7. Effect of welding speed for different vibration frequency on (a) Hardness (b) Deposition Rate

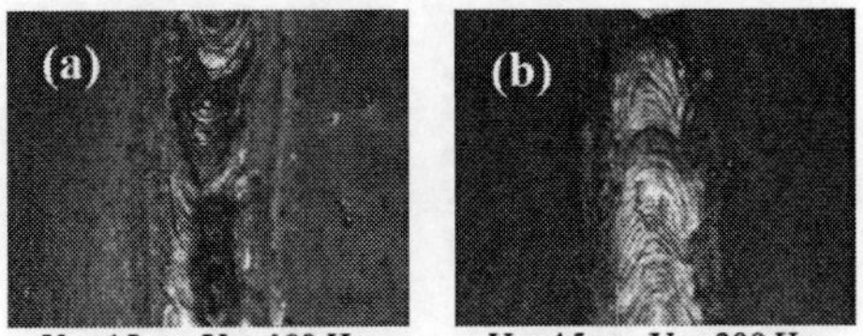

V_A=15μm, V_F=100 Hz, W_S = 2.0mm/sec, E_A=15^0 | V_A=15μm, V_F=200 Hz, W_S = 2.0mm/sec, E_A=15^0

Figure 6. Change of porosity in welding joint with the variation of frequency (a) high porosity (b) less porosity

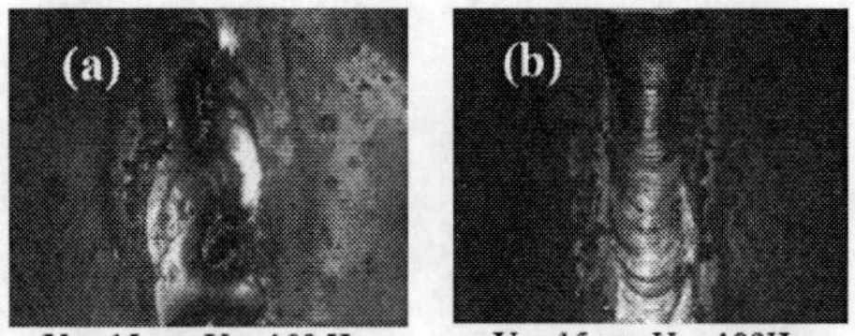

V_A=15μm, V_F=100 Hz, W_S = 5.0mm/sec, E_A=15^0 | V_A=15μm, V_F=100Hz W_S = 2.0mm/sec, E_A=15^0

Figure 8. Change of Porosity in welding joint with the variation welding speed (a) high porosity (b) less porosity

3.3 Effect of Welding Speed

Figure 7 (a) shows less welding speed is beneficial to the joining point because if welding speed is high then less amount of molten metal will enter into the joining point so obviously joining point will be weak. Here different coloured lines of amplitude show how hardness is varying with the decrease of welding speed. In figure 7 (a), there is black line which represents non-vibration line and it shows it is also decreasing with the increase of welding speed. Figure 8 (a) shows high welding speed produces welding joint with high amount of porosity. Moreover, figure 8 (b) less welding speed produces welding joint with less amount of porosity. Figure 7 (b) shows the amplitude is decreasing with the increase of welding speed that means deposition rate is decreasing.

3.4 Effect of Electrode Angle

Figure 9 (a) shows less electrode angle is good for the joining point because if electrode angle is high then molten metal will enter into the joining point in less amount so joining point will be weak. With high frequency, high amplitude, less welding speed and less electrode angle, molten metal will enter more into the joining point and release bubble from the joining point. Figure 10 (a) shows the porosity amount is higher when electrode angle is higher, which is not good for the welding joint. In figure 10 (b), the porosity amount is lower when electrode angle is lower so for better welding joint electrode angle should be maintained lower. Figure 9 (b) shows that with the decrease of electrode angle, having high vibration frequency and amplitude and less welding speed, and deposition rate gets higher. Because high frequency and amplitude provides less porosity and high penetration. As the porosity becomes lower in the molten material and metal is entering more into the joint.

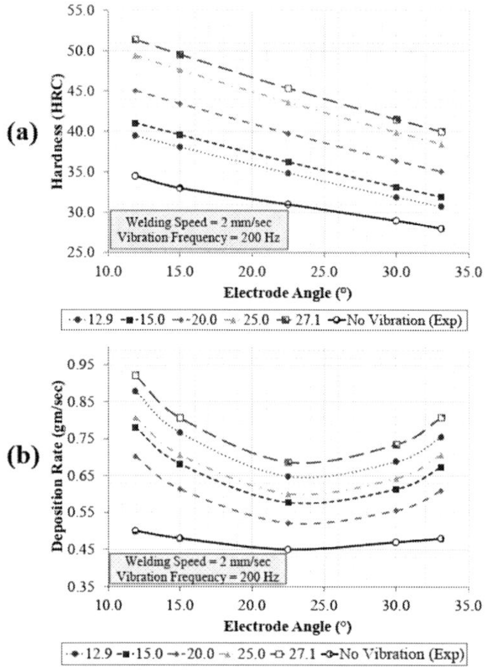

Figure 9. Effect of electrode angle for different vibration frequency on (a) Hardness (b) Deposition Rate

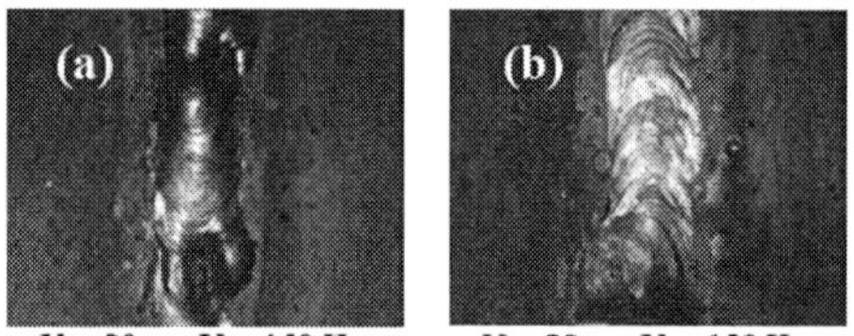

V_A=20μm, V_F=150 Hz, W_S = 3.5mm/sec, E_A=33.1^0 | V_A=20μm, V_F=150 Hz, W_S = 3.5mm/sec, E_A=11.9^0

Figure 10. Change of Porosity amount with the change of electrode angle (a) high porosity (b) less porosity

4 Conclusions

The input parameters- amplitude, frequency, welding speed and electrode angle have a very significant effect on welding process. These model equations have been used to find out the relation between input and output parameters through graphical analysis. Graphical analyze of input and output parameters shows that hardness of a welded joint increases when the amplitude, frequency, welding speed and electrode angle in a selected range. Hardness differs with the amount of porosity in welding section. If porosity amount is higher in welded section, hardness becomes low and if this porosity amount is lower, hardness of the joint gets higher. This study also shows the similar improvement phenomenon with the deposition rate, which gives an increase in production rate as well. The controlled input parameters in some specific ranges helps to get complete penetration in joining section, which provides the better joint than regular cases.

References

1. W. Wu, Micro- and Nano. Tech: Mat., Processes, Packaging, and Systems II **5650**, 438 (2005)
2. S. P. Tewari, ISIJ International **39**, 570 (1999)
3. S. P. Tewari, Shanker, Proc. IMechE B **207**, 173 (1993)
4. A. S. M. Y. Munsi, A. J. Waddell, C. A. Walker, Strain **35**,139 (1999)
5. S. P. Tewari, Thammasat Int. J. Sci. Technology **14**,17 (1999)
6. M. Fukuhisa, N. Hiroji, N. Kazuhiro, A. Ritsu, Transactions of JWRI **7**, 111 (1978)
7. A. Grill, Metall. Trans. B **12B**, 667 (1981)
8. S. Qingjie, L. Sanbao, Y. Chunli, F. Yangyang, Z. Guoqing, China Welding **17**, 52 (2008)

https://doi.org/10.1051/matecconf/201822104005

Influence of the Characteristics of a Workpiece on the Slicing Characteristics Including Tool Wear

Satoshi Sakamoto[1], Masaya Gemma[1], Yasuo Kondo[2], Kenji Yamaguchi[3], Mitsugu Yamaguchi[4], and Takao Yakou[1]

[1]*Yokohama National University, 79-2 Tokiwadai, Hodogaya-ku, Yokohama, Japan*
[2]*Yamagata University, 4-3-16 Jonan, Yonezawa, Japan*
[3]*Yonago National College of Technology, 4448 Hikona-cho, Yonago, Japan*
[4]*Salesian Polytechnic, 4-6-8 Oyamagaoka, Machida, Japan*

Abstract. Multi-wire saws with a diamond electrodeposited wire tool are widely used to slice hard and brittle materials. The properties of the materials significantly affect the saw's performance in terms of slicing quality, efficiency, and accuracy. In this study, the effects of the material properties of workpieces on wire tool performance and tool wear are described. The main conclusions are as follows: the higher the hardness and toughness of the workpiece material, the longer the slicing time and greater the damage to the wire tool. The brittleness of the workpiece adversely affects the thickness variation more than its hardness. In addition, the frequency of chipping mainly depends on the hardness of the workpiece, whereas the chipping size mainly depends on the toughness of the workpiece.

1. Introduction

Multi-wire sawing is an excellent method for slicing hard and brittle materials such as semiconductors and optical materials [1]. Recently, multi-wire sawing using a diamond electrodeposited wire tool has been adopted as the preferred slicing method for high-hardness materials [1]. In addition, new types of wire tools and slicing devices are being developed [2–4]. However, the occurrence of microcracks on the workpiece due to slicing adversely affects slicing efficiency and accuracy. Therefore, it is important to clarify the relationship between the workpiece and slicing characteristics.

We have previously clarified the influence of the mechanical properties of workpieces on the polishing and grooving characteristics [5, 6]. In this study, we describe the relationship between the mechanical properties of workpieces and slicing characteristics, including tool wear.

2. Experimental procedure

Figure 1 shows a schematic illustration of the experimental apparatus. The multi-wire saw used in this study is a wire reciprocating type. Therefore, one drum simultaneously controls the feeding and rewinding of a wire tool.

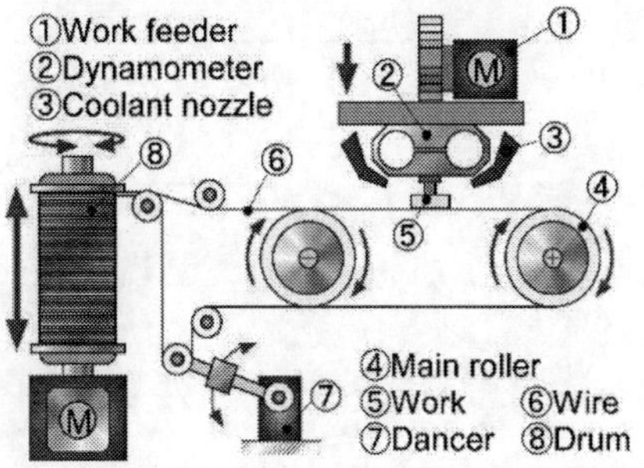

Figure 1. Schematic of the multi-wire saw device used in this study.

Table 1 shows the materials used and main experimental conditions. The multi-wire saw included a diamond electrodeposited wire tool with a core diameter of 100 μm, an outer diameter of 143 μm, and tension set at 14.6 N. Borosilicate glass (TEMPAX), silicon (single crystal, polycrystalline), and sapphire were used as workpieces with dimensions of 20 mm × 20 mm × 5 mm. A glycol-based water-soluble coolant supplied at 50 mL/min was used as the working fluid. The running speed of the wire tool was 100 m/min, and the feed speed of the workpiece was 0.8 mm/min. Table 2 shows the hardness and toughness value of the TEMPAX, silicon, and sapphire workpieces [7, 8]. Silicon is a numerical value of a single crystal.

Table 1. Experimental conditions and material used.

Wire tool	
Core diameter	100 [μm]
Abrasives	Diamond
Grain size	20–30 [μm]
Outer diameter	143 [μm]
Running speed	100 [m/min]
Tension	14.6 [N]
Usage length	250 [m]
Workpiece	
Material	Borosilicate glass (TEMPAX) Silicon (Mono-Si, Poly-Si) Sapphire
Dimension	20×20×5 [mm^3]
Feed speed	0.8 [mm/min]
Coolant	
Glycol-based water-soluble working fluid	
Supply amount 50 [mL/min]	

Table 2. Vickers hardness and fracture toughness of workpieces.

Workpiece	TEMPAX	Silicon (Mono-Si)	Sapphire
Vickers hardness [GPa]	6.2	10.0	23.0
Fracture toughness [MPa]	0.7	1.14	2.1

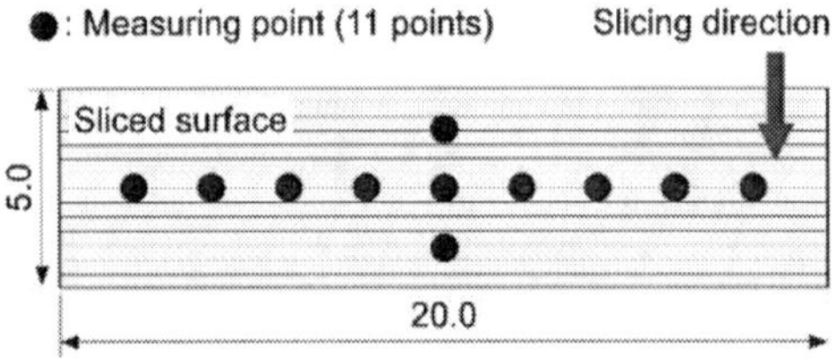

Figure 2. Thickness measuring points.

Figure 2 shows the measurement position of the wafer thickness. The thickness of the wafer was measured with a micrometer at 11 points, as shown in the figure.

3. Experimental results and discussion

3.1. Slicing time of workpieces

Figure 3 shows a comparison of the slicing time required until slicing is completed. Slicing time from the shortest to longest are ranked as follows: TEMPAX, silicon, and sapphire. This result is affected by the hardness and toughness of the materials, as shown in Table 2. Since sapphire had the highest hardness and toughness values among the three workpieces, slicing was more difficult and took much longer, whereas TEMPAX, which had the lowest hardness and toughness values, was easier to slice and took a much shorter time. No significant difference in slicing time was found between the single crystal silicon and polycrystalline silicon workpieces.

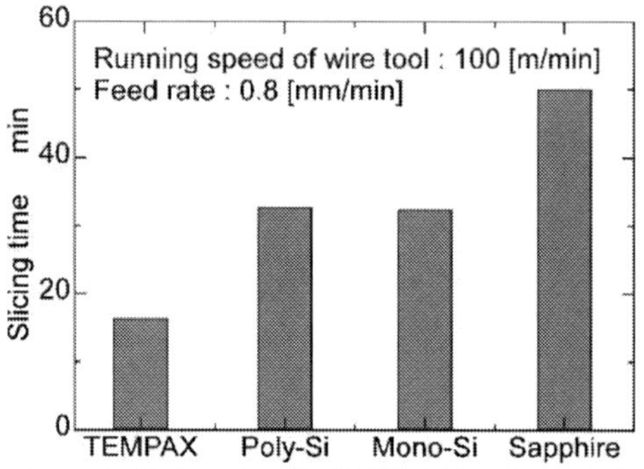

Figure 3. Slicing time for TEMPAX, silicon, and sapphire workpieces.

3.2. Thickness variation of TEMPAX, silicon, and sapphire wafers

Figure 4 shows a comparison of thickness variation of the TEMPAX, silicon, and sapphire workpiece. Sapphire has smaller thickness variation than the other workpieces. In general, the slicing of materials possessing a high value of hardness tends to cause the wire tool to stray. However, sapphire has both a high value of hardness and the highest value of toughness in the workpieces used in this study. Consequently, excessive microfracture of the workpiece and the wandering of the wire tool were suppressed. Thus, the slicing of sapphire requires higher slicing time, but the slicing accuracy is good. On comparing TEMPAX and sapphire, it is evident that the thickness variation of sapphire is smaller. In other words, the brittleness of the workpiece adversely affects the thickness variation more than its hardness.

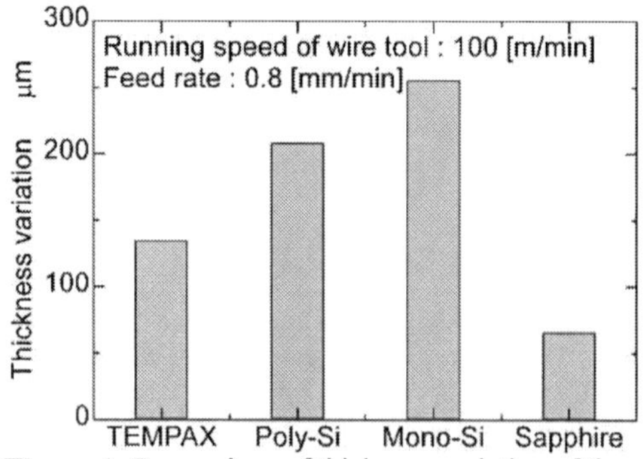

Figure 4. Comparison of thickness variation of the wafer.

3.3. Sliced surface of TEMPAX, silicon and sapphire wafers

Figure 5 contains images of the sliced surfaces for TEMPAX, silicon, and sapphire. Saw marks from the wire tool are apparent on the sliced surface of all workpieces. However, the sliced surface of sapphire shows slightly better properties. The toughness of the workpiece affects the properties of the sliced surface, i.e., a workpiece with higher toughness suffers less damage to its sliced surface than a workpiece with low toughness.

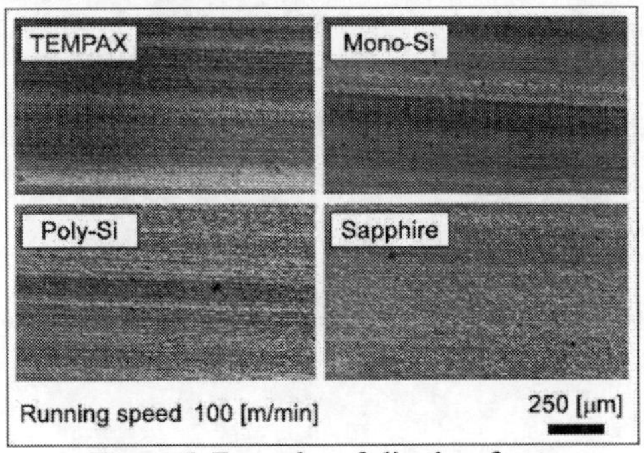

Figure 5. Examples of sliced surfaces.

Figure 6 shows the scanning electron microscopy (SEM) images of the TEMPAX, silicon, and sapphire workpieces along the edge of the wafer. The edge of the sapphire workpiece is relatively better than the edge of the other workpieces.

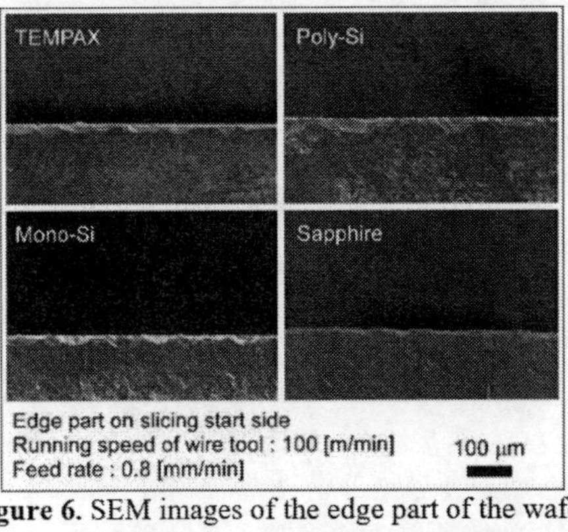

Figure 6. SEM images of the edge part of the wafer.

Figure 7 shows the chipping size and number of chipping that occurred at the edge of the wafer. The measurement distance of chipping is 600 μm. The indents that exceed the threshold are analyzed using the vector image analysis software and are measured as the number of chipping in this study. The threshold was set at 1.50 μm. The number of chipping on the edge of the sapphire and polycrystalline silicon workpieces is slightly larger. However, the size of chipping on the sapphire is very small. Conversely, the edge of TEMPAX has a smaller number of chipping, but its chipping size is large. The frequency of chipping mainly depends on the hardness of the workpiece, whereas the chipping size mainly depends on the toughness of the workpiece. In other words, the hard material tends to slightly increase the number of chips and the high toughness material tends to slightly reduce the chip size.

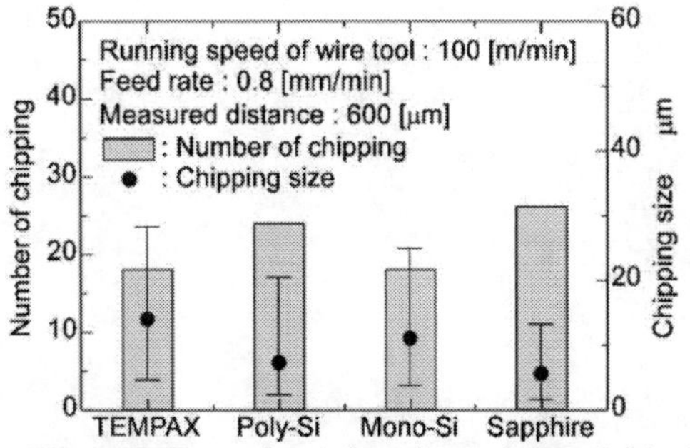

Figure 7. Comparison of number and size of chipping.

3.4. Influence on the damage of the wire tool

Figure 8 shows the wire tools after slicing the workpieces. Damage such as falling-off of abrasives occurred in all the wire tools for all workpieces. However, the wire tool used for slicing the TEMPAX appeared to be less damaged because the slicing time was shorter. On the contrary, the wire tool used to slice sapphire showed significant damage on both the surface and the core wire.

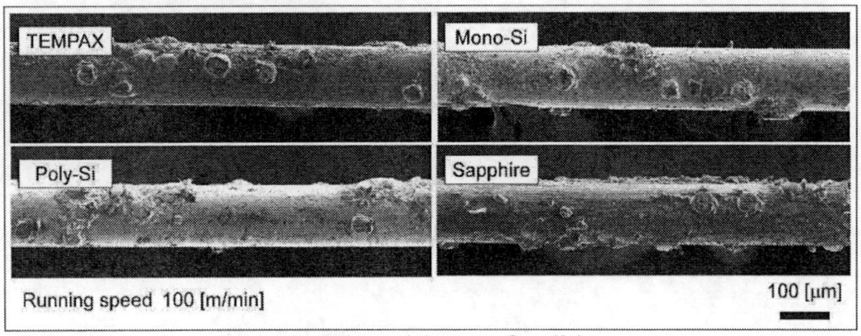

Figure 8. Wire tool damage after slicing.

4. Conclusions

We experimentally investigated the influence of the properties of workpieces on the slicing characteristics, including tool wear, of workpieces. The main conclusions obtained from this study are as follows. The hardness and brittle behavior of the work materials affect the slicing characteristics. Higher the hardness and toughness of the workpiece, longer is the slicing time required and greater is the damage to the tool. Especially, the wire tool used to slice sapphire with high hardness and toughness suffers significant damage. The brittleness of the workpiece adversely affects the wafer thickness variation more than its hardness. Therefore, in sapphire having high toughness, the wafer thickness variation is small. The workpiece with higher toughness suffers less damage to its sliced surface than the workpiece with low toughness. In addition, the hard material tends to slightly increase the number of chips and the high toughness material tends to slightly reduce the chip size.

Acknowledgments

We thank Instrumental Analysis Center of Yokohama National University for their support and cooperation. This work was supported by JSPS KAKENHI Grant Number 16K06005.

References

[1] H Wu 2016 *Precis. Eng.* **43** 1
[2] Kamiya O et al. 2010 *Int. J. Soc. Mater. Eng. Resour.* **17** 182
[3] Saito A et al. 2014 *Int. J. Soc. Mater. Eng. Resour.* **20** 181
[4] Suzuki T et al. 2017 *Precis. Eng.* **50** 32
[5] Sakamoto S et al. 2016 *Key Eng. Mater.* **703** 17
[6] Sakamoto S et al. 2017 *Mater. Sci. Eng.* **229** 012031
[7] Hayashi K et al. 1991 *J. of the Soc. of Mater. Sci.* **40** 405
[8] Ueda K et al. 1982 *Bull. of the Jpn. Inst. of Met.* **21** 225.

Extremely Thin Metal Foil Blades as Cutting Tools for Hard and Brittle Materials

Satoshi Sakamoto[1], Sanshiro Akaoka[1], Masaya Gemma[1], Yasuo Kondo[2], Kenji Yamaguchi[3], Mitsugu Yamaguchi[4] and Takao Yakou[1]

[1]*Yokohama National University, 79-2 Tokiwadai, Hodogaya-ku, Yokohama, Japan*
[2]*Yamagata University, 4-3-16 Jonan, Yonezawa, Japan*
[3]*Yonago National College of Technology, 4448 Hikona-cho, Yonago, Japan*
[4]*Salesian Polytechnic, 4-6-8 Oyamagaoka, Machida, Japan*

Abstract. The manufacturing costs of semiconductor products such as silicon wafers can be reduced by decreasing the kerf loss. In addition, a decrease in the kerf loss leads to an effective utilization of rare materials, which is environmentally beneficial from the viewpoint of saving resources. This study aims to reduce the kerf loss during slicing hard and brittle materials. Therefore, the possibility of using an extremely thin metal foil blade instead of a wire tool in slicing was examined. Initially, grooving characteristics using a metal foil blade (thickness: 50 μm or less) was investigated. The main conclusions are that grooving with a metal foil blade is possible and kerf loss can be reduced. The groove depth tends to increase as the machining time and particle size of abrasives increase. The groove width is smaller when a thin metal foil blade is used and vice versa. However, if the abrasive particle size is too large, grooving becomes impossible. Since the wear of metal foil blade increases with an increase in the particle size of the abrasive, it is necessary to use an abrasive with a particle size that is suitable for the thickness of the metal foil blade.

1. Introduction

Multi-wire sawing is an excellent slicing method for hard and brittle materials, such as silicon ingots, magnetic materials, ceramics, and sapphire [1]. In recent years, multi-wire sawing has emerged as the mainstream slicing method for ingots with large diameter. In addition, new types of wire tools and slicing devices are being developed [2–4]. Currently, when several thin layers are sliced from an ingot using a multi-wire saw, the volume of kerf loss is same or more than the volume of the wafers [5]. Large kerf loss during slicing not only increases the production cost of semiconductor products and optical parts but also is an important problem from the viewpoint of saving resource [6]. The diameters of the wire tool and particle size of the abrasives that are used for slicing greatly affect the volume of kerf loss. In other words, when the wire tool is thin and particle size is small, the kerf loss can be drastically reduced. However, the wire tool with a small diameter easily breaks during slicing. Therefore, it is very difficult to further reduce the diameter of the wire tool.

This study aims to drastically reduce the kerf loss by examining the possibility of a machining method that utilizes extremely thin metal foil blades instead of a wire tool. In this study, as a first step to realize slicing with reduced kerf loss, we discuss the fundamental grooving characteristics of using an extremely thin metal foil blade and fine abrasives.

2. Experimental procedure

Figure 1 shows a schematic illustration of the grooving experimental apparatus, and Figure 2 shows details of the machining section. The materials used and main experimental conditions are summarized in Table 1. The metal foil blade was fixed to the tool holder and pressed against the workpiece under a constant load (machining load). A reciprocating motion was applied to the metal foil blade and grooving experiments were performed. In the grooving experiments, 5 mm thick sheets of borosilicate glass (TEMPAX) were arranged side by side (Work [A] to [D] in Figure 2). Various measurements were made where work [B] and [C] were machined at a steady state. A 30 mm long metal foil blade made from alloy tool steel (SKS 81) with a thickness between 10 and 50 μm was used with a diamond paste of low viscosity, and the particle size of abrasives was between 1 and 6 μm. The Vickers hardness of the metal foil blade was adjusted to 780 HV by heat treatment. The protrusion of the metal foil blade from the holder was 2.5 mm for blades with a thickness between 10 and 30 μm, and 5 mm for blades with a thickness of 50 μm. The machining load was 0.1, 0.3, and 0.5 N for metal foil blades of thickness 10, 30, and 50 μm, respectively. The relative speed between the metal foil blade and the workpiece was 1212 mm/min, and the moving distance of the metal foil blade was 100 mm.

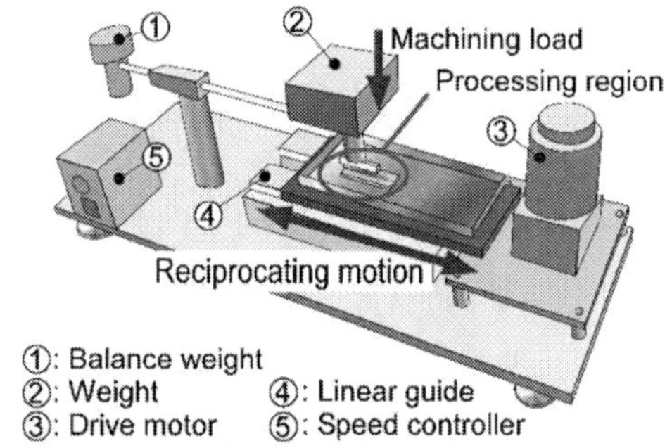

Figure 1. Schematic illustration of grooving experimental apparatus.

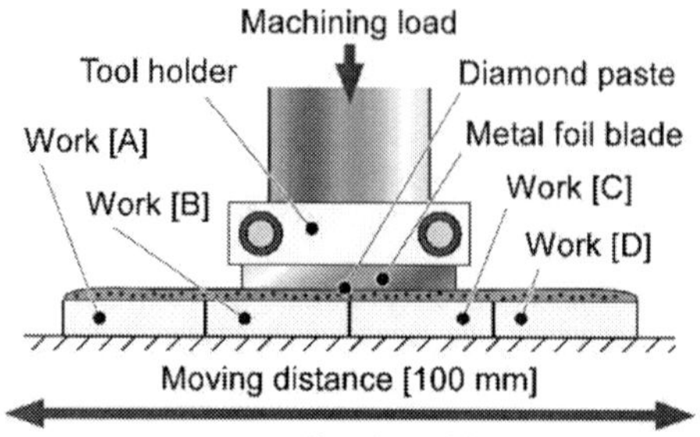

Figure 2. Details of machining section.

Table 1. Materials and experimental conditions of grooving experiments.

Blade material	Alloy tool steel (SKS81)
Dimensions	Length: 30 [mm], Thickness: 10, 30, and 50 [μm]
Protrusion	2.5, 5.0 [mm]
Abrasive	Diamond
Particle size	1, 3, and 6 [μm]
Work material	Borosilicate glass (TEMPAX)
Cross section	Rectangle (5 × 20 [mm])
Table speed	1212 [mm/min]
Moving distance	100 [mm]
Machining load	0.1, 0.3, and 0.5 [N]
Machining time	30, 60, 90, and 120 [min]

3. Experimental results and discussion

3.1. Example of a groove machined using a metal foil blade

Figure 3 shows an example of a groove machined using a metal foil blade. The thickness of the metal foil blade was 50 μm and a machining load of 0.5 N was applied for 120 minutes using an abrasive with a particle size of 1 μm. One side of the groove was relatively vertical while the other surface was slightly oblique. In this study, the metal foil blade was not tensioned and slight deflection of the metal foil blade due to the machining load was expected. Therefore, one side of the groove is inclined, but it is clear that a hard and brittle material like borosilicate glass can be machined using an extremely thin metal foil blade.

Figure 3. Example of groove machined using a metal foil (50 μm).

3.2. Influence on groove depth

Figure 4 shows the relation between machining time and groove depth. When the thickness of the metal foil was 10, 30, or 50 μm, the applied machining load was 0.1, 0.3, or 0.5 N, respectively. In all of these grooving experiments, the particle size of the abrasive was 1 μm. The groove depth tends to increase with an increase in the machining time. The machining load is several times different depending on the thickness of the metal foil blade. However, there is a slight difference in the depth of the groove, but there is no remarkable difference. This is simply explained because although the applied machining load was different for each blade, the actual machining pressure remained the same in all cases.

We also investigated the influence of the particle size of the abrasive on the groove depth using the 50 μm thick metal foil blade. Figure 5 shows the relation between the particle size and groove depth after a machining time of 120 minutes at an applied load of 0.5 N. The groove depth tends to increase linearly as the particle size increases.

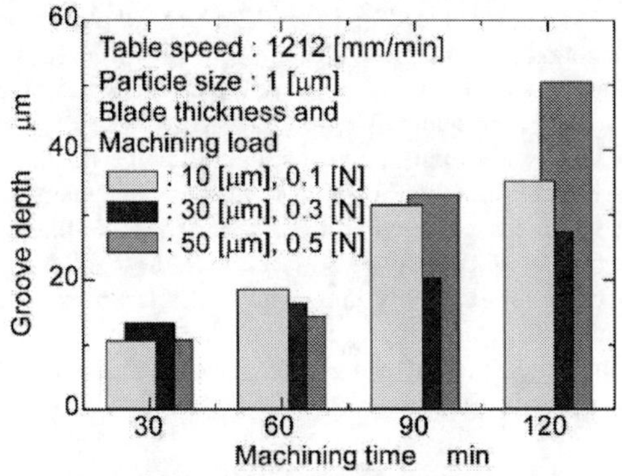

Figure 4. Relationship between machining time and groove depth as observed for three different blade thicknesses.

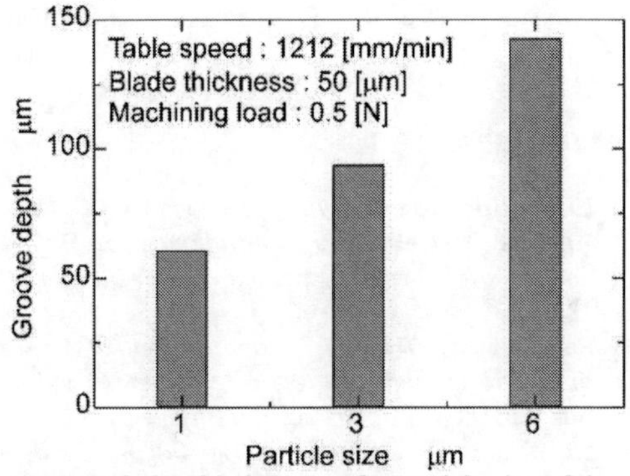

Figure 5. Relationship between the particle size of the abrasive and the groove depth manufactured from a blade with a thickness of 50 μm.

3.3. Influence on groove width

Figure 6 shows the influence of the thickness of the metal foil on the groove width (kerf width). Naturally, the groove width is directly related to the volume of kerf loss. The empirical formula for the width of a groove cut using a multi-wire saw is known [7]. Here, the groove width was calculated using the thickness of the metal foil blade instead of the diameter of the wire tool. Therefore, the calculated groove width is given by equation (1).

$$W = T + 3G, \tag{1}$$

where W is the calculated groove width, T is thickness of the metal foil, and G is particle size of the abrasive. W is called a calculated value in this study and is indicated by the dashed lines, as shown in Figure 6. It is clear from this figure that the groove width depends on the thickness of metal foil blade. In addition, the groove width is wider than the calculated value. Generally, the groove width in these experiments tends to be larger than the calculated value because it is influenced by the accuracy of the experimental apparatus.

A thinner blade is clearly more advantageous for narrowing the groove width. Therefore, the influence of the particle size on the groove width was also investigated using a thin metal foil blade having a thickness of 10 μm. Figure 7 shows the influence of the particle size on the groove width. The larger the diameter of the abrasive used, the wider the resulting groove width. However, grooving becomes impossible if the particle size is too large such as 6 μm. It appears that if the particle size of the abrasive is too large relative to the thickness of the metal foil blade, it becomes difficult to efficiently convey the abrasive to the machining section. Therefore, if an abrasive with large particle size is used, grooving becomes impossible. For efficient grooving, it is necessary to use the abrasive with the most suitable particle size in relation to the thickness of the metal foil blade that is being employed.

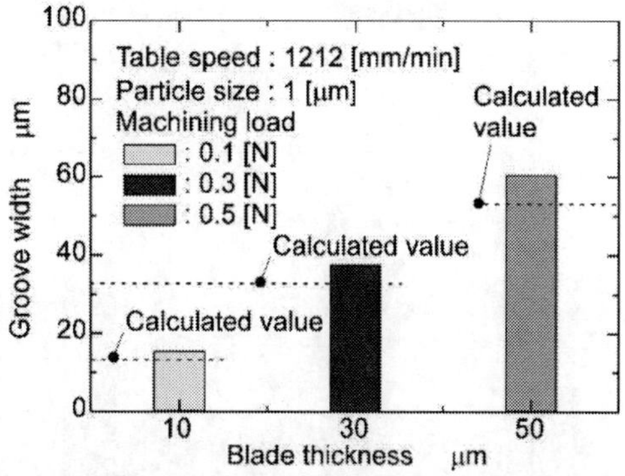

Figure 6. Influence of the thickness of the metal foil on the groove width using a 1μm particle abrasive.

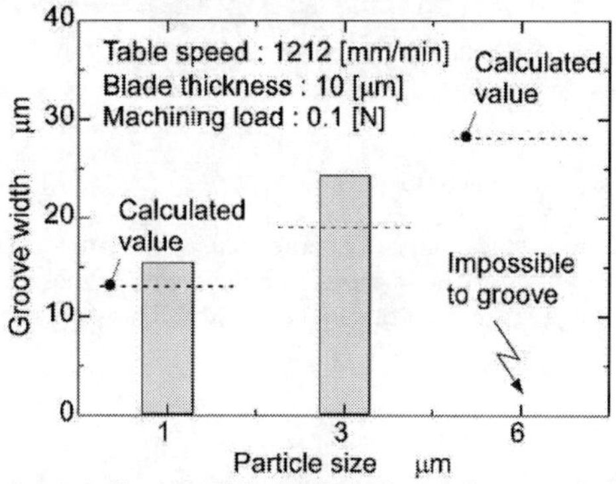

Figure 7. Influence of the particle size on the groove width cut using a 10 μm thick metal foil blade.

3.4. Influence on tool wear

SEM images and thickness analyses of the 50 μm metal foil blade after 60 minutes of machining are shown in Figure 8 and Figure 9, respectively. In Figure 8, the relation between the particle size of the abrasive and effect on the wear of the metal foil blade can be clearly seen. The thickness of the metal foil blade used was 50 μm. The wear width is the distance from the edge of the blade. The average value was adopted in this study, and it is clear that the larger the particle size, larger is the increase in wear width. Furthermore, the wear width of the metal foil blade is about 1.5 to 2 times the particle size of the abrasive used.

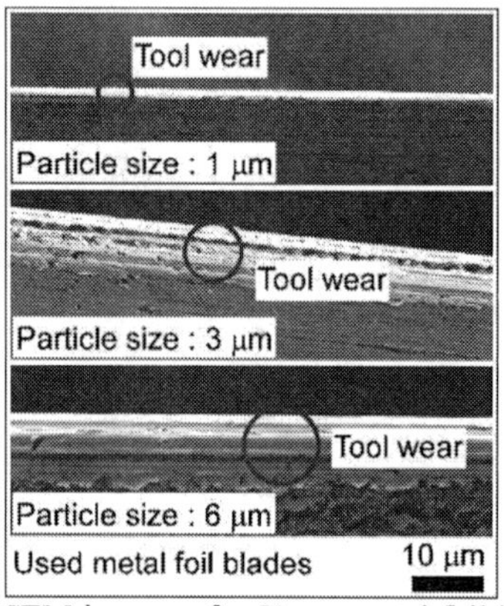

Figure 8. SEM images of a 50 μm metal foil blade after machining a groove using 1, 3, or 6 μm particle sized abrasives.

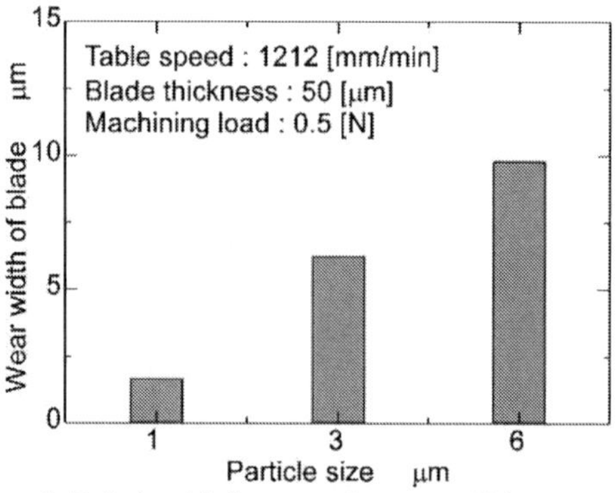

Figure 9. Relationship between the wear width on a 50 μm metal foil blade and the particle size of the abrasive used.

4. Conclusions

In this study, we experimentally investigated the possibility of using a metal foil blade with a thickness of 50 μm or less as a cutting tool for hard and brittle materials. We demonstrated that the precise machining of borosilicate glass using an extremely thin metal foil blade and fine abrasives is possible. The groove depth tends to increase with both the machining time and particle size of the abrasive. The width of the groove depends on the thickness of the metal foil blade. In other words, the groove width is less when a thinner metal foil blade is used, resulting in a reduced kerf loss. The larger the diameter of the abrasive used, the wider the groove width. However, using an abrasive with too large a particle size will make grooving impossible. Therefore, it is necessary to use an abrasive with a particle size suitable for the thickness of the metal foil blade. In addition, the wear of the metal foil blade increases as the particle size increases. In this study, we could reduce groove width (kerf width) to about 15 μm by using an extremely thin metal foil and fine abrasives. Therefore, metal foil processing can be expected to greatly reduce the manufacturing cost of semiconductor wafers and optical parts.

Acknowledgments

We thank the Instrumental Analysis Center of Yokohama National University for their support and cooperation. This work was supported by JSPS KAKENHI Grant Number 16K06005.

References

[1] H Wu 2016 *Precision Engineering* **43** 1–9
[2] Kamiya O et al. 2010 *International Journal of the Society of Materials Engineering for Resources* **17** 182–5
[3] Saito A et al. 2014 *International Journal of the Society of Materials Engineering for Resources* **20** 181–5.
[4] Suzuki T et al. 2017 *Precision Engineering* **50** 32–43
[5] Hoshiyama T 2009 *J. of the Japan Society for Abrasive Technology* **53** 655–8
[6] NEDO 2009 *Roadmap (PV2030+)* (Tokyo: New Energy and Industrial Technology Development Organization)
[7] Kinutani K and Kanamichi 1996 *Electronic Materials and Parts* **35** (Tokyo: Kogyo Chosakai Publishing Co., Ltd.) 29–33.

https://doi.org/10.1051/matecconf/201822104007

Non-linear Analysis of Vibrating Flip-flow Screens

Sanpeng Gong[1,2], Xinwen Wang[1] and Sebastian Oberst[2]

[1] *School of Chemical and Environmental Engineering, China University of Mining and Technology, Beijing, China*
[2] *Centre for Audio, Acoustics and Vibration, School of Software, University of Technology Sydney, Australia*

Abstract. Vibrating flip-flow screens provide an effective solution for the screening of highly viscous or fine materials. Apart from other factors, the vibration characteristics of the main and floating screen frames are largely responsible for the flip-flow screen's sifting performance and its processing capacity. In this paper, the vibration characteristics of a vibrating flip-flow screen with linear and nonlinear springs are compared. Analytical results highlight that increasing the relative amplitude and avoiding undesirable resonances of the main and the floating screen frames can be realised to improve the screen's performance. The materials on the screen panel have less an effect on the vibration characteristics of the vibrating flip-flow screen with nonlinear springs than using linear springs. Other design parameters which influence the performance of vibrating flip-flow screens are discussed.

1. Introduction

Dry screening of wet and fine materials is an on-going issue in processing minerals and ore such as gold, iron or coal. During traditional screening equipment frequency experience the problem of plugging, which is a blockage of the screen's grid with debris [1-2]. Vibrating flip-flow screens (VFFS) are used as machinery to tackle the problem of sifting the materials. A VFFS has elastic screen panels which are stretched and relaxed, thereby creating changing tension motion to efficiently sort loose materials. VFFS are widely used in screening processes of fine materials where it is superior to conventional vibrating screens [3-4].

Common influencing factors on the screening performance of VFFS include the rotational speed of the exciter, the inclination angle of the screen, the slack length of the screening panel among many other factors related to material itself. Derived from e.g. the rotational speed are the vibration characteristics of VFFS, which also largely affect the processing performance and capacity [5-6]. To achieve the best screen performance, an optimized design and control is required, which fundamentally needs an in-depth understanding of the vibrations.

Xiong [6] analysed the dynamics of a banana flip-flow screen with linear springs. However, in the actual situation, vibrating flip-flow screens with linear springs can't work at the ideal working area as the change of the system' natural frequency caused by material quantity (additional mass) or changing stiffness (pre-tension, tension). The need to design vibrating flip-flow screens which are efficient over a broad band of forcing frequencies led to the study nonlinear characteristics – considering nonlinear stiffness. This however, requires developing an analytical procedure, to study the system's steady state responses and stability behaviour which we are doing in the following.

2. Analytical model

The investigated VFFS is shown schematically in Fig.1(a). The main screen frame is supported by vibration isolation springs. The inserted floating screen frame is connected to the main screen frame by shear springs. Moreover, the VFFS is excited by an exciter shaft mounted to the main frame. The rotation of the eccentric blocks in the exciter causes a relative movement between the main and the floating frame, which harmonically slackens and stretches the screen panel.

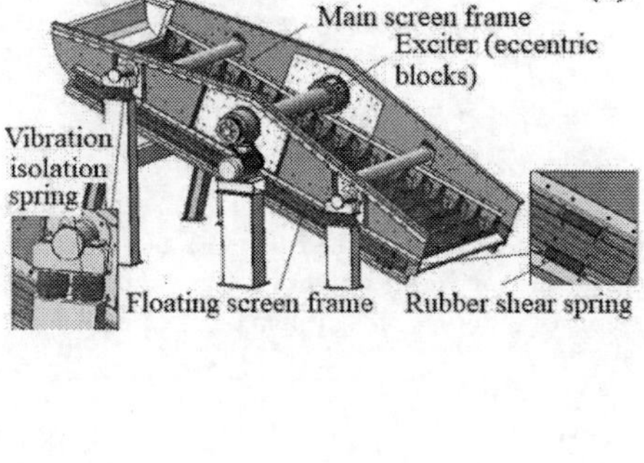

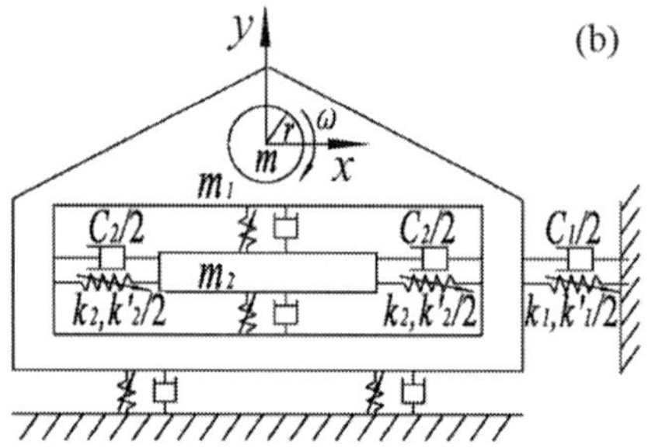

Figure 1. (a) Schematic and the (b) mathematical model of a vibrating flip-flow screen.

To investigate the dynamic of the vibrating flip-flow screen, a corresponding dynamic model is established, Fig.1(b). Here, m is the eccentric block mass of vibration exciter; m_1 and m_2 are the mass of main and floating screen frames respectively, r is the eccentricity; ω the angular velocity of m; k_1, k'_1 and c_1 - stiffness and damping coefficient of the vibration isolation springs, k_2, k'_2 and c_2 - stiffness and damping coefficients of the vibration shear spring. The x-axis is along the screen panel, the y-axis is perpendicular to the screen panel. The co-ordinate x_1 represents the displacement of the main screen fame's mass m_1 with respect to its foundation, while x_2 stands for the relative displacement of the floating screen frame's mass m_2 with respect to the mass m_1. The vibration of the vibrating flip-flow screen in y direction can be neglected, because that the main and floating screen frames will vibrate synchronously in y direction when VFFS works, which cause little effect on the performance of the VFFS. Therefore, in this paper, the dynamics of the VFFS is studied only considering x direction. The machine is excited by a harmonic forcing of the form $F = F_0 \cos \omega t$ ($F_0 = mr\omega^2$). The non-linear springs restoring forces and governing equations are written as [7-8]:

$f_i(x_i) = k_i x_i + k_i{}' x_i{}^3$, with $i = 1,2$, and (1)

$(m_1 + m_2)\ddot{x}_1 + m_2\ddot{x}_2 + c_1\dot{x}_1 + k_1 x_1 + k_1{}' x_1{}^3 = F_0 \cos \omega t$; and (2a)

$m_2\ddot{x}_1 + m_2\ddot{x}_2 + c_2\dot{x}_2 + k_2 x_2 + k_2{}' x_2{}^3 = 0.$ (2b)

Suppose that the characteristic length is described by $x_c = F_0/k_1$, then the displacements can be normalized using x_c, $y_i = x_i/x_c$ and the following dimensionless parameters are introduced:

$\tau = \omega t$, $\mu = m_2/m_1$, $\bar{\omega}_i{}^2 = k_i/m_i$, $\omega_i = \bar{\omega}_i/\omega$, $\eta_i = \omega_i{}^2$, $\zeta_i = c_i/2\sqrt{k_i m_i}$, $\varepsilon_i = 2\zeta_i\omega_i$, $\delta = \omega_2/\omega_1$, $\Omega = \omega/\bar{\omega}_1$, $\bar{\alpha}_1 = k_1' k_1 x_c^2/m_1{}^2\omega^4$, $\bar{\alpha}_2 = k_2' k_2 x_c^2/m_1 m_2 \omega^4$, $\alpha_i = \bar{\alpha}_i \Omega^2$, $\bar{\lambda} = F_0/m_1\bar{\omega}_1{}^2 x_c$, and $\lambda = \bar{\lambda}\omega_1{}^2$.

Important here are the dimensionless parameters $\bar{\alpha}_1, \bar{\alpha}_2$ and μ as they represent non-linear stiffness and mass ratio of m_2 and m_1 which we are going to vary in the following. The equations of motion of the system can be put in matrix form:

$$\mathbf{M}\ddot{\mathbf{Y}} + \mathbf{C}\dot{\mathbf{Y}} + \mathbf{K}\mathbf{Y} = \mathbf{F}, \quad (3)$$

with

$$\mathbf{Y} = \begin{bmatrix} y_1 \\ y_2 \end{bmatrix}, \ \mathbf{M} = \begin{bmatrix} 1+\mu & \mu \\ \mu & \mu \end{bmatrix}, \ \mathbf{C} = \begin{bmatrix} \varepsilon_1 & 0 \\ 0 & \mu\varepsilon_2 \end{bmatrix}, \mathbf{K} = \begin{bmatrix} \eta_1 & 0 \\ 0 & \eta_2 \end{bmatrix}, \ \mathbf{F} = \begin{bmatrix} \lambda\cos\tau - \alpha_1 y_1{}^3 \\ -\mu\alpha_2 y_2{}^3 \end{bmatrix}.$$

Applying the method of averaging [9-10] the steady state responses become of the form

$\mathbf{Y} = \mathbf{U}\cos\tau + \sin\tau$, and (4a)

$\dot{\mathbf{Y}}(\boldsymbol{\tau}) = -\mathbf{U}\sin\tau + \mathbf{V}\cos\tau$ (4b)

with the time dependence of $\mathbf{U} = [u_1(\tau),\ u_2(\tau)]^T$ and $\mathbf{V} = [v_1(\tau),\ v_2(\tau)]^T$ being assumed to be "slow" and of higher order. Differentiating the Eq. (4a) with respect to the time τ, we obtain

$\dot{\mathbf{Y}} = \dot{\mathbf{U}}\cos\tau - \mathbf{U}\sin\tau + \dot{\mathbf{V}}\sin\tau + \mathbf{V}\cos\tau$ (5)

Substituting Eq. (4b) into Eq. (5), one finds:

$\dot{\mathbf{U}}\cos\tau + \dot{\mathbf{V}}\sin\tau = 0.$ (6)

Also differentiating the Eq. (4b), we obtain

$$\ddot{\mathbf{Y}} = -\dot{\mathbf{U}}\sin\tau - \mathbf{U}\cos\tau + \dot{\mathbf{V}}\cos\tau - \mathbf{V}\sin\tau$$

Substituting the equations about $\mathbf{Y}$, $\dot{\mathbf{Y}}$ and $\ddot{\mathbf{Y}}$ into Eq. (3), the resulting equation becomes

$$(\mathbf{M}\dot{\mathbf{V}} - \mathbf{M}\mathbf{U} + \mathbf{C}\mathbf{V} + \mathbf{K}\mathbf{U})\cos\tau - (\mathbf{M}\dot{\mathbf{U}} + \mathbf{M}\mathbf{V} + \mathbf{C}\mathbf{U} - \mathbf{K}\mathbf{V})\sin\tau = \mathbf{F}(\boldsymbol{u}, \boldsymbol{v}, \boldsymbol{\tau}) \quad (7)$$

Then, multiplying Eq. (6) by the term $\mathbf{M}\cos\tau$, Eq. (7) is multiplied by $-\sin\tau$ and adding the two equations. The resulting equation is then integrated from 0 to 2π by assuming that $\mathbf{U}$ and $\mathbf{V}$ remain constant. The final outcome of these manipulations is

$$\mathbf{M}\dot{\mathbf{U}} = \frac{1}{2}(\mathbf{K} - \mathbf{M})\mathbf{V} - \frac{1}{2}\mathbf{C}\mathbf{U} + \frac{1}{2}\begin{pmatrix} \frac{3}{4}\alpha_1 v_1(u_1^2+v_1^2) \\ \frac{3}{4}\mu\alpha_2 v_2(u_2^2+v_2^2) \end{pmatrix} \quad (8)$$

Similarly, multiplying Eqs. (6) and (7) by either $\mathbf{M}\sin\tau$ or $\cos\tau$, respectively; adding both equations and integrating it from 0 to 2π, we obtain

$$\mathbf{M}\dot{\mathbf{V}} = \frac{1}{2}(\mathbf{M} - \mathbf{K})\mathbf{V} - \frac{1}{2}\mathbf{C}\mathbf{V} + \frac{1}{2}\begin{pmatrix} \lambda - \frac{3}{4}\alpha_1 u_1(u_1^2+v_1^2) \\ -\frac{3}{4}\mu\alpha_2 u_2(u_2^2+v_2^2) \end{pmatrix} \quad (9)$$

Eqs. (8) and (9) represent a system of first order, ordinary differential equations. For periodic steady vibration, the conditions are provided as

$\dot{\mathbf{U}} = \dot{\mathbf{V}} = \mathbf{0}.$ (10)

Substituting conditions (10) into Eqs. (8) and (9), a set of four coupled non-linear algebraic equations for u_1, v_1, u_2 and v_2 is obtained,

$$(\eta_1 - \mu - 1)v_1 - \mu v_2 - \varepsilon_1 u_1 + \frac{3}{4}\alpha_1 v_1(u_1^2 + v_1^2) = 0, \quad (11a)$$

$$-\mu v_1 + (\mu\eta_2 - \mu)v_2 - \mu\varepsilon_2 u_2 + \frac{3}{4}\mu\alpha_2 v_2(u_2^2 + v_2^2) = 0, \quad (11b)$$

$$(1 + \mu - \eta_1)u_1 - \mu u_2 - \varepsilon_1 v_1 + \lambda - \frac{3}{4}\alpha_1 u_1(u_1^2 + v_1^2) = 0, \text{ and} \quad (11c)$$

$$\mu u_1 + (\mu - \mu\eta_2)u_2 - \mu\varepsilon_2 v_2 - \frac{3}{4}\mu\alpha_2 u_2(u_2^2 + v_2^2) = 0. \quad (11d)$$

3. Numerical results and discussion

In all calculations, the non-dimensionless frequency Ω is taken as the independent variable. x_1 and x_2 denote the response amplitudes of the main screen frame and the relative displacement of main screen frame and floating screen frame. Both x_1 and x_2 can be expressed in terms of displacement and velocity using.

$x_i = x_c\sqrt{u_i{}^2 + v_i{}^2}$, with $i = 1,2$ (12a)

As spring stiffness and materials have greater effect on the performance of VFFS compared with other factors. Results are obtained by applying the above analysis, calculating the effect of nonlinear spring stiffness and the effect of the material on the steady state response of the system.

3.1. The effect of non-linear stiffness on the steady state response

Several groups of parameters are tried out to discuss the vibration characteristics of the system; the selection of the relative parameters is taken by an industrial flip-flow screen (m_1 = 4,482 kg, k_1 = 756.7 kN/m). The steady state response of the main screen frame (x_1) and relative displacement between the main and the floating screen frame (x_2) are shown in Figs. 2(a) and 2(b) considering different $\bar{\alpha}_1$ and $\bar{\alpha}_2$, which represent nonlinear stiffness of isolation vibration springs and shear springs, respectively.

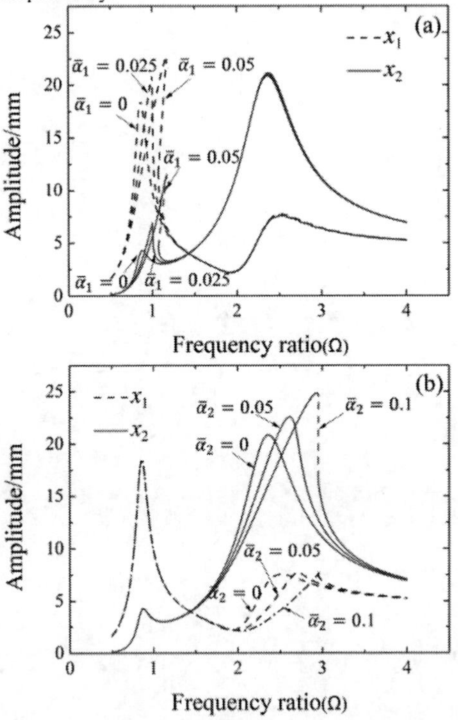

Figure 2. Steady-state response of the system: (a) u = 0.3, δ= 2, $\bar{\alpha}_2$= 0, $\zeta_1 = \zeta_2 = 0.1$ (b) u = 0.3, δ = 2, $\bar{\alpha}_1$= 0, $\zeta_1 = \zeta_2 = 0.1$

In Fig. 2(a), with the increase of the non-linear spring stiffness $\bar{\alpha}_1$ from zero to 0.05 in the first order resonance region, the peak values of x_1 and x_2 both increase, and the corresponding Ω of peak value also increases. This will cause an undesirable vibration mode in case the VFFS begins to work. In Fig. 2(b), increasing the non-linear spring stiffness $\bar{\alpha}_2$ from zero to 0.1 the peak value of x_2 increases from 20.85 to 24.58, and the peak value of x_1 remains almost constant. Because VFFS commonly work near the second resonance region, it needs to be tested whether the increase of $\bar{\alpha}_2$ could be beneficial to the performance and processing capacity of materials.

3.2. The effect of materials on vibration characteristic of the system

When the screening panel is loaded with material (2,045 kg) the mass-ratio changes from μ = 0.3 to μ = 0.38. The steady state responses of the VFFS with linear ($\bar{\alpha}_2 = 0$) and nonlinear ($\bar{\alpha}_2 = 0.1$) shear springs with and without material are shown in Figs. 3(a) and 3(b), respectively. In Fig. 3(a) it can be seen that the presence of materials has a large effect on the vibration characteristics of VFFS with linear shear spring; the large changes in x_1 and x_1 influences the stability of the system and requires to be monitored. Fig. 3(b) shows that an additional loading of material, which acts on the screen has a rather small effect on the nonlinear vibration characteristics when it works near the resonance area hence non-linear characteristics of shear springs could be beneficial to the stability of the system.

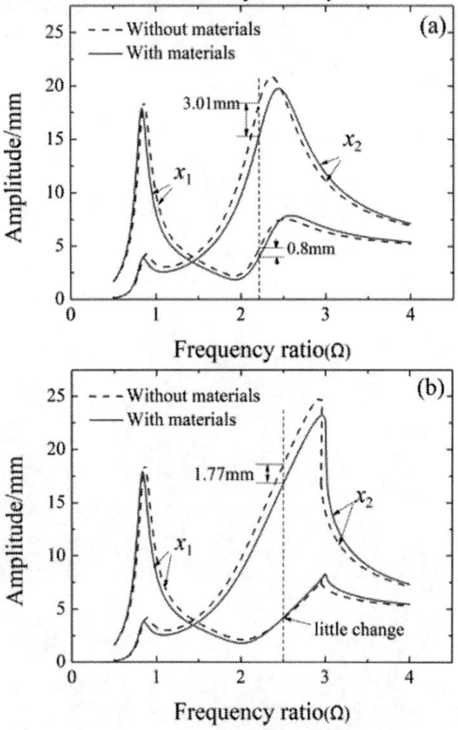

Figure. 3 Steady-state response of the system: (a) linear, and (b) nonlinear shear springs

4. Conclusion

Vibrating flip-flow screens can be effective solutions for the screening of highly viscous and fine materials. Our study employs an analytical procedure to determine the dynamic response of vibrating flip-flow screens by also considering their nonlinear shear spring characteristics. This analysis is applied to two cases, with the goal of identifying the effect of the system parameters on the steady-state responses. Results indicate that nonlinear springs in connection with properly system parameters would be beneficial for the vibrating flip-flow screens' performance. However, while the application of nonlinear springs could show to be useful over a wider range of forcing frequencies, the extension to a larger frequency band also requires to study whether stability issue can arise.

Further, a comparison with real experiments and an updated finite element model would be required in the future.

References

[1] Noble A and Luttrell G 2015 *Int. J. Min. Sci. Technol.* **25** 511-521
[2] Zheng G, Zhu J, Xia W and Liu S 2016 *Filtr. + Separation.* **53** 38-41
[3] Akbari H, Ackah L and Mohanty M 2017 *Int. J. Coal Prep. Util.* 1-23
[4] Guo Q and Gong G 2016 *XVIII Int. Coal Prep. Cong.* (Saint- Petersburg: Springer International Publishing Switzerland) P 913-918
[5] Peng L, Liu C, Hong H, Li J and Xia Y 2014 *J. China Coal Soc.* **39** 976-980
[6] Xiong X, Niu L, Gu C and Wang Y 2017 *J. Sound Vib.* **411** 108-128
[7] Zhang Z, Oberst S and Lai J 2016 *J. Sound Vib.* **377** 123-132
[8] Beltrán-Carbajal and Silva-Navarro G 2014 *J. Sound Vib.* **333** 3019-3030
[9] Zhu S, Zheng Y and Fu Y 2004 *J. Sound Vib.* **271** 15-24
[10] Natsiavas S 1992 *J. Sound Vib.* **156** 227-245.

Stress Analysis and Design of Main Reducer Housing Using Finite Element Method

S. Wang
School of Mechanical Science and Engineering, Jilin University, Chang Chun, 130022, China

Abstract. The main reducer housing takes over the shaft loads from gear engagement and transmits to other components, such as differential, semiaxle and driving wheels, so the main reducer housing with enough strength and stiffness is very important. Some factors preventing it from failure need to be taken into consideration when design it. To design a main reducer housing with better performance, in this paper, FEA (Finite Element Analysis) is used to analysis the main reducer housing and to find out some big stress regions. Then, some modifications are proposed to eliminate those big stress regions and obtain a reliable main reducer housing. During the analysis process, an annulus model is built and the reaction forces between the differential bearing seats and axle housing are calculated to determine whether they contact with each other. Finally, some design methods and improvements of the original design main reducer housing are proposed. And numerical comparison results of the stress distribution of the original and improved main reducer housing validate the effectiveness of the proposed methods and modifications in this paper. Those stress analysis and modifications in this paper are time-saving and money-saving before mass production.
Keywords: main reducer housing, tress analysis, FEA

1. Introduction

The main reducer housing, a basic structural component of axle assembly, is able to reduce the rotation speed and enhance the torque significantly[1]. It takes over the shaft loads from gear engagements and delivers those driving forces to other components, such as differential gears, semiaxis and driving wheels. The main reducer housing with enough strength and stiffness ensure more accurate gear engagements and better performance of transmission system, avoiding a disaster during work caused by failure. Therefore, making sure the main reducer housing has enough strength and stiffness is very important. The stress analysis is aim to find out some potential problem regions of the main reducer housing and make some improvements of its structure. Besides, the analysis results can be used to instruct the design of the main reducer housing, which may time-saving and money-saving before mass production. The main reducer housing in the axle assembly of a heavy commercial vehicle is shown in Figure 1.

Some studies about strength of the main reducer housing can be found, such as stress distribution of the main reducers housing using FEA [2 3 4]. The lightweight design about the main reducer housing is also studied [5]. But less study about the boundary conditions between the main reduce housing and axle can be found, which have significant influences on FEA.

To get more accurate stress distribution of the main reducer, FEA is carried out. To deal with the boundary conditions, an annulus model is built and reaction forces are calculated. Finally, some noticeable methods to design the main reducer housing are proposed according to the stress distribution. This paper is organized as follows. A full scale finite element model of the main reducer housing is built in section 2.1. Loads and boundary conditions of the main reducer housing are discussed and determined in section 2.2. In section 2.3, potential problem regions of the main reducer housing are found out. In section 3, some design enhancement solutions and modifications are put forward and applied to the main reducer housing. Comparisons between the main reducer housing stress distribution before and after improvement are also shown in this section. Finally, some conclusions about redesign of the main reducer housing are given out and verified in the end of this paper.

Figure 1. Typical axle assembly in heavy commercial vehicle.

2. Finite element analysis

2.1 FEA model

https://doi.org/10.1051/matecconf/201822104008

FEA is carried out with a full scale model of the main reducer housing. The FEA model is built with high-quality meshes in Figure 2 consisting of 437860 elements and 732016 notes. During the modeling process, meshes of the main reducer housing should be just refined in local regions with consideration of less calculating time, especially in the critical regions, which is called local refinement of meshes.

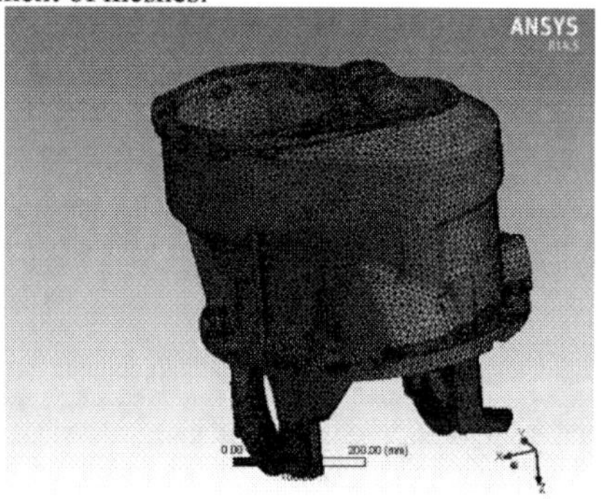

Figure 2. Discretized model of main reducer housing.

2.2 Loads and Boundary Conditions

The main reducer housing mounted on the axle housing through flange and bolts and it constrained in all the translational and rotational DOF's. Some small intervals between the axle housing and differential bearing seats exist. They may disappear when the main reducer housing experiences some deformations, causing the axle housing contacts with the differential bearing seats in nonlinear and preventing more deformation of the differential bearing seats and the main reducer housing. Figure 3 shows the details of those seven bearing loads and boundary conditions of the main reducer housing.

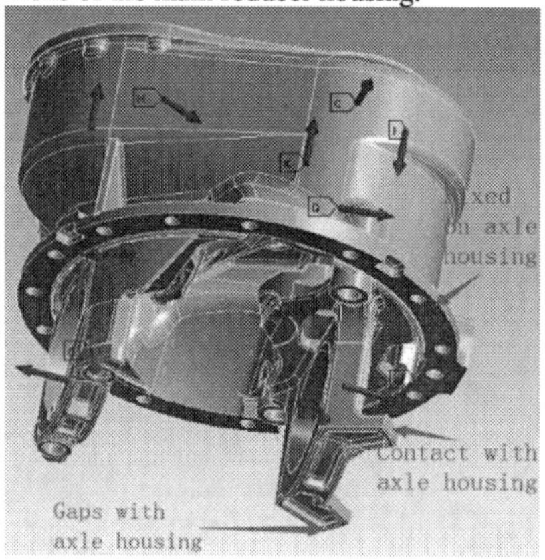

Figure 3. Boundary conditions and bearing loads of main reducer housing.

Two methods dealing with the boundary conditions of the axle housing and the differential bearing seats are pointed out in this section. The first one that an annulus model is built around differential bearing seats to simulate the axle housing and set intervals between them in Figure 4. The second one is that the reaction forces and deformations of those two differential bearing seats are calculated, which indicate whether intervals between them are equal to zero, which means the axle housing contacts with the differential bearing seats. These directions and magnitudes of reaction forces are shown in Figure 5.

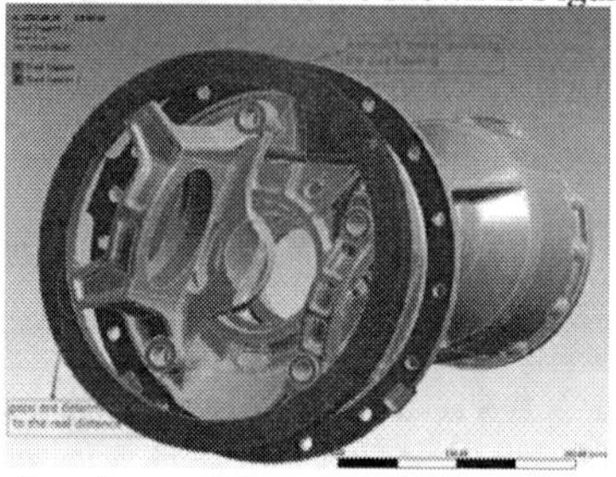

Figure 4. Annulus model around differential.

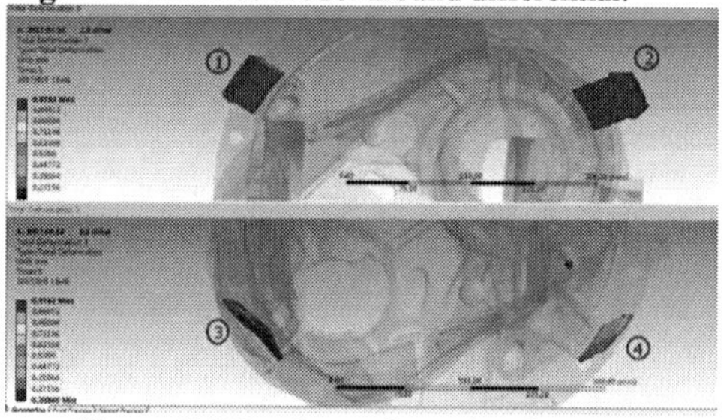

Figure 5. Directions of reaction forces.

From Figure 5, the directions of reaction forces numbered 1 and 2 are perpendicular to differential bearing seats which indicate the differential bearing seats numbered 1 and 2 contact with axle housing. The directions of reaction forces numbered 3 and 4 indicate intervals exist between them. According to the analysis of reaction forces, the differential bearing seats numbered 1 and 2 should be constrained in all translational and rotational DOF's, while those numbered 3 and 4 should be free in all directions during FEA.

2.3 Results of FEA

The material of the main reducer housing is ductile iron, yield strength is 310MPa, modulus of elasticity is 1.69×10^5 and poisson ratio is 0.257. According to the stress distribution, some big stress regions and low stress regions are pointed out. Figure 6(a) and Figure 6(b) show the stress concentration regions of the main reducer housing at the root of the differential bearing and guide bearing seats, while Figure 6(c) shows the low stress regions at the exterior stiffeners.

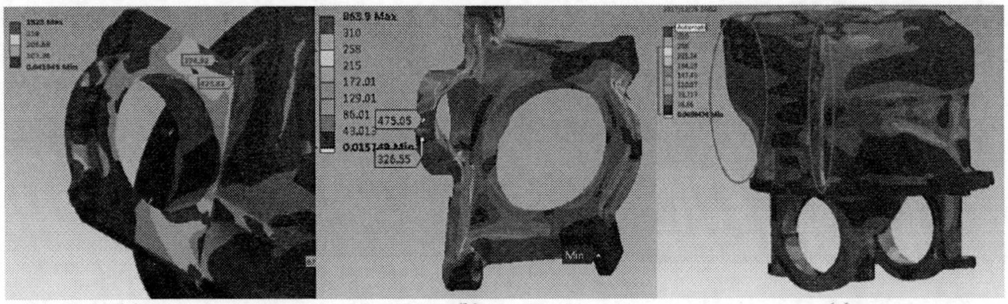

(a) (b) (c)

Figure 6. (a) Stress concentration at the root of differential bearing seat. (b) Stress concentration at the guide bearing seat. (c) Low stress region of exterior stiffeners.

3. Improved design of main reducer housing

3.1 Improved design of stress concentration regions

To get a better stress distribution, the stress concentration regions need to be eliminated. Increasing the thickness of structure with big stress regions is a common way but not a prefer solution, which may cause poor efficiency with mass increase during work. Given this, the roots of differential bearing seats are expanded as shown in Figure 7(a) compared with Figure6(a), and the guide bearing seat is thickened in Figure 7(b) compared with Figure6(b). Besides, the stiffeners are shortened for mass reduction in Figure 7(c) without big stress compared with Figure6(c). With these modifications, significant improvements appear in stress distribution of the main reducer housing. In order to further verify the effectiveness of these modifications, FEA is carried out again and the analysis results are shown below.

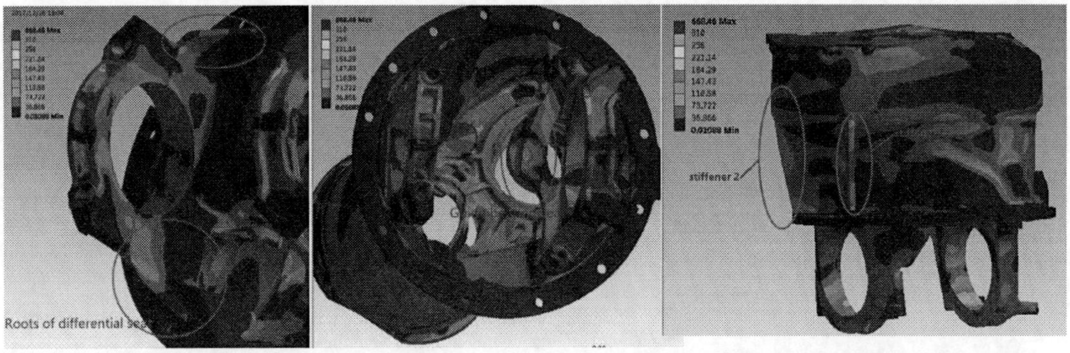

(a) (b) (c)

Figure 7. (a) Improvement of the differential bearing seats. (b) Improvement of the guide bearing seats. (c) Improvement of exterior stiffeners.

3.2 Comparisons of the original and improved main reducer housing

Combined with the improvement methods and FEA, better stress distribution of the main reducer housing are obtained. Figure 8 and Table 1 show stress distribution of local part of the main reducer housing and stress comparisons of the original and improved main reducer housing respectively. From those Figures, the strength of the main reducer housing really improve a lot than before, which can ensure safety in practice. The stress distribution of those circled areas in Figure 8 verify the validity of those improvement methods proposed in this paper.

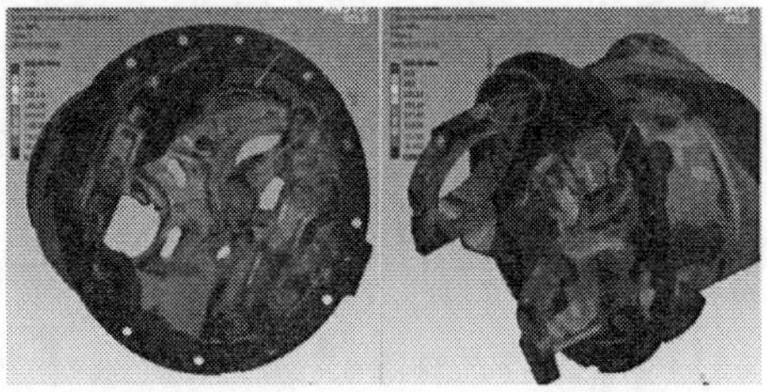

Figure 8. Modified points of main reducer housing.

Table 1. Stress comparisons between original and improved main reducer housing.

	A(MPa)	B(MPa)	C(MPa)	D(MPa)
Original	380	440	420	360
Improved	170	110	50	150
Improvement percentage	55%	75%	88%	58%

4. Conclusion

In order to find out big stress regions and to ensure enough strength of the main reducer housing during work, FEA method is carried out in this paper. During the analysis process, Local refinement of meshes are pointed out. Besides, an annulus model and reaction forces dealing with boundary conditions are also used to ensure more precise stress analysis. According to these analysis results of the main reducer housing, some potential problem regions are found out which is accordance with the practice. And some improvement methods aimed at those big stress regions are given out. Then, better stress distributions of improved main reducer housing are obtained. Finally, the stress comparisons between original and improved main reducer housing are given out and verify the effectiveness of those improvement methods proposed in this paper, and the improvement rates indicate low stress with better structural design.

References

[1] Hu F, Chen L Q, Qi B, 2012 Finite element analysis and optimum design of time-sharing 4WD vehicle main reducer. *J. Applied Mechanics & Materials* vol 246-247, ed F Y Cheng and Y Ma pp 692-696.

[2] Chen Y D, 2009 Finite element analysis of the main reducer housing of driving axle. *J. Manufacturing Informatization*. Vol 37. P 10.

[3] Zhou C, 2016 Finite element analysis on the support stiffness of final drive in vehicle drive axle. *J.Automotive Engineering*. Vol 38. No.8.

[4] Q. Q. Zhao. 2014 Finite element analysis of the main reducer housing of four wheel drive car. *J. Journal of Chifeng University (Natural Science Edition)*. Vol (30).No.4.

[5] Yang C, Qin F, Huang D, 2014 The lightweight design for main reducer shell based on topological optimization method. *J. Journal of Mechanical Transmission*.

To Determine the Effect of Dimpled Poppet Valve on the Flow in Engine Cylinder by Port Flow Simulation

Anirudh Singh[1], Apoorv Dubey[1] and Raj Kumar Singh[1]

[1]*Mechanical Engineering Department, Delhi Technological University, Delhi 42, India*

Abstract. Intake system is one of the crucial sub-systems in engine, the geometrical configuration of the inlet ports and the valves and their opening schedule creates organized motions in the cylinder, known as swirl (about the cylinder axis) and tumble (orthogonal to the cylinder axis) which help in proper mixing of fuel and air. Surface texturing of intake valve can improve the performance of engine drastically by generating more swirl, turbulent kinetic energy and reducing the coefficient of skin friction in the valve proximity area. Making dimples on intake valve is one of the many ways in which surface texturing can be done and its effectiveness can be easily measured by conducting a steady-state port flow simulation on ANSYS IC fluent solver using standard k-omega model. Port flow analysis was carried on dimpled intake valve at a lift of 5mm for 5 dimple ratio ranging from 0.1 to 0.3 in steps of 0.05. The dimple depth-over diameter ratio is a very crucial parameter in determining the optimum geometry of dimple on which best results are obtained. The results showed that the maximum mass flow rate is obtained at dimple ratio of 0.2 with an increase of 0.005% over the simple intake valve and a 38.4% decrease in the area weighted average skin friction coefficient in valve proximity area was seen in case of dimpled valve.

1 Introduction

The 4 stroke internal combustion engine consists of four cycles namely intake, compression, power stroke and exhaust. Out of these four processes, intake and compression strokes are the most important as these strokes have great effect on determining the pattern of air flow coming inside cylinder during intake stroke and the condition needed for the proper vaporization and mixing of fuel during fuel injection in compression stroke[1]. For maximizing the air passing into the cylinder, intake valve design has to be optimized. Hence optimum design of intake valve is crucial in engine design. Xiaofeng Yang, Zhaohui Chen, and Tang-Wei Kuo described that the development and optimization of intake system is to find the best trade-off between a) minimum flow restriction (discharge coefficient or volumetric efficiency) and b) an optimal in-cylinder charge motion (swirl or tumble ratios) [2]. Chenbo Ma, Jianjun Sun, Yan Wang , Bo Yu , Qiuping Yu and Qiaoan Tu stated surface texturing as a viable technique to significantly enhance the load-carrying capacity, friction coefficient, and wear resistance of vital components and to maximize the effect of surface texturing, many theoretical and experimental studies have been carried out that concentrate on the geometrical parameters of textures[3]. Etsion I. stated that to fully benefit from surface texturing, a proper optimization of the geometrical parameters, including the dimple depth-over diameter ratio (DOD) and area density, must be performed [4]. Therefore, the determination of the optimum dimple DOD ratio is very important. This concept of surface texturing was used to introduce dimples on the poppet valve and then measuring its effectiveness by conducting the Port flow simulation on ANSYS IC Fluent solver.

2 Flow in cylinder

All fluid flow is either laminar or turbulent, whether a flow is laminar or turbulent is usually determined by the Reynolds number. The flow in the cylinder is divided into several distinct phases. Firstly, the geometrical configuration of the intake system and the opening schedule of the inlet valves creates organized motions in the cylinder, known as swirl (about the cylinder axis) and tumble (orthogonal to the cylinder axis) and second, the jet itself is turbulent, and in addition much of the directed (non-turbulent) energy in the jet is converted to turbulence, resulting in a very high turbulence level during the inlet stroke[5]. The flow through the inlet valve forms a hollow conical jet, which is roughly equal to the angle of the valve seat.

An air flow bench is a device used for testing the internal aerodynamic qualities of an engine component and is used primarily for testing the intake and exhaust ports of cylinder heads of internal combustion engines. A flow bench consists of an air pump, a metering element, temperature and pressure measuring instruments such as manometers, and various controls. The test piece is connected in series with the pump and measuring element and air is pumped through the whole system. As the volume flow rate through the metering element is known which is equal to the flow through the test piece, the mass flow rate can be calculated using the known pressure and temperature data to calculate air densities, and then multiplied by the volume flow rate. Most flow testing is done between 10 and 28 inches of water pressure (2.5 to 7 kilopascals).

Nur Adila Mohamad Shafie et al.[6] showed that the minimum and maximum deviation as small as 1.6% and 5.1% respectively were obtained when they compared the results obtained by port flow simulation and those obtained from engine flow bench test(experimental test) which shows that port flow simulation can accurately predict the breathing parameter of engine.

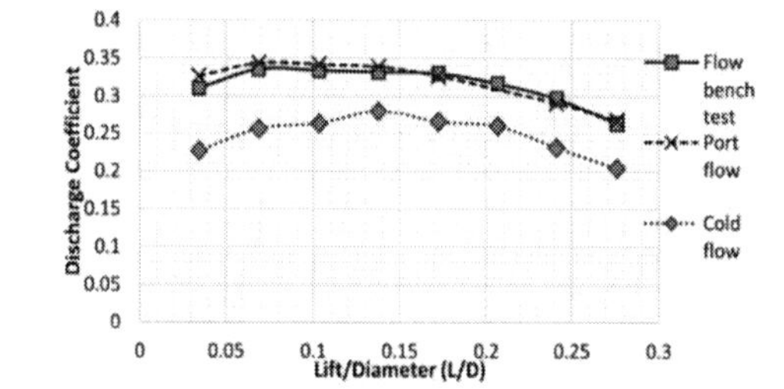

Figure 1. Deviation in results obtained from Port flow simulation and engine flow bench (experimental test) [6]

3 Methodology

A simplified design of port has been used for conducting this simulation, the design of intake plenum, outplenum, cylinder and port is kept same for all the cases in order to compare the effect of introducing dimples on the poppet valve over the simple valve. In case of the dimpled valves, two sets of 20 equidistant dimples were introduced on the valve. The print diameter of the dimples is kept at a constant of 1.2mm and the depth of the dimple is varied, governed by the dimple ratio $h/_D$ =c (here D=print diameter, h=depth, c=constant ranging from 0.1 to 0.3 in steps of 0.05). The port flow analysis in all the cases is carried out at an intake valve lift of 5mm and the exhaust port is deactivated by defining the exhaust valve lift to be 0mm[7].

The placement of dimples on poppet valve can be seen in fig. 2 for more better understanding of the geometry on which analysis is being carried upon, their configuration as well as their shape have a great effect on various parameters like drag, skin friction coefficient, standton number and reynolds number of the flow[8].

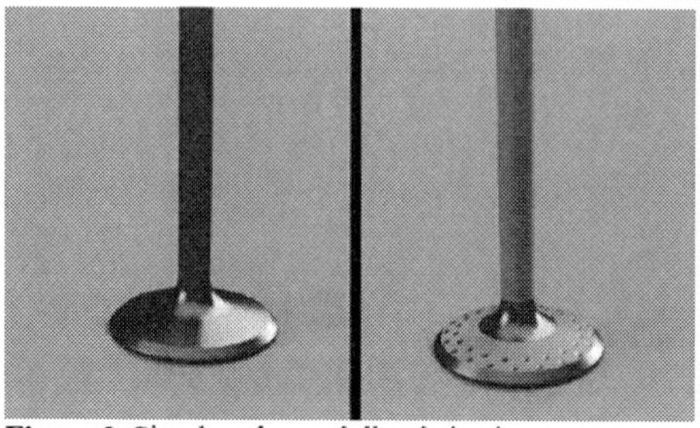

Figure 2. Simple valve and dimpled valve

3.1 Decomposition and meshing

The assembled model is then defined in the IC Input Manager in the design modeler and decomposed into the intake, exhaust ports and the cylinder. Four post planes (swirl planes) were added at a distance of 15mm, 30mm, 45mm and 60mm from the reference point for analyzing the swirl and velocity distribution due to introduction of dimples on poppet valve. Values of various parameters inserted in the input manager are shown in the Fig. 3.

The decomposed geometry is then meshed using a hybrid mesh topology. The maximum mesh size is reduced to 2.5mm, the number of inflation layers is increased to 12 and the chamber growth rate is reduced to 1.12. The size of the elements in the valve proximity area is kept at 0.2mm to capture the properties of flow precisely in that area. The inflation layers are essential for capturing the features of the boundary layers perfectly as prismatic elements provide good alignment with the flow near wall boundaries and thus special emphasis is laid on them. For regions other than the boundary layers, tetrahedral elements are predominant. The values of various mesh parameters are shown in Table 1.

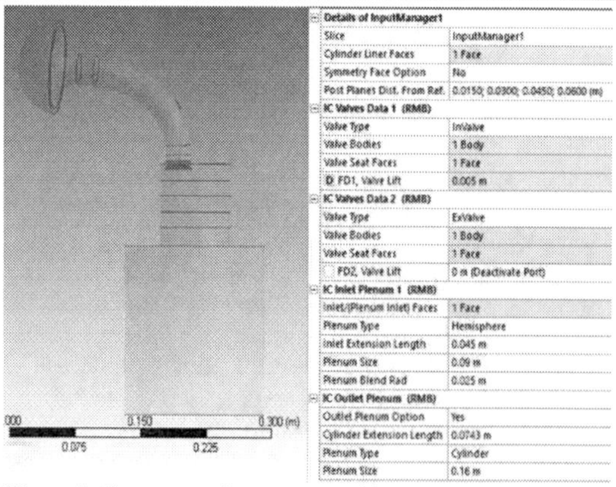

Figure 3. Decomposed geometry and input parameters

Table 1. IC Port Mesh Parameters

S.no	Mesh Parameters	Input value
1	Mesh type	Hybrid
2	Reference size(mm)	0.993
3	Maximum mesh size(mm)	2.5
4	Curvature normal angle(degree)	18
5	Number of cell across gap	5
6	Number of inflation layer	12
7	Chamber size(mm)	1.2
8	Chamber growth rate	1.12
9	Valve proximity face size(mm)	0.2

3.2 Solver Setup

For the purpose of simulation, standard k-omega model is used and the boundary conditions applied on various surfaces is shown in Table 2. An area weighted average surface monitor was set on the ice-int-chamber2-outplenum and a convergence monitor was applied on the plane with convergence criteria of $1*10^{-5}$ and the numbers of iterations to be ignored was set to 600. Different zones on which boundary condition is applied can be seen in Fig. 4.

Table 2. Boundary conditions

S.no	Zones	Type	Temperature	Gauge Pressure(Pa)
1	ice-slipwall-outplenum	wall	300 K	X
2	ice-slipwall-inplenum	wall	300K	X
3	ice-outlet	pressure outlet	300K	5000
4	ice-inlet-inplenum1	pressure inlet	300K	0

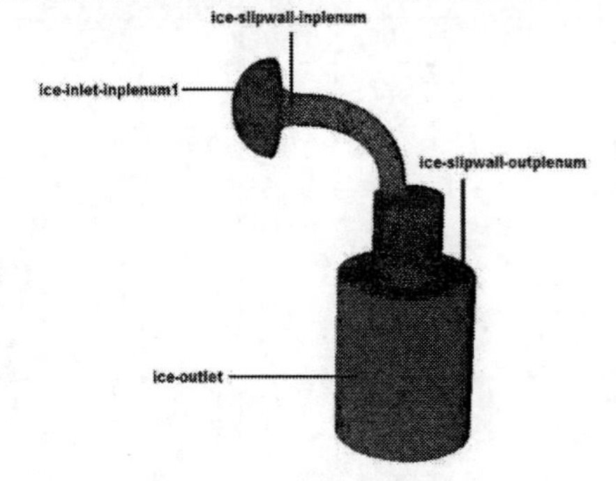

Figure 4. Different Zones on which boundary condition is applied

Once the solution had converged, obtained results were analyzed for all the case.

4 Result And Discussion

To compare the results obtained from port flow simulation conducted with simple intake poppet valve to that obtained from dimpled intake poppet valve with various dimple ratios, some key parameters were focused upon. In total four major parameters were studied to determine the effect of introduction of dimples on the in-cylinder flow. These parameters are as follows:

4.1 Mass Flow Rate

Mass flow rate is one of the many parameters that is given major focus upon as it determines the ability of intake port by which the engine can move the charge of air into and out of the cylinder. The results obtained after port flow simulation for dimple ratio from 0.1-0.3 showed that we get maximum mass flow rate of 34.663g/s at the dimple ratio of 0.2. Fig. 5 shows the variation of mass flow rate (g/s) to the dimple ratio (h/D). The curve shows that the mass flow rate increases for dimple ratio from 0.1 to maximum at 0.2 and the suddenly decreases for dimple ratio from 0.2 to 0.3. Thus it can be concluded that there is an optimum valve of dimple ratio that will increase our mass flow rate for the same port design. The mass flow rate obtained for the simple valve was 34.661g/s. Thus there is a small increase in mass flow rate of 0.005%.The value of mass flow rate obtained for other dimple ratios can be seen in Table 3.

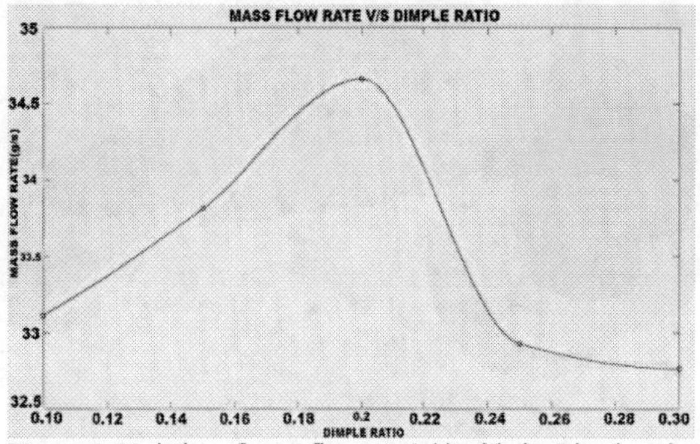

Figure 5. Variation of mass flow rate (g/s) with the Dimple ratio (h/D)

4.2 Turbulent Kinetic Energy

Mohd Aizad Sazrul Sabrudin, Mohd Farid Muhamad Said and Zulkarnain Abdul Latiff tried to study the effect of asymmetric intake valve lift configuration towards in-cylinder air flow behaviour by conducting a steady state port flow simulation and described that Turbulent kinetic energy (TKE) measurement on swirl monitoring planes is one of the methods that can be used to visualize the flow mixture behaviour inside the combustion chamber [9].

Instead of focussing on maximum or minimum turbulent kinetic energy (m^2/s^2) obtained on the swirl plane, focus on the area weighted average turbulent kinetic energy on the first swirl plane (SWIRL PLANE-1) was laid. Table 3 shows the value of area weighted average turbulent kinetic energy obtained on the first swirl plane monitor formed during the geometry decomposition process.

4.3 Swirl

There are a number of reasons for inducing swirl. High turbulence levels at ignition produce higher effective flame speeds, and more reliable combustion at very lean ratios, or with exhaust gas recirculation. At normal air/fuel ratios, the higher speed will allow the flame front to reach the end gas before the chemical reaction resulting in auto ignition has time to take place, permitting higher compression ratios without knock, the idea of swirl or tumble is to encapsulate some momentum of the inlet valve jet in the organized motion (swirl or tumble), which is less dissipative than the turbulence. The value of area weighted average swirl obtained on all four swirl planes for all the cases is shown in Table 3.

4.4 Coefficient of Skin Friction in Valve Proximity Area

According to Superflow Technologies Flowbench Operators' Manual and study conducted by Abdul Rahim Ismail et al.[10,11], source of flow losses in the port are wall friction, bend at valve guide, contraction at push-rod, expansion behind valve guide, expansion in 25 and 30 degrees, bend at the exit of the valve and expansion exiting valve. These sources of flow losses can be seen in Fig. 6.

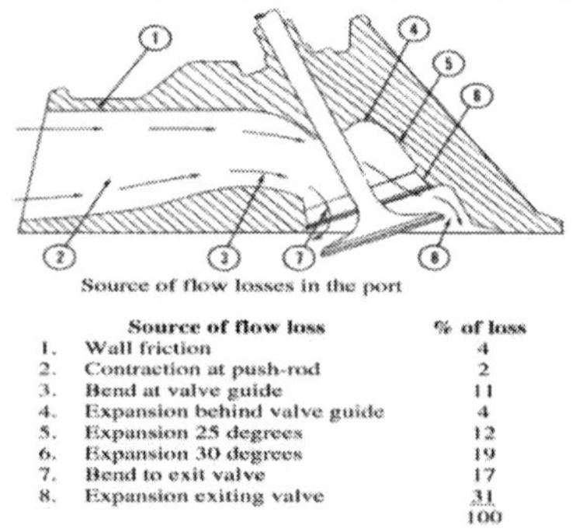

Figure 6. Sources of flow losses in the port [10,11]

The area weighted average of skin friction coefficient in the valve proximity area was analyzed and it can be seen from Table 3. that for all values of dimple ratio(0.1-0.3), the area weighted average of skin friction coefficient in valve proximity area was less than that in case of simple intake poppet valve. Also, it can be seen from Fig.7 that as the dimple ratio is increased from 0.1 to 0.3 there is an increase in the coefficient of skin friction in the valve proximity area.

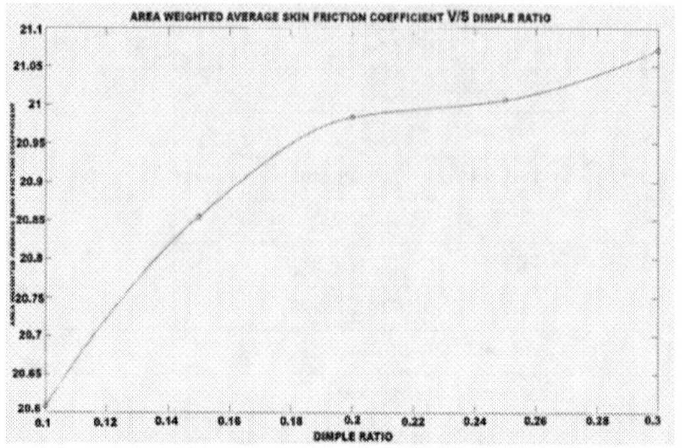

Figure 7. Variation of area-weighted average skin friction coefficient with dimple ratio

The lowest value of area weighted average skin friction coefficient(20.608) is obtained with a dimple ratio of 0.1, while the value of area weighted average skin friction coefficient obtained in case of simple valve is 33.507.Thus the skin friction coefficient of valve is decreased by 38.4% due to introduction of dimples.

The velocity distribution on all the swirl planes as well as cutplane for all the cases can be seen in Fig. 8. Major focus was given upon these four parameters to evaluate the effectiveness of the dimples created on the valve.

Table 3. Values of various parameters obtained for all the cases

Parameter	**Simple Valve**	**Dimpled Valve**				
		0.1	0.15	0.2	0.25	0.3
Mass flow rate(g/s)	34.661	33.111	33.812	34.663	32.927	32.765
Turbulent kinetic energy(m^2/s^2)	48.41	53.25	52.33	51.56	58.75	57.57
Skin friction coefficient	33.507	20.608	20.854	20.983	21.006	21.071
Swirl on Swirl plane-1	-0.02	0.007	0.022	0.023	0.015	0.036
Swirl on Swirl plane-2	-0.05	0.013	0.041	0.029	0.044	0.035
Swirl on Swirl plane-3	-0.04	0.0137	0.07	0.036	0.019	0.078
Swirl on Swirl plane-4	-0.01	0.046	0.018	0.055	0.011	0.12

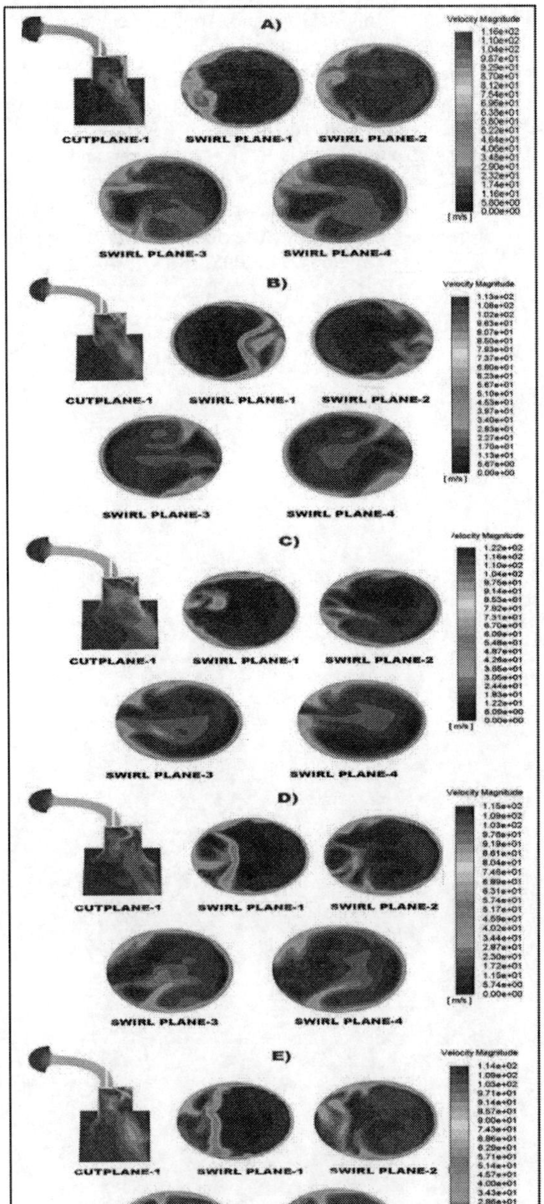

Figure 8. Velocity distribution on Swirl planes and Cutplane: A) simple valve, B) dimple ratio-0.1, C) dimple ratio-0.15, D) dimple ratio-0.2, E) dimple ratio-0.25, F) dimple ratio-0.3

5 Conclusion

In this work, the effect of introduction of dimples on the intake valve on the various performance characteristics of the engine was studied. The port flow simulation was carried with gauge outlet pressure of 5000Pa and the intake valve was given a lift of 5mm. Port flow analysis was carried on dimpled intake valve at a lift of 5mm for 5 dimple ratio ranging from 0.1 to 0.3 in steps of 0.05. The main conclusions are as follows:

1. The results obtained after port flow simulation for dimple ratio from 0.1-0.3 showed that we get maximum mass flow rate of 34.663g/s at the dimple ratio of 0.2. The mass flow rate increases for dimple ratio from 0.1 to maximum at 0.2 and the suddenly decreases for dimple ratio from 0.2 to 0.3.
2. The lowest value of area weighted average skin friction coefficient(20.608) is obtained with a dimple ratio of 0.1, while the value of area weighted average skin friction coefficient obtained in case of simple valve is 33.507.Thus the skin friction coefficient of valve is decreased by 38.4% due to introduction of dimples.
3. The value of area weighted Turbulent Kinetic energy (m^2/s^2) on SWIRL PLANE-1 was found to be greater in case of dimpled valve for every dimple ratio over the normal valve.
4. Higher value of swirl obtained on each swirl plane in case of dimpled valve shows the ability of the engine to mix air and fuel properly and effectively, that will enable a controlled as well as high speed flame front which will produce more power for the same quantity of the provided fuel.

References

[1] Heywood JB Internal Combustion Engine Fundamentals. Mcgraw-Hill; 1989
[2] Yang X, Chen Z, Kuo T Pitfalls for accurate steady-state port flow simulations. Journal of Engineering for Gas Turbines and Power 2013;135(6):061601–061601
[3] Ma C, Sun J, Wang Y, et al. On the optimum dimple depth-over-diameter ratio for textured surfaces. Adv. Mech Eng 2017; 9: 1–8.
[4] Etsion I, Kligerman Y and Halperin G. Analytical and experimental investigation of laser-textured mechanical seal faces. Tribol T 1999; 42: 511–516.
[5] FLOW IN THE CYLINDER by John L. Lumley, Cornell University (excerpt from his book: Engines – an introduction. Cambridge University Press, 1999)
[6] Nur Adila Mohamad Shafie, Mohd Farid Muhamad Said, Azhar Abdul Aziz, Zulkarnain Abdul Latiff, Ahmad Kamal Mat Yamin and Noreffendy Tamaldin MATEC Web Conf., 90 (2017) 01067, DOI: https://doi.org/10.1051/matecconf/20179001067

[7] Winroth, P.M., Ford, C.L. & Alfredsson, P.H. Exp Fluids (2018) 59: 24. DOI: https://doi.org/10.1007/s00348-017-2478-8

[8] Leontiev A I, Kiselev N A, Vinogradov Y A, Strongin M M, Zditovets A G and Burtsev S A 2017 Experimental investigation of heat transfer and drag on surfaces coated with dimples of different shape Int. J. Therm. Sci.118152–67

[9] Mohd Aizad Sazrul Sabrudin, Mohd Farid Muhamad Said and Zulkarnain Abdul Latiff. Effects of Asymmetric Intake Valve Lift Configuration Towards In-Cylinder Air Flow Behaviour. ARPN Journal of Engineering and Applied Sciences; VOL. 12, NO. 7, APRIL 2017

[10] SuperFlow Technologies Group, 2004. SF-1020 Flowbench Operators' Manual, SuperFlow Corporation, Colorado Springs, USA.

[11] Abdul Rahim Ismail, Rosli Abu Bakar and Semin, An Investigation of Valve Lift Effect on Air Flow and Coefficient of Discharge of Four Stroke Engines Based on Experiment,American Journal of Applied Sciences 5 (8): 963-971, 2008, ISSN 1546-9239.

Human Ergonomics Study in Microgravity Environment

Pengyan Liu [1], Dong Zhou[1], Long Xue[1] and Yuan Li[1]

[1]*Beihang University, Beijing, China*

Abstract. In recent decades, human exploration of space has gradually deepened, and more and more countries and regions have successfully sent astronauts into space. How to complete the space experiment as efficiently and safely as possible in the shortest time with limited resources has become an important issue in the field of aerospace technology. Taking full account of such matter, we consider that ergonomics in design stage is one of the most effective ways to solve this problem. The microgravity environment is the biggest difference between ground and space. Therefore, the study of ergonomics under the microgravity environment is of great significance. This article deeply analyses and summarizes the physiological differences between human and normal gravity under microgravity environment. From the ergonomics point of view, the requirements for spacecraft design and space mission planning are put forward for the reference of engineers and scholars.

1 Introduction

With the improvement of aerospace science and technology, the United States, Russia, China and many other countries have successfully sent humans into space. Since the beginning of the establishment of space stations by humans, the astronauts' length of stay in space has also been continuously refreshed. To ensure the safety of astronauts and improve work efficiency, spacecraft design must fully consider ergonomics.

Ergonomics take the human-oriented principle, focusing on human physical and psychological factors, research on the relationship between people, machines and the working environment to determine the best combination [1]. Unlike the operation task in ground conditions, the astronauts are in the microgravity environment during orbital operations, and their visual, operational, cognitive, and strength are different from those in the 1G gravity environment. Research shows that the application of human factors engineering plays an important role in space safety, especially in space operations and the human factor [2]. On the other hand, astronauts' work in space is tough, therefore they need to complete as many experiments as possible within a limited time quickly as they can. In addition, unlike the feature that the support equipment on the ground can support personnel movement for a long time, the spacecraft transportation capacity is limited and the protection resources are very expensive. Therefore, consider ergonomics fully in the design process of spacecraft is of great significance to ensure personnel safety, shorten task time, increase task completion rate, and reduce maintenance costs.

This article first elaborates in Chapter 1 the importance of considering ergonomics in the design of spacecraft systems. Then, on the basis of large amount of literature, the results of the research on the effect of microgravity environment on human physiology, cognition and accessibility are introduced in Chapters 2, 3, and 4 respectively. In Chapter 5, the physiology of the astronaut is summarized, and some suggestions on the design of the spacecraft and the space mission are put forward.

2 Effects of microgravity on human physiology

2.1 Influence of microgravity on human body

Long-term weightlessness will affect the size and posture of the human body. As the astronauts have been living under gravity on the ground, their body will experience a downward force. This force has always existed on the earth, and people's height generally does not change in a short time. However, under the condition of microgravity, when the body's downward gravity is lost, the spine will be lengthened, the upper body's size will also increase, and the overall height will increase by about 3%.

The neutral posture under microgravity will change, with the plantar flexion of the ankle (bending towards the sole of the foot), the flexion of the hip and knee joints, slight hip abduction, slight flexion of the trunk forward, and movement of the head and limbs towards the midpoint of the range of motion, shoulders, arms, and elbows lift up and the neck bends forward, as shown in Figure 1 [3-6].

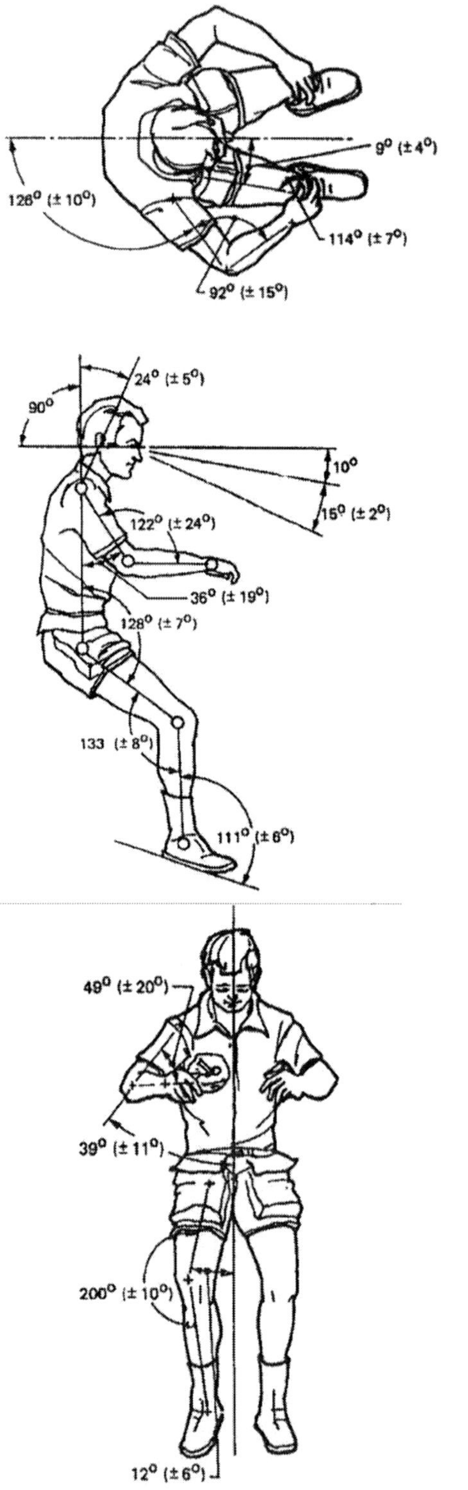

Figure 1. The neutral posture under microgravity

2.2 The effect of microgravity on skeletal muscle

In the microgravity environment, skeletal muscles suffer from problems such as atrophy and bone loss, resulting in decreased muscle strength. The entire center of mass of the human body changes, as well as the ability to exercise. These changes will have a great impact on the accomplishment of the astronauts' tasks in the space station, especially for work that requires strength. Biomedical research shows that the microgravity environment can lead to a decrease in the number of muscle proteins and can easily lead to functional changes such as muscle fatigue [7]. At the same time, the muscle weight loss of the lower extremities will significantly exceed the muscle weight loss of the upper extremities. Animal experiments have shown that the rat's hind-limb soleus muscle has a 23 to 40% loss in weight, whereas the forelimb muscle's weight loss is only 12 to 24% [8].

2.3 Bone and Mineral Metabolism

When people are in a microgravity environment for a long time, their bones will grow due to the lack of weight, and their height will increase by about 5cm. Loss of weight will lead to lower bone density in the lower body, and the skull bone density will increase [9]. This phenomenon is related to the tendency that body fluid concentrates on the head. Because of the change of bone metabolism, a large amount of calcium will be lost [10]. According to the data of Skylab, on-track astronauts lose 0.6% of the total calcium per month on average [11, 12].

3 The influence of microgravity on cognition

Cognitive ability is controlled by the human brain and central nervous system, including human emotions, memory functions, language, thinking and other psychological states. The microgravity environment can lead to a decline in the ability of movement, reaction, and vision which closely related to mission operations [13]. For example, the response time for sound, vision, and selection under microgravity will increase by 100ms, 120ms, and 1000ms, respectively. This change can lead to problems such as spatial disorientation, posture illusion and ataxia, as well as lengthening the task completion time [14].

3.1. Spatial orientation and object recognition

Under the microgravity condition, the downward gravitational sensation has disappeared, and thus the vestibular system that controls the body's adaptation to the earth's gravitational sensations has also changed. The most important directional sensory vestibular system is the most affected. The vestibular system under microgravity can cause eye movement disorders, and decreased ability of posture control and coordinate movement. The absence of vertical gravity and downward inertial forces will cause the misjudgement on own body position and movement illusions. The human body loses its long-term dependent gravity-referenced coordinate system under microgravity. Therefore, it will lead to wrong judgment of the object's distance [15].

3.2 Motor perception

The astronaut's cognitive mode of moving objects under microgravity will change. On the Earth, because of a force toward the center of the earth, people are used to using this downward force as the reference for motion. Under microgravity, due to the loss of the habitual coordinate system, the astronaut's vestibular system and somatosensory sensation will make mistakes when judging moving objects [16]. During the first 2-3 days of a spaceflight, 60-80% of the astronauts will be affected by space motion sickness, which is a typical motion sickness produced during adaptation of the vestibular system to the microgravity environment [17]. At the same time, due to changes in the angle of view, the ability to determine the movement of objects will be biased. The astronauts usually establish reference frames in the space station with self-centeredness, which may lead to obstacles in motor perception [18, 19].

3.3 Coordinated movement

Coordinated movement relies on information input from systems such as vestibular systems, somatic sensation, and vision which may work together to complete the coordinated movement of the body. When the astronauts are in a microgravity environment for a long time, these coordinated motion information control systems will undergo significant changes. Since there is no downward gravity, the control of the posture in the space station can be difficult which requires the external forces to control [20]. The correlation between different parts of the movement exercise ability and gravity are different. Among them, the movement is the biggest changed part. The astronauts are walking with their feet on the ground, while floating laterally under microgravity. The adjustment of posture is even more difficult than on the ground and requires handrails. The overall coordination ability of people will decline, and the speed of action will be affected [21].

4 The effect of microgravity on accessibility to operation

Microgravity can affect the visual reachability of operational accessibility. During the operation of the mission, the astronaut's line of sight must stay on the operation. If the perspective is affected, the task will be delayed and the corresponding action will be slow. Under the condition of microgravity, the human body is completely relaxed.

Under the condition of 1G, the best viewing zone of the upper and lower viewing zones is located at 15°-45° below the horizontal line of sight, and the effective viewing zone is located at 0°-75° below the horizontal line of sight, as shown in FIG. 2 . Under the microgravity condition, the layout design should be based on the change of the main line of sight falling by 15° in the neutral posture state, as shown in FIG. 3. Regardless of 1G conditions or under weightless conditions, the best view zone for the left and right viewing zones is the left and right 15° viewing zone, and the effective viewing zone is the left and right 95° viewing zone.

The perspective of the operation determines whether the action can be carried out normally, because the astronaut needs to focus when conducting the action and stay focused [22]. As microgravity changes the astronaut's normal physiology perspective, and visibility is an important factor in the human-computer interaction process, changes in the perspective of the activity will cause difficulties for the astronauts and increase the task completion time and reduce work efficiency.

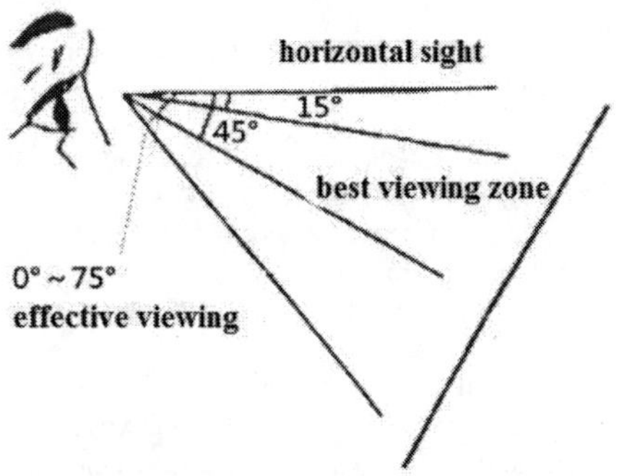

Figure 2. The upper and lower view in 1G condition

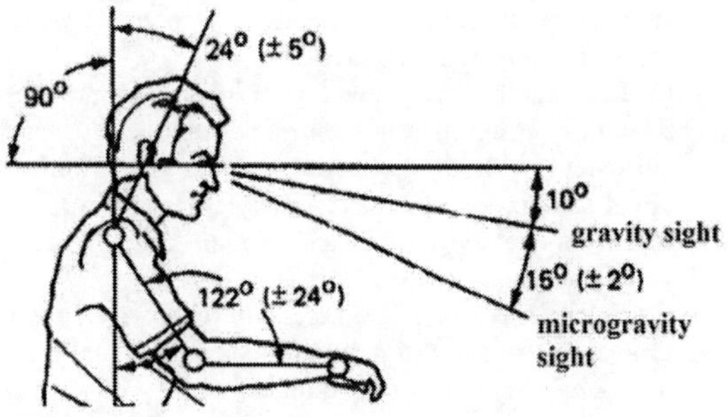

Figure 3. Changes in the upper and lower viewing zones in a microgravity environment

5. Conclusion

This article mainly summarizes the influence of microgravity environment on the astronaut's physiology, cognition and accessibility. Astronauts are the main factor of space activities, and changes in astronauts' physiology affects the completion of space missions. For example, the changes of cognition will affect the astronauts' decisions and prolong the reaction time, thereby increasing the difficulty of the task and the duration of the task. The influence of visibility puts forward new design requirements for the placement of experimental instruments and operating interfaces.

5.1 Influence of Microgravity on Human Physiology

The limitations of astronauts under microgravity are mainly reflected in the following aspects:
(1) The decline in muscle power;
(2) People's various types of information processing mechanisms, such as tactile, pain, visual, etc., receive slower responses;
(3) Decreased spatial orientation of the human body;
(4) The person's movement speed slows down and depends on assistance;
(5) Changes in the human field of vision and optimal operating space;
(6) The overall judgment ability is reduced.

5.2 Ergonomics Requirements in Microgravity Environment

Based on the physiological changes of humans in the microgravity environment analyzed in the previous article, the space environment poses the following requirements for the ergonomic design of spacecraft:
(1) Minimize work that requires long-term execution and greater strength;
(2) Minimize work that requires long-term attention and work that cannot be suspended halfway;
(3) Increase the positioning and orientation signs in the spacecraft to help the astronauts carry out spatial orientation;
(4) Set foot limiters, hip limiters and handrails with suitable position, size, and quantity to assist astronaut activities and operating facilities;
(5) Reasonably arrange the layout of each component and laboratory equipment in the design process in order to be consistent with the astronauts' vision and comfortable operation space in the microgravity environment;
(6) Minimize the occurrence of multitasking in a short period of time;
(7) Set up a reasonable emergency plan for emergencies to avoid astronauts to judge by themself, so as to reduce the probability of error occurrence.

5.3 Conclusion

This paper analyzes the influence of the microgravity environment on the astronauts' physiology and operational activities, and obtains the differences from the tasks performed on the ground in the 1G environment. It provides suggestions and references for ergonomics design of spacecraft and space station in microgravity environment.

References

1. M. Oshima, Ergonomics, *Iyo Denshi to Seitai Kogaku Japanese Journal of Medical Electronics & Biological Engineering*, 10.6:556 (1972)
2. A. R. Brody, Space operations and the human factor, *Aerosp Am*, 31.10:18-21(1993)
3. V. V. Bogomolov, A.I. Grigoriev, I.B. Kozlovskaya, The Russian experience in medical care and health maintenance of the International Space Station crews, *Acta Astronautica*, 60(4–7):237-246 (2007)
4. S. Glasauer, H. Mittelstaedt, Perception of spatial orientation in microgravity, *Brain Research Reviews*, 28(1–2):185-193 (1998)
5. G. Leone, The effect of gravity on human recognition of disoriented objects, *Brain Research Brain Research Reviews*, 28(1–2):203-214 (1998)
6. P. P. Di, M. V. Narici, Muscles in microgravity: from fibres to human motion, *Journal of Biomechanics*, 36(3), 403-412, (2003)
7. G.B. Prange, L.A. Kallenberg, M.J. Jannink, et al, Influence of gravity compensation on muscle activity during reach and retrieval in healthy elderly, *J Electromyogr Kinesiol*, 19(2):e40-e49 (2009)
8. A. Ishihara, F. Nagatomo, M. Terada, et al, Effects of microgravity on the mouse triceps brachii muscle, *Muscle & Nerve*, 52(1):63-68, (2015).
9. P. Ghosh, J. N. Stabley, B.J. Behnke, et al, Effects of spaceflight on the murine mandible: Possible factors mediating skeletal changes in non-weight bearing bones of the head, *Bone*, 83:156-161, (2016).
10. S. M. Smith, M. Heer, L. C. Shackelford , et al, Bone metabolism and renal stone risk during international space station missions, *Bone*, 81:712-720, (2015).
11. M.F. Reschke, J.J. Bloomberg, D.L.Harm D L, et al, Posture, locomotion, spatial orientation, and motion sickness as a function of space flight, *Brain Research Brain Research Reviews*, 28(1–2):102-117 (1998)
12. J. Wei, The hierarchical characteristics of physiological effects of weightlessness, *Aerospace Medicine and Medical Engineering*, (2):134-139, (1994), (in chinese)
13. K. Koga, Gravity cue has implicit effects on human behaviour, *Aviat Space Environ Med*, 71(9):78-86. (2000)
14. J. Mcintyre, A. Berthoz, F. Lacquaniti, Reference frames and internal models for visuo-manual coordination: what can we learn from microgravity experiments, *Brain Research Brain Research Reviews*, 28(1-2):143 (1998)
15. T. Pozzo, C. Papaxanthis, P. Stapley, A Berthoz, The sensorimotor and cognitive integration of gravity, *Brain Research Reviews*, 28(1–2):92-101 (1998)
16. G.D. Gironimo, C.D. Martino, A. Lanzotti, et al. Improving MTM-UAS to predetermine automotive maintenance times, *International Journal on Interactive Design & Manufacturing*, 6(4):265-273 (2012)
17. W. Chen, J. G. Chao, J. K. Wang, et al., Subjective Vertical Conflict Theory and Space Motion Sickness, *Aerospace Medicine & Human Performance*, 87(2), (2016).
18. J.R. Lackner, P. Dizio, Motor function in microgravity: movement in weightlessness, *Current Opinion in Neurobiology*, 6(6):744-750 (1996)
19. D. Manzey, B. Lorenz, Mental performance during short-term and long-term spaceflight, *Brain Research Reviews*, 28(1–2):215-221 (1998)

20. B. Yang, T. Hu, Research and analysis of space station microgravity environment, *Manned Space Flight*, (2):178-183 (2014), (in chinese)
21. B. Fowler, D. Manzey , Summary of research issues in monitoring of mental and perceptual-motor performance and stress in space, *Aviation Space & Environmental Medicine*, 71(9 Suppl):76-7.a (2000)
22. G. Clément, Alteration of eye movements and motion perception in microgravity, *Brain Research Reviews*, 28(1–2):161-172 (1998)

Increase of Critical Mach Number by Local Linearization Method and Airfoil Construction at Subsonic and Transonic Regimes

Ali. S. Bazgir[1], Sergey. A. Takovitskii [2]

[1]*Moscow Institute of Physics and Technology, Department of Aeromechanics and Flight Engineering140180, Gagarina Street, 16, Zhukovsky, Russia.*
[2]*Central Aerohydrodynamic Institute (TsAGI), Zhukovsky Str., Zhukovsky, 140180, Moscow Region, Russia.*

Abstract. A symmetrical airfoil has been constructed by local linearization method. A single-point objective function is defined to check the convergence of the method. As an example, the nose and tail zone of supercritical airfoil is fixed and a flat line is placed between them. The optimizable element of the airfoil contour was conjoined with the nose and tail elements of fixed shape at the sections with coordinates x_{s1}= 0.11 and x_{s2}= 0.66, respectively. The optimizable part of airfoil (the fixed chord line) is divided into N=55 segments. The convergence of this method has been shown with the airfoil constructed with higher critical Mach number rather than the initial airfoil. Finally, this airfoil has been compared with the supercritical airfoil NASA SC (2)-0012 at $M_\infty = 0.76$. At the second part, several airfoils have been constructed and simulated over different Subsonic and Transonic Mach numbers. Finally, the drag coefficient on constructed airfoils have been compared with supercritical airfoil.

1 Introduction

Unlike supersonic flows in the range of subsonic velocities, there are no local models connecting the pressure on the body surface with the local geometrical parameters and the conditions of the incoming flow. However, successful examples of applying local linearization in small shape variations based on connecting the pressure and contour curvature [1, 2] are known. The variation method, which uses the relationship between variations of pressure and an external normal vector of the optimizable element, was implemented in Ref. [3].

For small deformation of the shape the local pressure coefficient change was estimated by the relation with the element length and the variation of the second derivative of the ordinate y" at subsonic Mach numbers [4]. The same can be estimated for supersonic Mach numbers by the relation of the local pressure coefficient with the element length and the variation of the first derivative of the ordinate y'[5].

2 Local linearization method

Use Maximum Critical Mach number, M_{cr}, achieves when pressure coefficient on the optimizable part of the airfoil has a constant value ($C_p = constant$). In this case, we have to put M=1 on the optimizable part of the airfoil and get a desired value of pressure coefficient ($C_{p_{desired}}$) by:

$$C_p = \frac{2\left(\frac{P}{P_\infty}-1\right)}{\gamma M_\infty^2} \quad (1)$$

The optimizable part of the airfoil is divided to several segments and the Mach number is averaged on these elements based on the data of the numerical flow field modeling:

$$M_n = \left[\frac{2}{\gamma-1} \times \left(\frac{(1+\frac{\gamma-1}{2}\times M_\infty^2)}{(\frac{\gamma}{2}\times M_\infty^2 \times C_{p_n}+1)^{\left(1-\frac{1}{\gamma}\right)}} - 1\right)\right]^{0.5} \quad (2)$$

The second order derivative is determined by finite difference approximation through the ordinate values of the three nodal points. A similar approach was implemented for supersonic flows when the linear dependency of the change in pressure on the inclination angle of the element, which is specified by the ordinates of two adjacent points, was used [5]:

$$d_{y_n} = \frac{\pi}{8} \times (x_n - x_{n-2}) \times |1 - M_{n-1}^2|^{0.5} \times (Cp_{desired} - Cp_{n-1}) + 2 \times d_{y_{n-1}} - d_{y_{n-2}} \quad (3)$$

$$d_{y_n} = \frac{(x_n - x_{n-2})}{2} \times |1 - M_{n-1}^2|^{0.5} \times (Cp_{desired} - Cp_{n-1}) + dy_{n-1} \text{ for } M > 1 \quad (4)$$

We will use these formulas to connect the changes in the gas-dynamic functions and geometrical parameters in the local analysis. To get new vertical position of each nodal, the following formula is used:

$$y_{new_n} = y_n + d_{y_n} k \quad (5)$$

Where y_n and y_{new_n} are the initial airfoil vertical coordinates and new vertical coordinates after applying the linearization of flow parameters d_{y_n} at each local segment.

k is an additional constant value which can be chosen different numbers during the optimization process.

The proposed linear dependence of the surface pressure on the geometrical parameters leads to the quadratic approximation of the objective function and determines the shape variations, which are directed to its reduction:

$$OF = \left[\frac{\sum_i^n (Cp_i - Cp)^2 l_i}{\sum l_i}\right]^{0.5} \quad (6)$$

where l is the length of the contour element, and the summation is done along all elements of the convex part of the contour. Cp is desired value of the mean pressure coefficient.

The described model gives a sufficiently rough evaluation of the aerodynamic load change. However, at every iterative step the proposed shape variations are checked within the framework of the numerical computation. In addition, as long as the optimization procedure converges, the variations will decrease and the accuracy of the approximation will increase. In the presented investigation, numerical simulation of the flow over constructed airfoil is performed based on solving Naiver-Stokes equations.

2.1. Critical Mach number increase

The nose and tail zone of supercritical airfoil NASA SC (2)-0012 is fixed and a flat line is placed between them. The optimizable part of the airfoil contour was conjoined with the nose and tail elements of fixed shape at the sections with coordinates $x_{s1} = 0.11$ and $x_{s2} = 0.66$, respectively. The optimizable part of airfoil (the fixed chord line) is divided into N=55 segments. The nose element of the contour was specified by N=13 segments and the tail element was specified by N=35 segments. At Mach condition of $M_\infty = 0.76$, Reynolds number based on reference wind speed and airfoil length is approximately 1.1×10^6. The calculated value of desired pressure coefficient is $C_{p_{desired}} = -0.5576$.

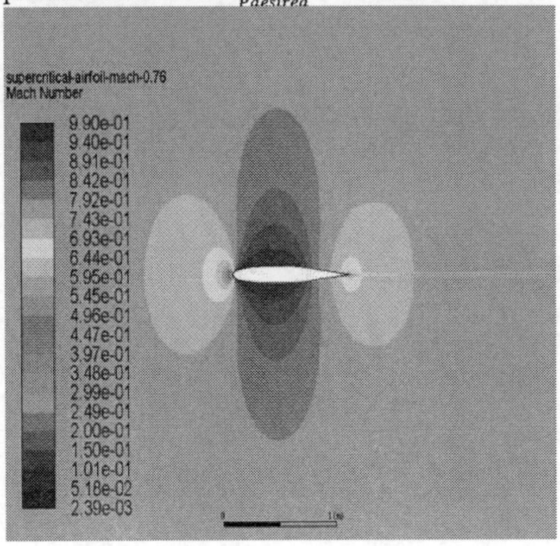

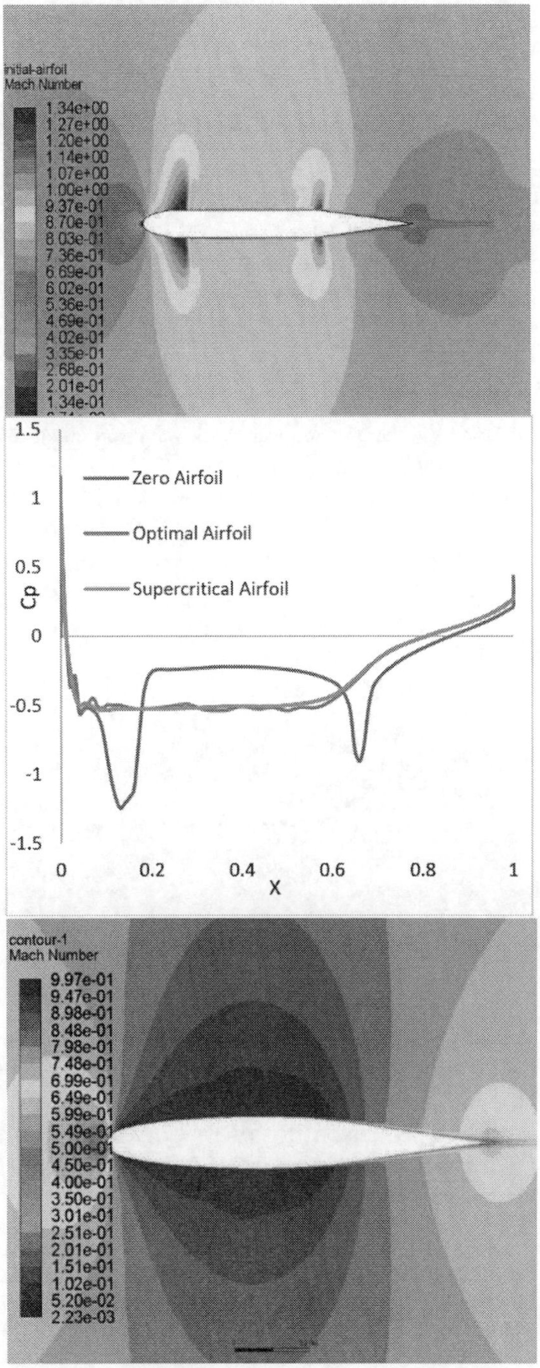

Figure 1. Comparison of the results of initial and optimized airfoil by local linearization.

A The presented figures (fig. 1) compare the pressure coefficient distribution on the optimal airfoil from local linearization, initial airfoil and supercritical airfoil. There are two huge pressure drop in the fixed edges of the initial airfoil where the minimum pressure coefficient reaches to approximately -1.24. It is because the local velocity of the flow increases hugely and the maximum local Mach number on the airfoil's surface reaches to 1.34 in these zones. However, the final optimal airfoil by local linearization method has approximately smooth distribution like the supercritical airfoil. While the shock waves and pressure jumps disappear on the airfoil surface, the overall drag on the airfoil reduces and so Critical Mach number increases.

The optimal airfoil from local linearization method achieves after five cycle of optimization. The deviation of the pressure coefficient value from the desired constant pressure coefficient on the optimizable part of the airfoil is minimum. Maximum local Mach number on the optimal airfoil surfaces reaches approximately to M=0.997. In fact, due to the constant pressure distribution on the optimizable part of the airfoil, the value of the Critical Mach number is delayed and there is no local shock waves on the airfoil surface at $M_\infty = 0.76$. Results from Fig.2 demonstrates the accuracy of this method in which it shows the convergence of the objective function after few cycles of optimization.

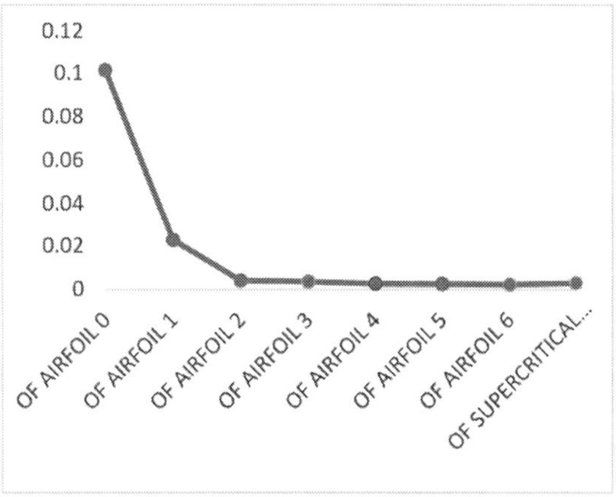

Figure 2. Minimization of objective value during the optimization process.

3 Drag force mitigation for the Mach number regimes more than critical Mach number by several methods

The tail and nose of the supercritical airfoil fixed in two points and for the central part of the airfoil (fig. 3), three methods are used to create new airfoil. The first method is to use local linearization method for construction of the central part. The second method is to use an elliptic function to create an elliptical line and then fixe the elliptic line with the nose and tail of the airfoil. Third method is consisted from two inclined flat segments at the central part.

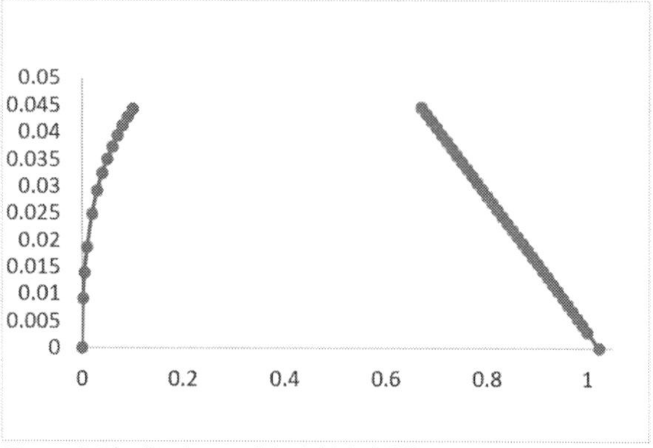

Figure 3. The fixed part of the initial airfoil.

3.1. Local linearization method

The Optimal airfoil from local linearization method of the previous part has been chosen. This airfoil, which was optimal airfoil at $M_\infty = 0.76$, has been simulated over different Mach numbers.

3.2. Elliptical airfoil

P To obtain the airfoil whole shape, the fixed flat line is replaced by Elliptic shape over different x_c(corresponding longitude of the maximum ordinate) and constant value of $y_c = 0.06$ (maximum ordinate). The used elliptical equation is as:

$$y = \sqrt{1 - \left(\frac{x - x_c}{a}\right)^2} \times b + y_c \quad (7)$$

Fig.4 indicates the drag coefficient at different constructed elliptical airfoils in which the middle of the airfoils is constructed by elliptic equations. The airfoil was fixed in zero angle of attack under farfield Mach number of M_∞ =0.84. As it is indicated from the results, the airfoil with elliptic central line of x_m=0.425 has the minimum value of drag force coefficient. As a result, this airfoil is the optimal airfoil of elliptical airfoil. This airfoil is simulated over different Mach numbers $M_\infty = 0.7, 0.72, 0.74, 0.76, 0.78, 0.8, 0.82, 0.84, 0.86, 0.88$ *and* 0.9

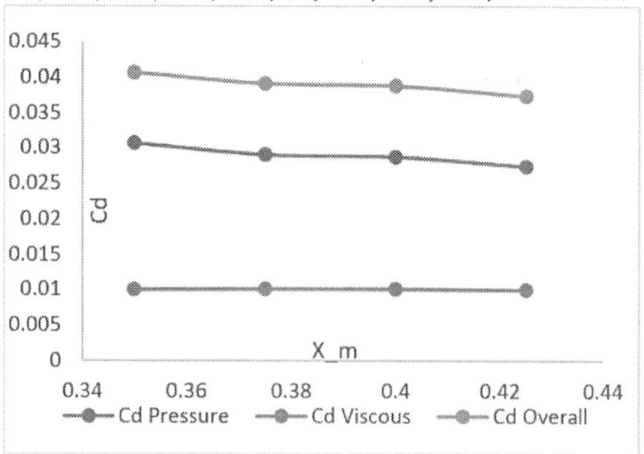

Figure 4. Drag coefficient on the constructed elliptical airfoils.

3.3. Airfoil Consisted from Two Inclined Lines

This airfoil is a combined airfoil from supercritical airfoil and classic diamond shape of supersonic airfoils. The nose and tail part of the supercritical airfoil is fixed and two flat inclined lines are fixed to the maximum ordinate of the airfoil.

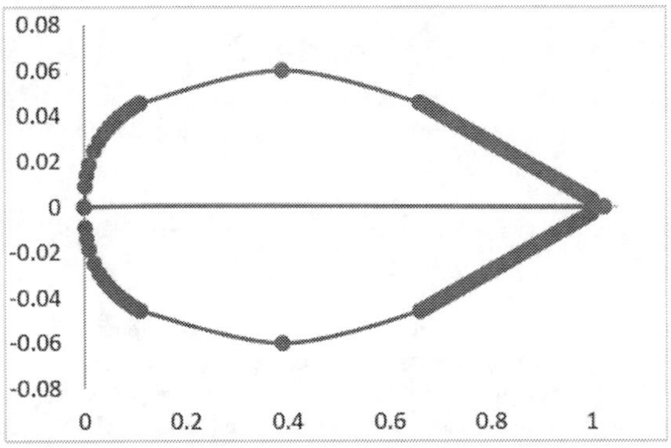

Figure 5. Geometry of the constructed airfoil.

3.4 Comparison and Results

Fig.6 indicates graphs of overall drag coefficient over different Mach numbers for the constructed airfoils. From the overall drag coefficient, it can be observed that the supercritical airfoil, airfoil consisted from Elliptical equation and constructed airfoil from local linearization have the same drag at low subsonic regimes but the supercritical one has a bit lower drag coefficient at high subsonic flows. However, the lowest drag coefficient at high subsonic flows belongs to the airfoil consisted from two inclined flat segments. For example, at $M_\infty = 0.86$ the relative reduction of drag exceeds 20% and 25% rather than supercritical and optimal airfoil from local linearization airfoils, respectively. This percentage a bit reduces at the higher Mach numbers.

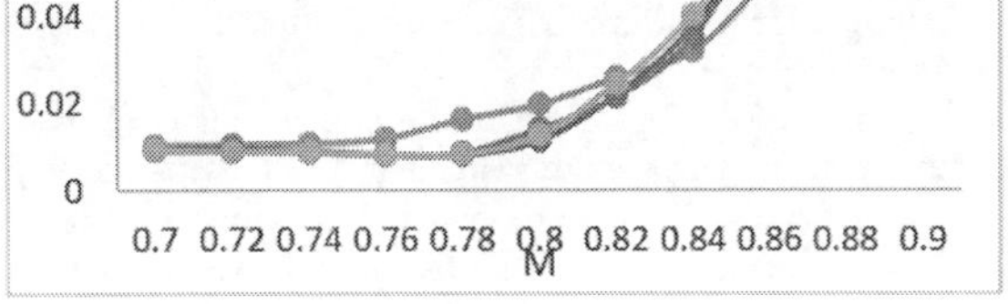

Figure 6. Drag coefficient at different Mach numbers over the constructed airfoils.

4 Conclusion

The first part of this research was aimed to increase M_{Cr}. The results from an airfoil construction example shows that Local linearization provides fast convergence because of the fact that it handles many parameters. In fact, it is shown the possibility to construct the central part of the supercritical airfoil on condition of flat pressure distribution.

The second part of this research was about Drag Mitigation of $Mach\ numbers > M_{Cr}$. The constructed airfoils were:

A) Local Linearization B) Supercritical Airfoil
C) Elliptical Airfoil D) Airfoil Consisted from Inclined Segments

The supercritical airfoil, Elliptical airfoil and optimal airfoil from local linearization had better performance at low Subsonic but Airfoil Consisted from Inclined Segments had the minimum Drag at high subsonic and transonic speeds. For example, at M=0.86 the relative reduction of drag exceeds 20% and 25% rather than Supercritical and optimal airfoil of local linearization method, respectively.

References

1. R.L. Barger. and C.W. Brooks, "A streamline curvature method for design of supercritical and subcritical airfoils", NASA TN D-7770, (1974).
2. W.H. Davis, "Technique for developing design tools from the analysis methods of computational aerodynamics", *AIAA Paper 79-1529*, (1979).
3. M.A. Brutyan and S.V. Lyapunov, "Optimization of shape of symmetrical flat bodies with the purpose of increase of the critical Mach number", *Uch. Zap. TsAGI*, 12(5):10–22, (1981).
4. Sergey Takovitskii, "Aerodynamic shapes optimization on the base of method of local linearization", 7th european conference for aeronautics and space sciences (eucass), eucass 2017-95.
5. S. A. Takovitskii, "Analytical Solution in the Problem of Constructing an Airfoil with Minimum Wave Drag", Fluid Dynamics, (2003).

Optimal Aerodynamic Shape Optimization of a Paraglider Airfoil Based on the Sharknose Concept

Amir A. Abdelqodus[1,a] and Innokentiy A. Kursakov[1,b]

[1]*Moscow Institute of Physics and Technology, 140180, Zhukovsky, Russia*

Abstract. The effects of classical air inlet configurations on the aerodynamic performance of a paraglider airfoil are firstly presented. Followed by conducting gradient-based aerodynamic shape optimization of the baseline airfoil used in the investigation of the classical air inlets. Different air inlet configurations are introduced to the optimized profile, and there effects on the aerodynamic performance are then compared to the initial and optimized airfoils. The Reynolds averaged Navier-Stokes equations (RANS) are solved for the flow field around the closed and open airfoils. The canopy is assumed to be smooth, rigid and impermeable. Results are focused on both of lift and drag coefficients for performance analysis and on the internal pressure coefficient which can be critical for a real flexible wing regarding the risk of collapse.

1 Introduction

Maintaining a good internal pressure at a high angle of attack offers, among other things, longer brake travel and increased stall resistance at lower speeds. Therefore one needs to place the air intake at or close to the stagnation point in order to get higher internal pressure than the external pressure at every point on the airfoil surface. Unfortunately the stagnation point is not fixed, but it varies with the angle of attack.

One way to fix this, is to design the *air inlet* or simply *inlet*, with the novel SharkNose shape, which was developed and patented by the company "Ozone" [1]. It adds a concave part in the usual range of the stagnation point. This concave part will greatly reduce the range size of the stagnation point, and shall be considered as an area where the airflow slows down. It produces the opposite effect to a venturi by presenting a larger area where the air will flow slower, knowing that as the airflow slows down in a given area, the closer to unity the pressure coefficient, C_p, of this area will be. The symmetrical shape, added to a rounded profile enables satisfactory behavior at both low and high angles of attack without added drag. Moreover, with the stagnation being less mobile, it is possible to reduce the size of the intake and therefore to obtain an even more pressure in front of it.

The airfoil presented in this paper is a custom paraglider airfoil, which has a chord length of 3m and flies at a cruise speed about 10m/s. The angle of attack varies from relatively low to high values, representing accelerated and normal flights, respectively. The main objectives are to quantify the influence of different classical inlets on the aerodynamic performance, and find the optimal aerodynamic design shape of the inlet to reduce the decrease in performance due to the classical inlets.

2 Numerical methodology

The open-source SU2 code [2, 3] is used to perform the flow analysis, and gradient-based optimal shape design for the paraglider airfoils. The 2-D simulation of the cross section of the airflow around a paraglider airfoil was obtained by solving the steady-state RANS over a C-topology flow domain. Domain boundaries for this case are placed 150 chord lengths from the airfoil surface to minimize the effect of the far-field boundaries on the solution. The flow is considered incompressible with constant properties. For the free stream Reynolds Number of 2.086×10^6, the turbulent Reynolds stresses were taken into account by choosing the Spalart-Allmaras (S-A) [4] turbulence model. Mesh spacing at the airfoil boundary 2×10^{-6} to ensure $y+ < 1$ everywhere over the airfoil surface. Due to practical constraints, the interior of the paraglider is solved using the same turbulence model, regardless of the low Reynolds number in this area for which this model may not be well adapted.

The closed and paraglider airfoil geometries were imported into the mesh generator software Pointwise®. Over the closed airfoils, two-dimensional discretized volumes consisting of quadrilateral elements conforming to the airfoil surface are generated. For the open airfoils, the discretized volume consists of hybrid meshes conforming to the airfoil surface. Structured meshes are used to analyze the boundary layer in the proximity of the canopy and the outer domain. While unstructured meshes are created for the interior domain of the open airfoils with mostly quadrilateral cells. The interior mesh

[a] Corresponding Author: a.amir@phystech.edu [b] i.kursakov@phystech.edu

characteristics are adapted to the node distribution of the external airfoil surface's mesh.

Roe's second-order upwind scheme is used to calculate the convective fluxes. Venkatakrishnan's limiter is applied to the primitive variables. Turbulent variables for the S-A model are convected using a second-order scalar upwind method, and the viscous fluxes are calculated using the corrected average-gradient method. Implicit, local time-stepping is used to converge the problem to the steady-state solution, and the linear system is solved using the GMRES method [5] with a maximum error tolerance of $O(10^{-6})$ for each nonlinear iteration of the flow solver. Characteristic-based far-field boundary conditions are applied to the outer domain boundaries, and an adiabatic wall boundary condition is enforced on the airfoil surface.

3 Classical inlet configurations

Two inlet configurations will be presented in this section. The first inlet is created using a cut in the pressure side of the airfoil. The second inlet is an offset inlet applied to the baseline airfoil with the same width as the first inlet. Both inlet shapes and sizes are used in industry.

A near view of the leading edge of the baseline airfoil with the plain cut, or inlet 1 for short, and with offset inlet, or inlet 2 are shown in Fig. 1a and 1b, respectively, with cut dimensions of 3%c width each. A view of the mesh around the whole closed and open airfoils with inlet 2 are shown in Fig. 2a and 2b, respectively.

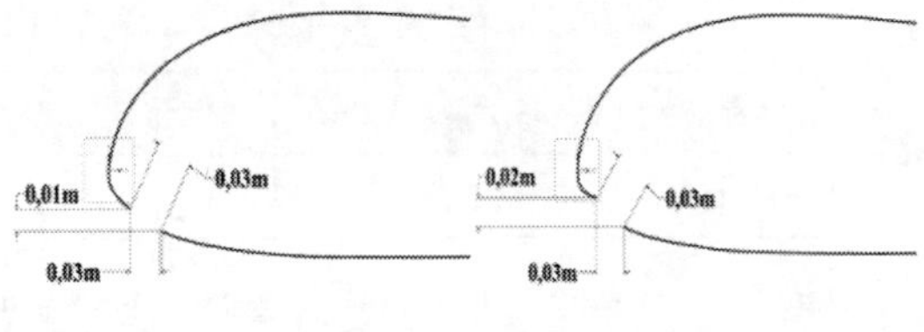

(a) (b)

Figure 1. Baseline airfoil with inlet (a) 1 and (b) 2 configurations.

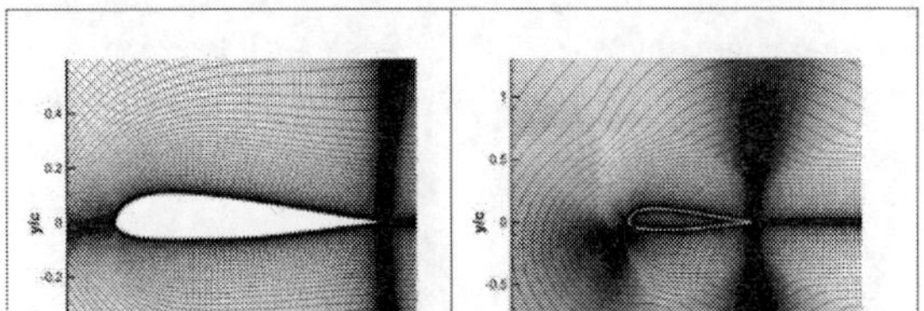

(a) (b)

Figure 2. Baseline (a) closed airfoil and (b) open with inlet 2 (airfoil in white).

The effects of the two inlet configurations on the aerodynamic performance compared to the closed airfoil are displayed in Fig. 3.

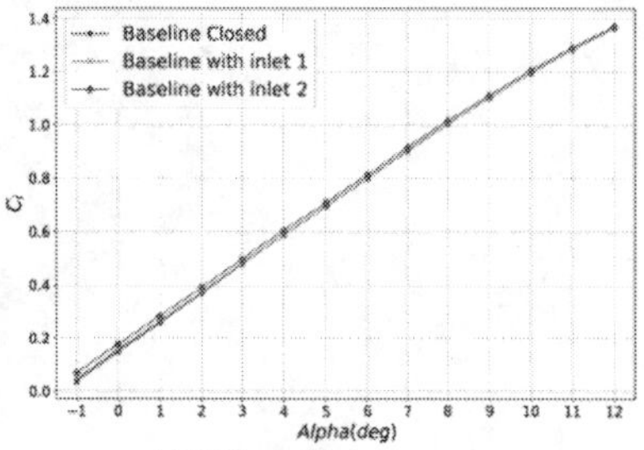

(a) Lift coefficient curves

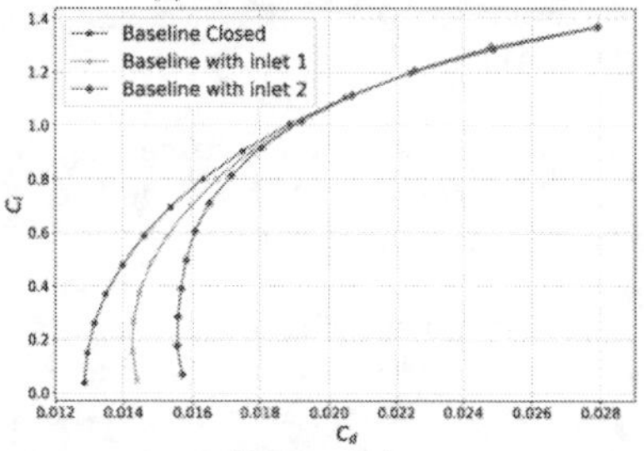

(b) Drag polar

Figure 3. The effects of the two classical inlets on the aerodynamic performance.

The increase in drag moving on from the closed, to inlet 1 then further increase with inlet 2 is due to increasing the size of the separation bubble near the opening on the lower edge for both inlets, increasing the thickness of the boundary layer than for the baseline airfoil. Additionally, a second separation bubble appears on the upper edge near the opening of inlet 2 at low angles of attack. Whereas at higher angles of attack, this effect becomes less prominent, where the curves converge with increasing angles of attack. The flow has a better orientation with the inlet, thus, the air flows directly through it, and the additional separation bubble on the upper edge of inlet 2 vanishes. This is illustrated through the streamlines for inlet 1 and 2 in Fig. 4 and 5, respectively, showing also the movement of the stagnation point position for both inlets, at angles of attack of 3° and 9°, representing the accelerated and normal flights, respectively.

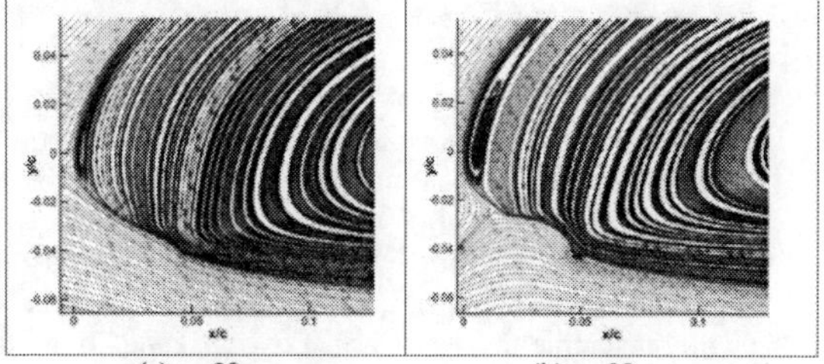

(a) α=3° (b) α=9°

Figure 4. Streamlines inside and around airfoil with inlet 1 (airfoil edge in red).

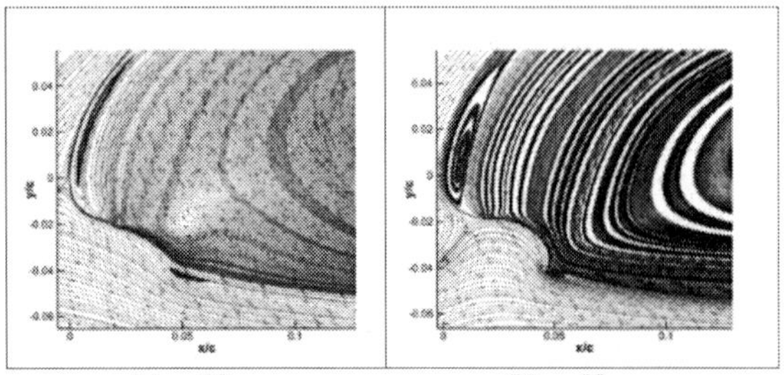
(a) α=3° (b) α=9°

Figure 5. Streamlines inside and around airfoil with inlet 2 (airfoil edge in red).

The internal pressure of the baseline airfoil due to inlet 1 and 2 are shown in Fig. 6. The loss in internal pressure with inlet 1 at low angles of attack, is due to positioning the inlet downstream, increasing its distance from the stagnation point (Fig. 4a), with no offset in the lower edge. As a consequence, when the flow reaches the air inlet, it has a larger distance to accelerate and has become tangent to the canopy. Resulting in a reduction in the deflection of the fluid penetrating the canopy.

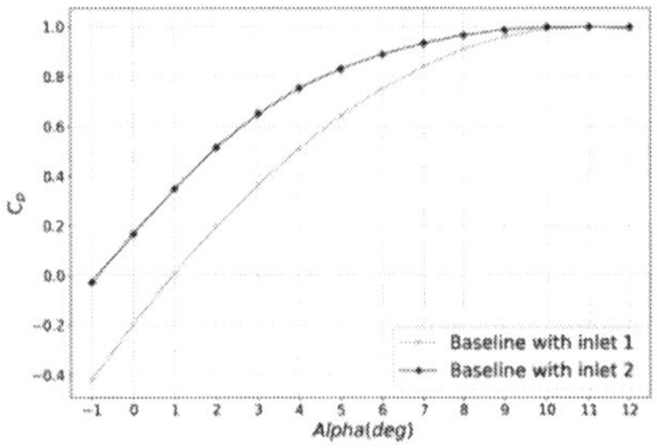

Figure 6. Internal pressure of the baseline airfoil due to classical inlets 1 and 2.

4 Airfoil optimization

The methodology and results for a constrained optimal shape design of a two-dimensional airfoil at viscous turbulent flow are presented here. The optimization goal is to increase the surface static pressure on the inlet position of a modified baseline airfoil, over the atmospheric pressure at 0° angle of attack, to ensure that the internal pressure of the paraglider will always satisfy the stability condition of the paraglider over all the range of flight. The maximum thickness is constrained to be greater than or equal to the initial airfoil maximum thickness which is equal to 0.164c.

Firstly, the closed baseline airfoil geometry is modified using the open-source FreeCAD package to roughly resemble the desired SharkNose shape, to aid the optimizer in reaching the optimal required shape of the inlet area. Figure 7a shows the modification which is applied on the lower surface near the leading of the baseline geometry. The resulting surface pressure coefficient compared with the baseline Spirit airfoil is shown in Fig. 7b at 0° angle of attack, achieving a surface pressure higher than the atmospheric pressure at a low angle of attack.

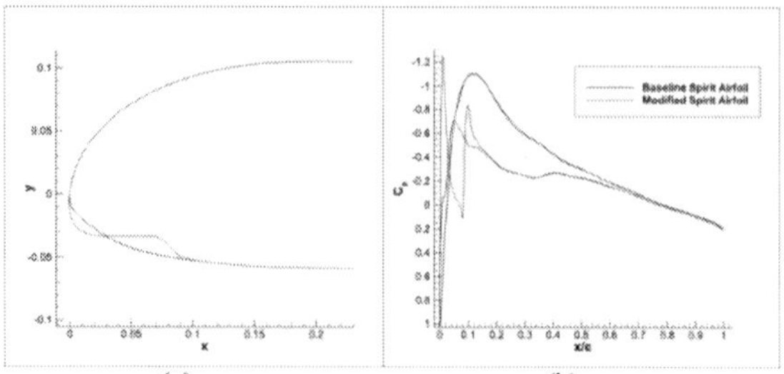

(a) (b)

Figure 7. Modified vs Baseline airfoils.

The 2D Free-Form Deformation (FFD) [6] control point is employed as the parameterization method and the FFD box has a rectangular shape and is defined by four boundaries. which are listed in Table 1. In order to properly capture the curvature of the inlet area, a value of 0.01c is chosen for the uniform spacing of the control points on the surface of the FFD box. Accordingly, the FFD is created with control point dimensions of 102 by 1 in the x-, and y-directions, respectively. Only the lower surface of the airfoil is allowed to deform, and only the control points surrounding the inlet area on the lower surface of FFD box are set as design variables, and the rest are held fixed. Hence, the number of design variable are 9. The lower and upper bounds for each design variable is set to 0.05c. To ensure that the deformed geometry does not contain any jumps or kinks. The optimization tolerance is set to $1\mathrm{x}10^{-6}$ for the Karush-Kuhn-Tucker (KKT) condition [7, 8].

Table 1. FFD box definition.

	B_{upper}	B_{lower}	B_{left}	B_{right}
FFD control point	0.2	-0.2	-0.01	1.01

Finally, a redesign of the airfoil is performed using the gradient information obtained from the adjoint formulation [2, 9]. Figure 8 displays the final deformed airfoil shape. Upon completion, the average surface pressure on the lower side was successfully increased from -10.19 to -9.88, which is a 3% increase, and the maximum thickness of the final airfoil design met the constraint. The convergence histories of the objective function are plotted in Fig. 9.

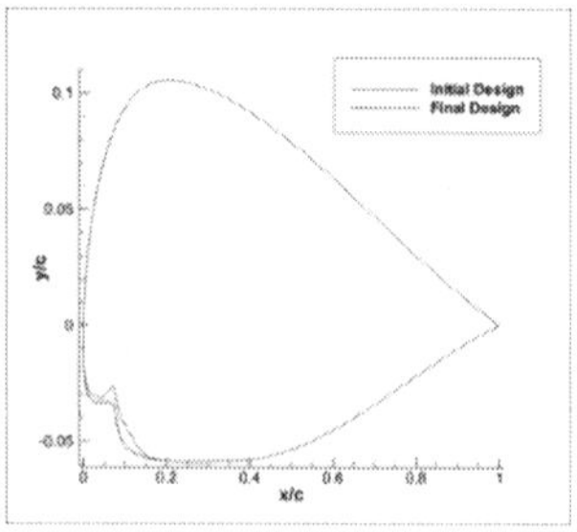

Figure 8. Deformation comparison for the initial and optimized airfoil.

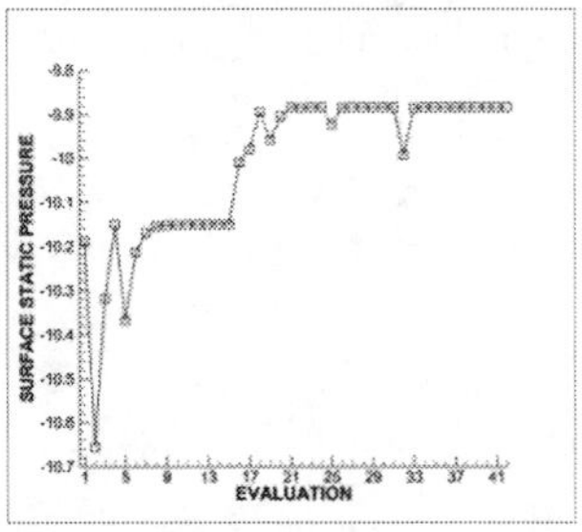

Figure 9. Convergence histories of the average surface static pressure for the modified baseline airfoil optimization.

The C_p distributions of the initial and final designs (Fig. 10) show the achievement of the desired higher C_p, with smoothening the pressure distribution the at the optimized inlet area, which would also decrease the flow separation near the inlet.

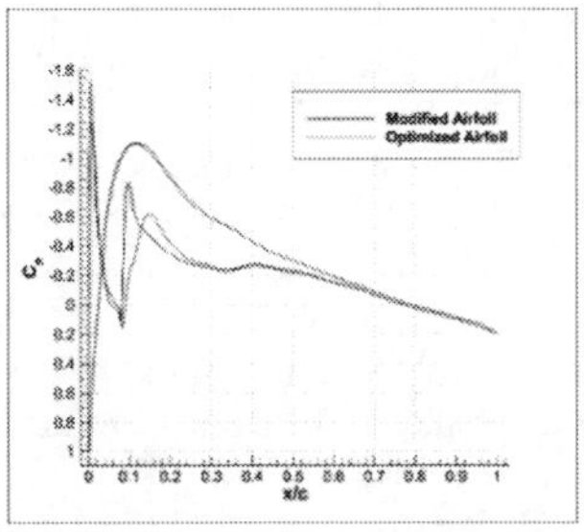

Figure 10. Surface pressure coefficient comparison for the initial and optimized airfoils at α=0°.

5 Optimized air inlet configurations

The width dimensions of the air inlets are chosen to be greater than or equal 1%c to represent sport paragliders, as illustrated in Fig. 11. .

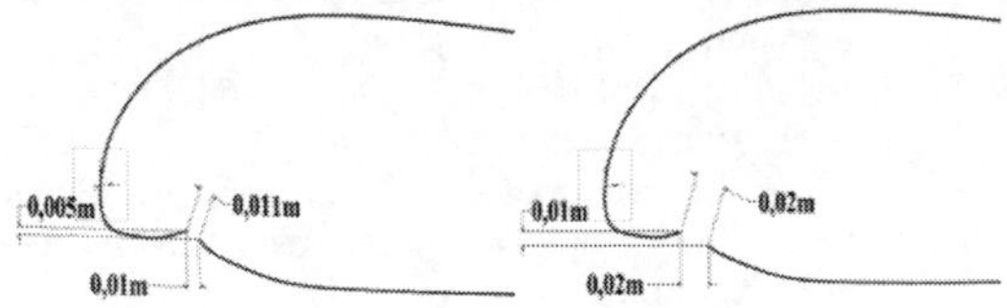

(a) (b)

Figure 11. The optimized airfoil with inlet (a) 1 and (b) 2 configurations.

The grids used in this case have twice the tangential grid nodes conforming to the airfoil surface, which were used in the classical open airfoils case with the same number of nodes in the normal direction. Thus, the optimized airfoil is re-meshed to match the open airfoils' grids.

The effects of the two inlet configurations on the aerodynamic performance compared to the closed optimized airfoil are displayed in Fig. 12. It is shown that inlet 1 has the lesser effect on the degradation of the aerodynamic performance as expected, owing to its smaller width. For small air inlet widths, there is no difference in the slope between the optimized closed airfoil and the open airfoils. The lift curve of the paraglider airfoil with inlet 1 is almost the same as that of the baseline closed airfoil, but inlet 2 has slightly lower lift. The adverse effects on drag due to the openings are very small compared to the effects of the two classical inlets that were discussed earlier. Separation bubbles only appear on the external surface of the upper edge of the inlet. No separation occurs on the lower edge of the inlet due to its concave shape, reducing the opening effects on the rise in drag, opposing the investigated classical inlets. Inlet 2 has a slight increase in drag over inlet 1, which is probably due to the increase of the inlet width, obscuring the flow. Even though, the separation bubbles on the external side of the upper edge of the inlets are slightly higher for the inlet with the smaller width. This is shown in Fig. 13, and Fig. 14 for inlets 1 and 2, respectively, through the streamlines at 3° and 9° angles of attack. Thus, the main contributor to the overall increase in drag, is the increase in inlet width, owing to the concave shape of the inlet.

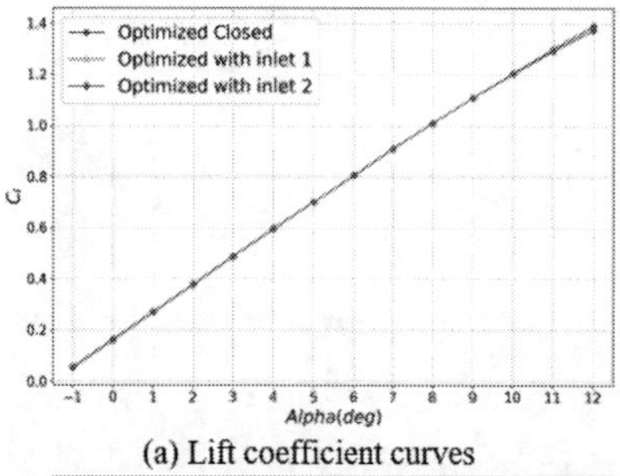

(a) Lift coefficient curves

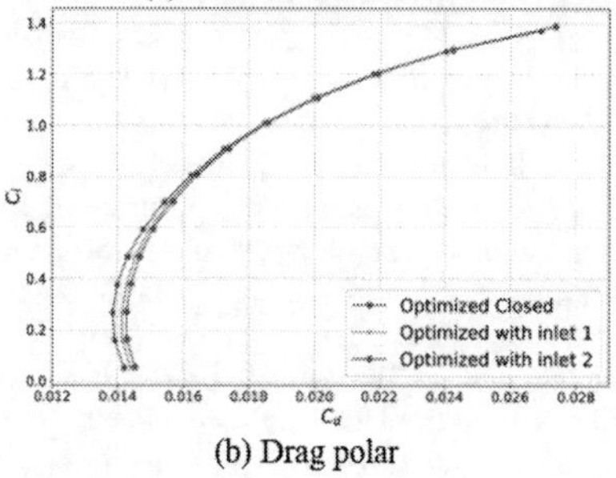

(b) Drag polar

Figure 12. The effects of the optimized inlets on the aerodynamic performance.

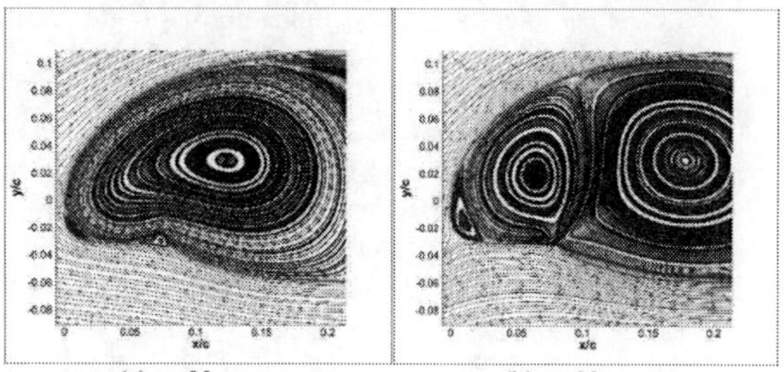

(a) α=3° (b) α=9°

Figure 13. Streamlines inside and around the optimized airfoil with inlet 1 (airfoil edge in red).

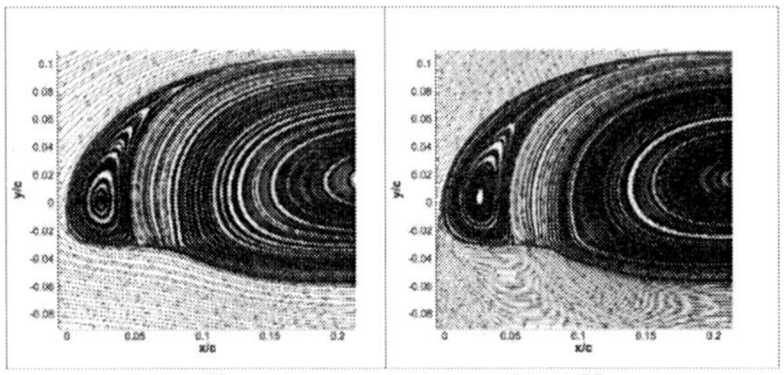
(a) α=3° (b) α=9°

Figure 14. Streamlines inside and around the optimized airfoil with inlet 2 (airfoil edge in red).

The internal pressure of the optimized inlets are shown in Fig. 15. The results are close to each other, but inlet 1 achieved a slightly higher internal pressure over the flight range than inlet 2.

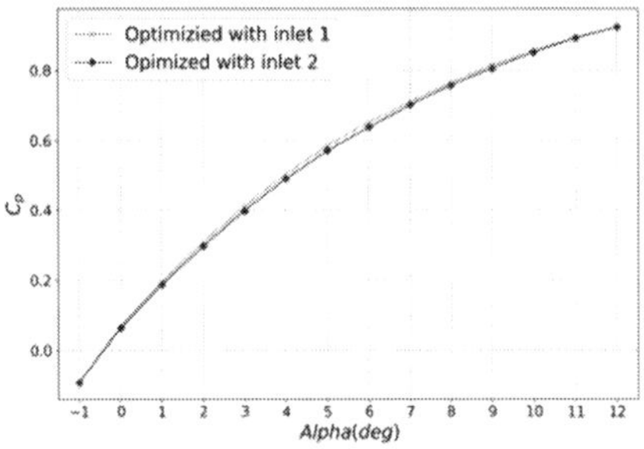

Figure 15. Internal pressure comparison of the baseline airfoil with inlet 1 and 2.

6 Best inlet configuration

The classical inlet with an offset proved to have the best inflation at the whole range of flight, and has also achieved a slight increase in lift over its baseline closed profile. However, this came with the price of having the highest additional parasite drag of all the studied inlet configurations. On the other hand, both openings that were introduced to the optimized airfoil, showed balanced effects on the overall performance, with full inflation starting from 0° angle of attack. Achieving a higher internal pressure than the surrounding external pressure at every point, with less adverse effects on drag and almost the same lift characteristics compared to its closed geometry. This resulted in higher glide ratio than the classical inlet by 2 counts in the normal flight regime. Additionally, the concave shape of the optimized airfoil eliminated the separation bubbles appearing on the external lower edge of both classical inlets, which not only decreased the effects of the inlet on drag by 20 drag counts compared to the classical inlet, but also it would improve recovery in case of stalling of the paraglider. As a matter of fact, the effects of an opening on the lift and drag characteristics, are less for an airfoil with the SharkNose concept, rather than introducing an inlet to a classical airfoil, as was concluded from this study, and previous studies that investigated different classical inlet configurations numerically or experimentally [10-12] to cite a few. Finally, the optimized airfoil with 1%c has a slightly better overall paragliding performance than that with 2%c . Therefore, it is chosen as the optimal air inlet configuration for the paraglider airfoil, based on this study.

References

1. P. Frederic, A. Luc, and D. David, Ozone parapente, Institut National De La Propriété Industrielle (INPI), French patent no. 2 972 422, (2011).
2. F. Palacios, M. R. Colonno, A.C. Aranake, A. Campos, S. R. Copeland, T. Economon, A. Lonkar, T. W. Lukaczyk, T. W. R. Taylor, and J. Alonso. Stanford University Unstructured (SU2): An open-source integrated computational environment for multi-physics simulation and design, (2013).
3. T. Economon, F. Palacios, S. R. Copeland, T.W. Lukaczyk, and J. Alonso. SU2: An Open-Source Suite for Multiphysics Simulation and Design. AIAA Journal, **54**:1-19, (2015).
4. P. Spalart and S. Allmaras, "A one-equation turbulence model for aerodynamic flows," in 30th Aerospace Sciences Meeting and Exhibit, (1992).
5. Y. Saad and M. H. Schultz. GMRES: A Generalized Minimal Residual Algorithm for Solving Non-symmetric Linear Systems. SIAM J. Sci. Stat. Comput., **7**:856-869, (1986).
6. T. W. Sederberg and S. R. Parry. Free-form deformation of solid geometric models. In Proceedings of the 13th annual conference on Computer graphics and interactive techniques, **20**:151-160, (1986). ISBN 0897911962. doi: 10.1145/15922.15903.
7. W. Karush. Minima of functions of several variables with inequalities as side conditions. Ph.d. thesis, (1939).
8. H. W. Kuhn and A.W Tucker. Nonlinear Programming. Proceedings of the Second Symposium on Mathematical Statistics and Probability, (1951). ISSN 01605682. doi: 10.1007/BF01582292.
9. T. Albring, M. Sagebaum, and N. R. Gauger. Efficient Aerodynamic Design using the Discrete Adjoint Method in SU2. AIAA 2016-3518.
10. G.M. Ware and H. J.L., Wind Tunnel Investigations of Ram-Air Inflated All- Flexible Wings of Aspect ratio 1.0 to 3.0. NASA TMSX-1923 (1969).
11. J. Ross, "Computational Aerodynamics in the Design and Analysis of Ram-Air-Inflated Wings," in RAeS/AIAA 12th Aerodynamic Decelerator Systems Technology Conference, AIAA Paper 93-1228 (Reston, VA, 1993).
12. H. Belloc, V. Chapin, F. Manara, F. Sgarbossa, and A. M. Forsting. Influence of the air inlet configuration on the performances of a paraglider open airfoil. International Journal of Aerodynamics **5**:83-104 (2016).

Numerical Simulation of the Motion of an Unmanned Aerial Vehicle

Il'ya O. Akimov[1], Vsevolod V. Koryanov[1]

[1]*Bauman Moscow State Technical University, 105005 2-nd Baumanskaya, str., 5, b1, Russia*

Abstract. Unmanned aerial vehicles are used for research in many areas: photography and video shooting and so on. The development of unmanned aerial vehicles is directly related to the development of airspace. Today, a mathematical model is required that would describe the movement of such an aircraft with the purpose of predicting, correcting and optimizing it. The paper presents the results of a study of the controlled motion of an unmanned multi-rotor aircraft using the example of a quadrocopter. The study included the development of a law governing the apparatus and its modeling in the form of a software package. The structure of the autopilot, its main contours and parameters of these circuits are considered. After determining the necessary characteristics of the autopilot, modeling of the controlled motion of the quadrocopter in the execution environment was carried out.

1 Introduction

The aim of the presented work is the development of the law governing the multi-rotor unmanned aerial vehicle by the example of a quadrocopter providing motion to a given point, as well as the definition of the motion model implementing this law. The structure of the autopilot, the basic contours of the control system of the flight of the quadrocopter and the parameters of these circuits are considered [1, 2].

After determining the necessary system characteristics, simulated Quadrocopter controlled motion in the Common Language Runtime (CLR) runtime in the Microsoft Visual Studio shell in C # is performed.

Solved in the process of research tasks:

- formation of the shape of the autopilot of the quadrocopter;
- development of a quadrocopter control law;
- creating a mathematical model of the movement of the device;
- correction of the motion model taking into account external disturbance - wind load;
- development of a software package that implements the movement model.

2 Model description

Quadrocopter is an unmanned multi-rotor aircraft with four rotors [3]. In the figure 1:

- B_i - the beams of the frame of the apparatus on which the engines are fixed;
- D_i - carrying DC motors;
- v_i - air propellers;
- F_i - traction forces created by propellers;

The distance from the center of mass of the device to each of the four engines is the same and equal to r. The following assumptions are accepted: the device is symmetrical about the x and z axes, its frame and screws have absolute stiffness, each of the engines is at the end of the rod.

The frame of the device, formed by four rays, is the main support plane of the quadrocopter. Its angular position relative to the coordinate system is described by three angles - ψ_k, ϑ_k, γ_k, which determine the rotation around the axes. The spatial position of the center of mass of the body is given by the linear coordinates x_k, y_k, and z_k.

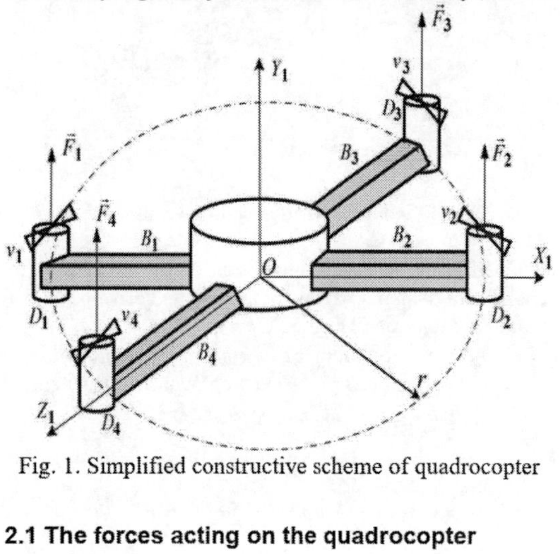

Fig. 1. Simplified constructive scheme of quadrocopter

2.1 The forces acting on the quadrocopter

Basic coordinate systems in which the motion model is considered:

- Start (CCS) $O_CX_CY_CZ_C$ - fixed coordinate system, the beginning of the starting point of the vehicle movement. The motion in it is described by the height above the surface (y = Ly = H), the longitudinal distance (Lx) and the lateral deviation (Lz);
- Normal (NSC) $OX_GY_GZ_G$ is a mobile non-rotating coordinate system, the origin is located in the center of mass of the device, the axes are aligned with the axes of the starting system.
- Basic linked (BSC) $OX_1Y_1Z_1$ is a moving rotating coordinate system, the position relative to the normal system is determined by the Euler angles ϑ_k, γ_k and ψ_k.

In Figure 2 shows the forces acting on the device, the corresponding notations are introduced [4]:

- $\vec{F}_T^{Y_1} = F_T^{Y_1}\vec{y}_1^{\,0}$ – vector of total thrust along the axis OY_1;
- $F_T^{X_G}, F_T^{Y_G}, F_T^{Z_G}$ – its projection on the axis of the normal coordinate system;
- $\vec{V}_k$ – velocity vector of the apparatus relative to the starting coordinate system;
- $\vec{F}_\Sigma^V = (\vec{G} + \vec{F}_T)$ – is the vector of the total force applied to the apparatus;
- $\vec{G}$ – vector of gravity acting on the quadrocopter.

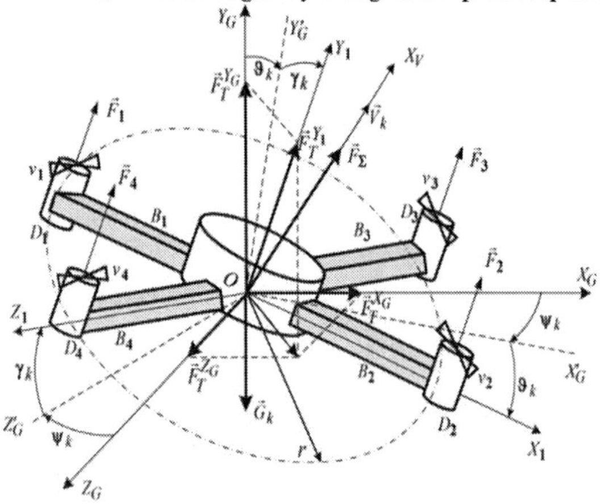

Fig. 2. Scheme of forces acting on the device

In the projections on the axes of the corresponding coordinate systems, these forces [5] make it possible to obtain the relations given below.

- Dynamic equations of motion of the center of mass of the apparatus in the starting coordinate system:

$$\begin{cases} -m_k v_k \frac{d\theta_k}{dt} = \sum F^{X_G} = F_T^{X_G}; \\ m_k \frac{dv_k}{dt} = \sum F^{Y_G} = F_T^{Y_G} - G; \\ m_k v_k \frac{d\varphi_k}{dt}\sin\theta_k = \sum F^{Z_G} = F_T^{Z_G}; \end{cases} \quad (1)$$

- Kinematic equations of motion of the center of mass of the apparatus in the starting coordinate system:

$$\begin{cases} \dot{x}_k = v_k^{X_G} = v_k \cos\varphi_k \cos\theta_k; \\ \dot{y}_k = v_k^{Y_G} = v_k \sin\theta_k; \\ \dot{z}_k = v_k^{Z_G} = -v_k \sin\varphi_k \cos\theta_k; \end{cases} \quad (2)$$

To determine the position of the apparatus in the starting coordinate system, the following relations were used:

$$\begin{cases} x_k^C(t) = x_k^C(0) + \int_0^t v_k \cos\varphi_k \cos\theta_k \, dt; \\ y_k^C(t) = y_k^C(0) + \int_0^t v_k \sin\theta_k \, dt; \\ z_k^C(t) = z_k^C(0) - \int_0^t v_k \sin\varphi_k \cos\theta_k \, dt. \end{cases} \quad (3)$$

- Dynamic equations of motion of the device around the center of mass in the base coupled coordinate system:

$$\begin{cases} \sum M^{X_G} = J_x \frac{dw_x}{dt} + (J_z - J_y) w_y w_z; \\ \sum M^{Y_G} = J_y \frac{dw_y}{dt} + (J_x - J_z) w_x w_z; \\ \sum M^{Z_G} = J_z \frac{dw_z}{dt} + (J_y - J_x) w_y w_x, \end{cases} \quad (4)$$

where J_i, w_i – moments of inertia and angular velocities of the apparatus relative to the axes of the base coordinate system.

- Kinematic equations of motion of the apparatus around the center of mass in the base coupled coordinate system:

$$\begin{cases} \dot{\psi}_k = \frac{1}{\cos\vartheta_k}(w_y \cos\gamma_k - w_z \sin\gamma_k); \\ \dot{\gamma}_k = w_x - \tan\vartheta_k (w_y \cos\gamma_k - w_z \sin\gamma_k); \\ \dot{\vartheta}_k = w_y \sin\gamma_k + w_z \cos\gamma_k. \end{cases} \quad (5)$$

2.2 The moments and forces taken into account when designing a mathematical model

The forces forming the linear motion of the center of mass of an unmanned aerial vehicle, as well as the moments that form the rotational movement of the apparatus around the center of mass, are created by rotating the carrier propellers installed on the rotors of DC motors, and the moments of forces depend on the mutual relations between the thrust forces of each of the screws. Achieving the specified angular velocities of the screws is the task of speed stabilizing drives for the respective engines.

The following notations are accepted:

- k_{vz} – coefficient of viscosity of air;
- c_{mi} – coefficient of electromechanical gain;
- c_{ei} - electronic gain factor;
- R_{ai}, L_{ai} – electrical parameters of the motor winding;
- J_{ni} – the moment of inertia of the rotating parts;
- $T_{ai} = L_{ai}/R_{ai}$ – electromagnetic time constant of the motor ;
- M_{dvi} – engine torque ;
- $M_{i\Sigma}$ – torque applied to the screw ;
- M_{pi}^{Y1} – the reaction time of the engine ;
- k_{di}, k_{tgi} – scale factors of sensors ;
- k_{kori}, T_{kori} – the gain and the time constant of the correcting link.

In order to meet the imposed conditions for creating a mathematical model of the movement of the device, it is necessary to take into account the following moments of forces acting on the device:

- the moment of forces $\vec{M}_{\Delta F_{21}}^{Z_1}$ around the axis OZ_1, created by the difference in thrust of the engines 1 and 2.
- the moment of forces $\vec{M}_{\Delta F_{34}}^{X_1}$ around the axis OX_1, created by the difference of the thrust of the engines 3 and 4.
- the moment of forces $\vec{M}_{p\Sigma}^{Y_1}$ around the axis OY_1, created by the difference in the reaction moments of the engines.

Motors 1 and 3 rotate clockwise, and 2 and 4 - against. Then:

$\vec{M}_{p\Sigma}^{Y_1} = (\vec{M}_{p1}^{Y_1} + \vec{M}_{p3}^{Y_1}) - (\vec{M}_{p2}^{Y_1} + \vec{M}_{p4}^{Y_1})$. Then you can record the transfer function of the drive for stabilizing the engine speed:

$$M_{pi}^{Y1}(s) = Q_i \frac{T_{vzi}S + 1}{a_{i3}S^3 + a_{i2}S^2 + a_{i1}S + 1}\varphi_{vizad}(s) \qquad (6)$$

where: $Q_i = \frac{k_{kori}k_{ui}k_{vz}}{c_{ei}K_{pri}}$; $K_{pri} = 1 + \frac{k_{kori}(k_{ui}+k_{tgi})}{c_{ei}} + \frac{T_{mi}}{T_{vzi}}$.

In the flight control system of the device, the control actions are aimed at changing the parameters $F_i^{Y_1}$ and $M_{pi}^{Y_1}$. The adjustable coordinates are x_k, y_k, and z_k. The adjustable angles are ψ_k, ϑ_k, γ_k. The intermediate angles are θ_k and φ_k.

Measured parameters that are used in the process of controlling the quadrocopter: $\dot{x}_k$, $\dot{y}_k$, $\dot{z}_k$, and w_x, w_y and w_z. They are available for measurement by sensors located on the machine itself.

The executive body responsible for the formation of force impacts in accordance with the commands of the control system is a system of four drives for stabilizing the speed of DC motors [8].

2.3 Numerical simulation of the obtained system

Numerical simulation of the controlled motion of a multi-rotor aircraft used the Common Language Runtime (CLR) runtime in the Microsoft Visual Studio shell in C #.

There are several restrictions:

- since the speed of rotation of the screws does not change instantaneously and can not be too high due to physical reasons, the maximum speed (490 r / s) and the acceleration (500 r / s2) of this parameter are set;
- each of the engines is considered a material point fixed to the weightless beam of the frame, and the apparatus itself is a ball, according to which moments of inertia were calculated by the formulas:

$$J_x = \frac{2mR_k^2}{5} + 2r^2 m_d, \qquad (7)$$

$$J_z = \frac{2mR_k^2}{5} + 2r^2 m_d, \qquad (8)$$

$$J_y = \frac{2mR_k^2}{5} + 4r^2 m_d. \qquad (9)$$

- the Euler method with a base step of 0.0001 seconds was adopted as the integration method;
- basic accuracy of arrival to the point - 0,1 m;
- the maximum permissible speed at which the device is considered to arrive at the point is 0.1 m / s
- atmosphere model complies with GOST 4401-81
- some of the parameters (step, initial spatial position, specified spatial position, etc.) should be able to change in the user interface;
- it is necessary to output the basic flight characteristics to a text file after the simulation is over.

3 Model analysis

A number of tests of the obtained model [6, 7] were carried out. In the first one, a flight with a flight program is considered that assumes control of the motion along the OYG axis (from the point with coordinates 0.10.17 to the point with coordinates 0.25.17). The resulting flight parameters are shown in Figures 3-6. At the same time, no movement is observed along the OX_G and OZ_G axes, which allows to conclude that the autopilot is operating correctly when controlling the movement along the OY_G axis.

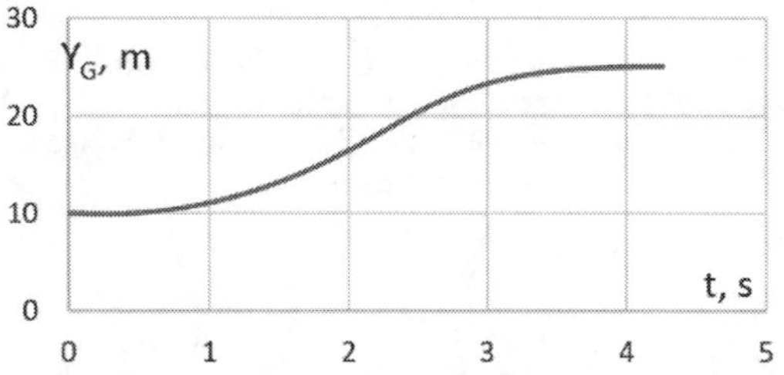

Fig. 3. Dependence of the Y_G coordinate on time with vertical flight of the device.

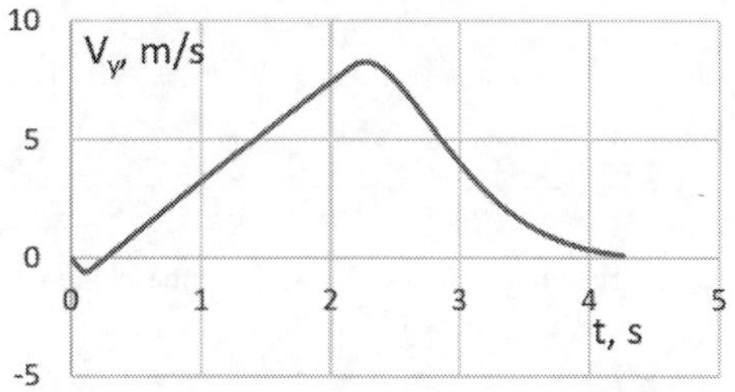

Fig. 4. Dependence of speed Vy on time with vertical flight of the device.

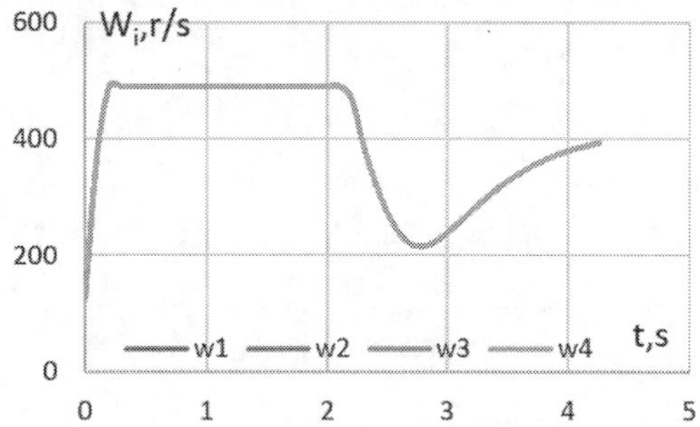

Fig. 5. Dependence of rotational speeds of screws on time

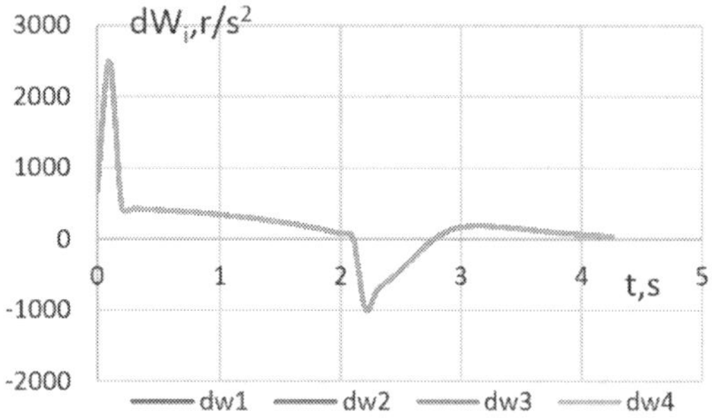

Fig. 6. Dependence of screw rotational acceleration on time

The following test provided for the change of all spatial coordinates of the quadrocopter during the execution of the flight program (from the point with coordinates 5,3,16 to the point with the coordinates 48,35,81). The resulting flight parameters are shown in Fig. 7-10. The change in the parameters corresponds to the expected ones, which made it possible to proceed to modeling taking into account external disturbances (wind load).

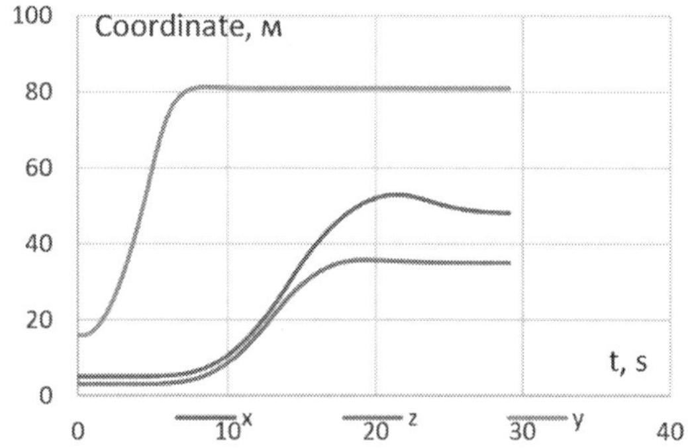

Fig. 7. Dependence of coordinates on time in the space flight of the device

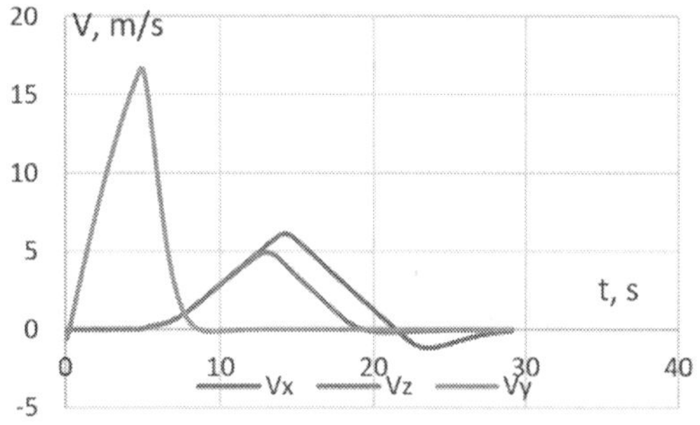

Fig. 8. The dependence of velocities on time in the spatial flight of the apparatus

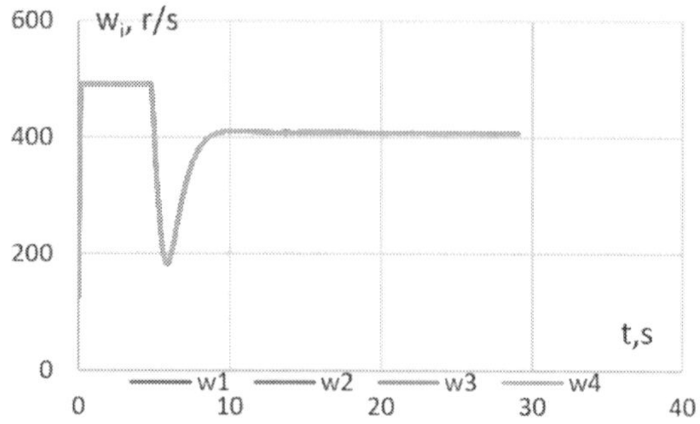

Fig. 9. Dependence of the rotational speed of the screws on time in the space flight of the apparatus

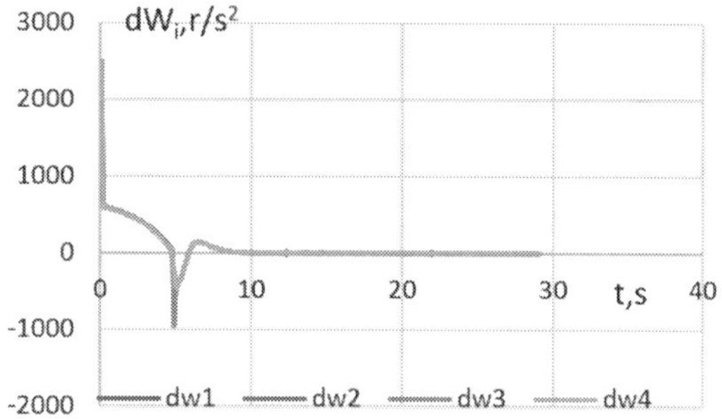

Fig. 10. Dependence of the acceleration of rotation of the screws on time in the spatial flight of the apparatus

The model has the option of taking into account the effect of the wind constantly acting on the apparatus. The wind effect was determined by changing the speed of the quadrocopter, acting continuously along one of the axes.

For clarity, a flight program similar to the previous point was specified, but a constant counter-impact velocity along the X axis in the size of 0.5 m/s was added. The flight parameters under the condition of an external disturbance are shown in Fig. 11-14.

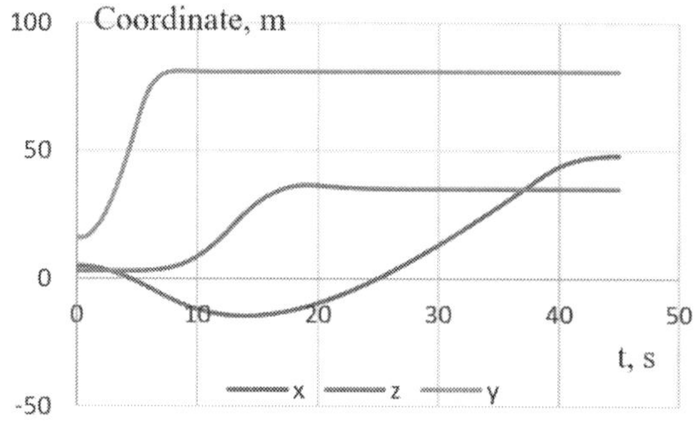

Fig. 11. Dependence of coordinates on time in the space flight of the apparatus in the presence of disturbances (wind load)

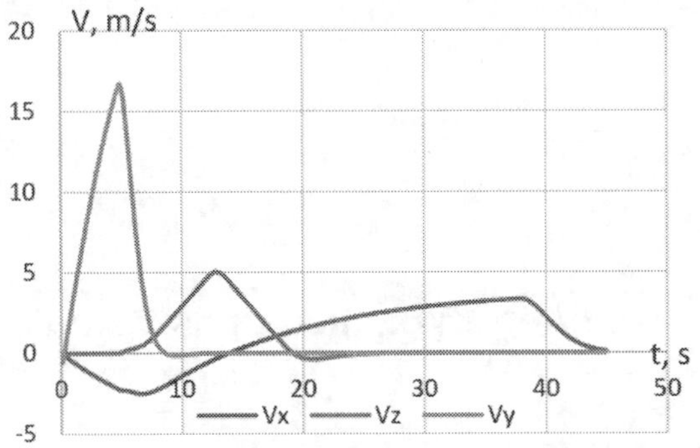

Fig. 12. The dependence of speed on time in the space flight of the apparatus in the presence of disturbances (wind load)

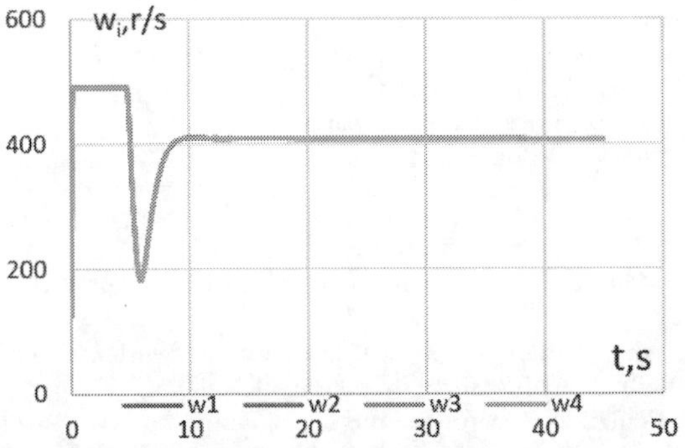

Fig. 13. Dependence of the rotational speed of the screws on time for a space flight of the apparatus in the presence of disturbances (wind load)

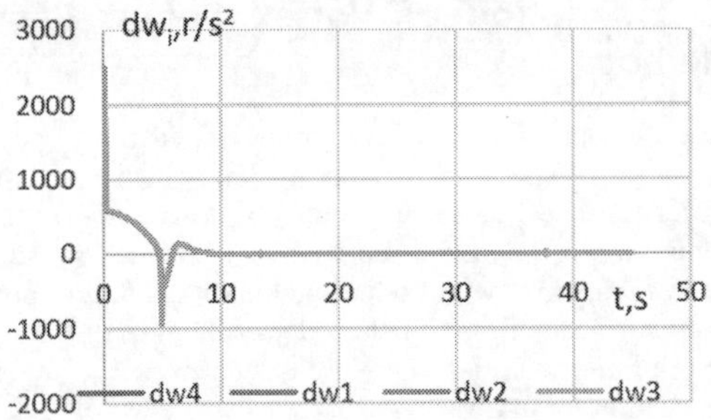

Fig. 14. Dependence of the rotor rotation acceleration on time in the space flight of the apparatus in the presence of disturbances (wind load)

4 Conclusion

In the process of performing the presented work, all the tasks were fulfilled:

- the shape of the autopilot of the quadrocopter is formed;
- the law of quadrocopter control has been developed;
- mathematical model of the movement of the apparatus for the case of an absolutely rigid design was created;
- the model of movement is corrected taking into account external disturbance - wind load.

Software tests with various programs and flight conditions were carried out, including taking into account external disturbance, the results of tests in the form of graphical dependencies are presented. Analysis of the flight simulation results allows us to conclude that the autopilot of the quadrocopter developed in the work does not require significant processing for operation under wind load conditions. This work is useful in minimizing the cost of designing a quadrocopter, and the use of a mathematical model is advisable when working on the on-board control system of the apparatus.

References

1. Eswarmurthi Gopalakrishnan. Quadcopter flight mechanics model and control algorithms. Supervisor: Prof. Dr. Martin Hromčík. Prague, May 2017.
2. Francesco Sabatino - Quadrotor control: modeling, nonlinearcontrol design, and simulation. Master's Degree Project. Stockholm, Sweden June 2015.
3. Ogoltsov I.I., Rozhnin N.B., Sheval V.V. Development of a mathematical model for the spatial flight of a quadrocopter // Proceedings of the MAI. Issue number 83, 05.10.2015.
4. Guryanov A.E. Modeling of quadrocopter control. Electronic scientific and technical journal "Engineering Herald". №8, 2014, pp. 522-534
5. Savitsky A.V., Pavlovsky V.E. Quadrocopter model and neural network control algorithm // Preprints of IPM. M.V. Keldysh. 2017. №77. 20 p.
6. Teppo Luukkonen - Modelling and control of quadcopter - School of Science. Independent research project in applied mathematics. Espoo, August 22, 2011.
7. Internet resource: Modeling Vehicle Dynamics, information as of 04/20/18 (http://charlestytler.com/quadcopter-equations-motion/).
8. Zsurzsan, Adriana Gabriela; Brogaard, Rune Yding; Andersen, Nils Axel; Ravn, Ole. General model and control of an n rotor helicopter. Published in: Journal of Physics: Conference Series (Online). 2015.

Vortex Loops Based Method for Subsonic Aerodynamic Loads Calculation

Georgy A. Shcheglov[1] and Sergey A. Dergachev[1]

[1]*Bauman Moscow State Technical University, Aerospace Department, 105005 2-nd Baumanskaya st., 5/1, Moscow, Russia*

Abstract. The new modification of vortex method is considered. It is based on closed vortex loops that can be considered as the separate vortex "superelement". It is shown that such approach has number of advantages for vortex wakes evolution simulation. The loops generation algorithm permits to simulate the separation zone due to the self-organization of the loops. In this case, there is no need to determine the location of the vortex sheet separation zones. The developed software package can be used in the calculation of aerodynamic loads on unmanned aerial vehicles, as well as in solving the aeroelasticity problems of structural elements of these aircraft.

1 Introduction

Vortex elements methods are highly efficient for calculation of unsteady aerodynamic loads acting on aircraft moving at low subsonic speed, because they require significantly less computational resources in comparison with grid-based methods [1, 2]. In these methods Lagrangian particles for vorticity transportation are used. Such particles are called vortex elements.

There are number of known models of vortex elements for flow simulation around 3D bodies: closed vortex framework, vortex filament, vorton, vortex dipole, vortex fragmenton, etc. Each of them has some advantages and restrictions. In "classical" vortex element methods, for example, in the discrete vortex method [3], vorticity is concentrated in vortex framework segments and it is absent outside them in the flow. However, this method requires number of empirical models to determine the location of the vortex sheet separation lines. In case of vortex methods with separated vortex particles (vortons, vortex blobs, etc.) flow separation zones are formed "naturally" due to vorticity flux approach – vortex elements are generated on the whole body surface and these elements self-organization in the flow [4, 5]. The main part of the vorticity is concentrated in neighbourhood of the vortex elements themselves, however, there is distributed non-zero "additional" vorticity in the flow domain, according to the Helmholtz theorems. Its intensity vanishes far from vortex elements [6]. This additional vorticity may cause significant errors at computing aerodynamic loads, acting the body.

Some interesting approach has been proposed in computer graphics for 3D smoke dynamics simulation, which implies vorticity flux simulation through vortex loops generation on the body surface [7]. The application of this approach to calculation of unsteady aerodynamic loads acting on bluff bodies showed that the obtained results are in good agreement with known experimental data [8].

The aim of this paper is application of vortex element method modification based on closed vortex loops to calculation of aerodynamic loads, acting on wing of the finite span and to simulation of vortex wake behind wing-fuselage layout.

2 Description of Vortex Loops Based Method

The incompressible 3D flow around fixed rigid body is considered. The flow of the media, which has small viscosity and constant density, described by the continuity equation and momentum equation (Navier — Stokes equation) with no-slip boundary condition on the body surface and perturbation decay boundary condition on infinity. Initial conditions correspond to the circulation-free flow.

In the developed modification of vortex method the vortex wake is considered to consist of K closed vortex filaments – vortex loops of the same circulation Γ. The loop with index k is simulated by the polygonal vortex filament with N_k vertices. These vertices are considered as the Lagrangian markers $\mathbf{r}_{k,i}$. We assume the loop legs between two neighboring vertices to be rectilinear line segments, which are described by vectors

$$\Delta\mathbf{r}_{k,i}=\mathbf{r}_{k,i+1} - \mathbf{r}_{k,i}.$$

They induce the velocities $\Gamma\ \mathbf{v}_{k,i}\ (\mathbf{r})$, which can be calculated analytically as vortex fragmentons influences, calculated by using Biot — Savart law and regularized by introducing of small smoothing radius ε [6] and finally give contribution to the velocity field $\mathbf{V}_\Omega$, generated by vortex wake:

$$\mathbf{V}_\Omega = \Gamma\ \Sigma^{K}_{l=1}\ \Sigma^{Nk}_{m=1}\ \mathbf{v}_{l,m}\ (\mathbf{r})$$

Markers move in the flow along the streamlines, i.e. their motion is described by ODE:

$$d\mathbf{r}_{k,i}/dt = \mathbf{V}\,(\mathbf{r}_{k,i},t), \qquad (1)$$

$$\mathbf{r}_{k,i}(0) = \mathbf{r}_{k,i}{}^{0},\ k=1,\ldots,K,\ i=1,\ldots,N_k$$

Numerical integration of system (1) is carried out by using numerical method of the first order of accuracy (explicit Euler method) with constant step Δt. Initial conditions for markers positions in (1) are parameters of the loops at the time of their generation on the body surface.

It is well-known from physical point of view, that in order to take into account the presence of the body in the flow, it is possible to replace it with the vortex sheet of unknown intensity, placed on the body surface, which generates the velocity field $\mathbf{V}_\gamma(\mathbf{r})$. From mathematical point of view, the velocity $\mathbf{V}_\gamma(\mathbf{r})$ potential can be expressed through unknown double layer potential density g($\mathbf{r}$). This expression coincides with the Biot — Savart law for incompressible flows [3].

For a body having a closed surface, the calculation of the no-sleep boundary condition subject to creation of a vorticity on the whole streamlined surface can be reduced to the no-through condition on body surface $\mathbf{r}_B$ with unit normal vector $\mathbf{n}_B$ [9]

$$V_n = \mathbf{V_B}\, \mathbf{r}_B = 0, \qquad (2)$$

For boundary condition (2) satisfaction linear system

$$\mathbf{S}\ \mathbf{g} = \mathbf{V}_n, \qquad (3)$$

should be solved for unknown circulations g_j of vortex fragmenton frameworks, placed on M body surface panels. For solving of (3) the regularization variable and additional condition is used: algebraic sum of vortex frameworks circulations should be equal to zero.

The summary velocity field $\mathbf{V}(\mathbf{r})$ is the superposition of the velocity field, generated by surface panels $\mathbf{V}_\gamma(\mathbf{r})$, incident flow velocity $\mathbf{V}_\infty$ and velocity field, generated by vorticity inside the flow domain $\mathbf{V}_\Omega(\mathbf{r})$. Finally, the velocity of the i-th marker of the k-th loop has the form

$$\mathbf{V}\,(\mathbf{r}_{k,i},t) = \mathbf{V}_\infty + \mathbf{V}_\Omega(\mathbf{r}_{k,i}) + \mathbf{V}_\gamma(\mathbf{r}_{k,i})$$

Vortex loops generation algorithm consists of three main operations. Firstly, maximal and minimal values of the double layer intensity for all the panels vertices on the body surface g_{max}=max(g_j), and g_{min}=min(g_j), j=1,…,M should be found. Secondary shuold be construct the level lines of the potential density, which correspond to potential values, which differ from one to another on value Γ. These level lines determine the initial shape of the vortex loops. For the given value of the potential density there can be one or more closed level lines, which correspond to separate vortex loops. The number of level lines values is determined as the integer part of expression

$$N_q=[(\,g_{max} - g_{min})/\,\Gamma]-1.$$

Then the generated vortex loops become part of vortex wake. The vortex loops, generated on the surface, should be shifted from the surface in normal direction on small constant distance Δ. It should be noted, that the contribution of the vortex loop to the velocity field $\mathbf{V}_\Omega$ is not considered at the time step immediately after its generation, and for velocity field influence of the corresponding panels on the body surface (which give contribution to $\mathbf{V}_\gamma$) is taken into account.

When vortex elements move in the flow, some marker positions can intersect the body surface, mainly due to numerical errors in velocity field reconstruction and vorticity motion equations integration. In such cases, the loop legs, which intersect the body surface, is replaced with some other loop legs, which lay on the body surface and provide the shortest way. The Dijkstra's algorithm is used for such procedure [10].

At every time step several procedures for smoothing of loops geometry, loop segments length alignment and loops reconnection are used [11].

It is necessary to control the angle values between the neighboring line segments in vortex loops:

$$\Psi_{k,i} = \arccos[(\Delta\mathbf{r}_{k,i},\ \Delta\mathbf{r}_{k,i-1})/(|\ \Delta\mathbf{r}_{k,i}\ |\ |\ \Delta\mathbf{r}_{k,i-1}|)].$$

If for some vortex loop vertex $\Psi_{k,i} < \varphi$, where φ is a given constant, then vertex position should be corrected.

After angle correction, when the vortex loops are smooth enough, it is necessary to "rediscretize" them in order to provide nearly the same length h for all the segments. In this procedure cubic spline interpolation is used in order to reconstruct smooth shape of the vortex filaments.

Vortex loops reconnection is carried out according to the following algorithm. For each marker $\mathbf{r}_{k,i}$ the nearest marker $\mathbf{r}_{l,j}$ is searched (j=1,…,N_l, l=1,…,K), for which the conditions are satisfied:

$$|\ \mathbf{r}_{l,j} - \mathbf{r}_{k,i}| < \mu,\ \ \arccos[(D_{k,i},\ D_{l,j})/(|\ D_{k,i}\ |\ |\ D_{l,j}\ |)] > \varphi, \ (4)$$

$$D_{l,j} = \mathbf{r}_{l,j+1} - \mathbf{r}_{l,j-1}\ \ D_{k,i} = \mathbf{r}_{k,i+1} - \mathbf{r}_{k,i-1}.$$

where μ>0 is some given value.

From all the marker pairs, for which conditions (4) are satisfied, that pair is selected, for which $|\ \mathbf{r}_{l,j} - \mathbf{r}_{k,i}|$ has minimal value, and the reconnected is performed for the coppesponding legs. In this procedure either two vortex loops are formed from one, or, vice verca, two loops are merged into one. Then for the newly formed vortex loop (or loops) the reconnection algorithm is repeated.

Pressure field at every time step is being calculated from the positions of loops by using the analog of the Cauchy – Lagrange integral [12].

$$p\,(\mathbf{r},t) = p_\infty + \rho_\infty\,[V_\infty^2/2+V(\mathbf{r},t)^2/2 - \Sigma^{K}_{i=1}\,(A_i\,g_i)/(4\,\pi\,\Delta t) +$$

$$+\, \Sigma^{K}_{l=1}\, \Sigma^{Nk}_{m=1}\, \Gamma\, \mathbf{v}_{l,m}\,(\mathbf{r})\, \mathbf{V}\,(\mathbf{r}_{l,m},t)]$$

where A_i is the solid angle under which the panel from the point of pressure calculation is visible.

Aerodynamic loads can be found by integration of the pressure distribution over the surface of the body.

3 Simulations Results

To implement the vortex element method modification, the C++ program was developed, which uses MPI technology for parallel computations. Calculations were performed on a cluster with 16 cores (were used 3 GHz Intel I7 processors). The execution time of one calculation did not exceed 1000 seconds.

For calculation of unsteady aerodynamic loads acting on wing of the finite span the model of wing with foil NACA- 2217, chord b=0,1 and wingspan elongation 5,0 is considered. The numbers of panels on the streamlined surface are M=5212. Other parameters were chosen as: the smoothing radius of the vortex element ε=0,002, time integration step Δt=0,00219; the constant circulation of the vortex loop Γ=0.00526; nominal length of loop segment h=0,001; loop legs smoothing angle threshold φ =160°; loops shifting distance Δ=0,0001; loops reconnection distance μ=0,04; upstream velocity V_∞=1,0.

The shape of a vortex loops in the wake behind the wing after calculating of 500 time steps is shown in Figure 1. Thin lines denotes the vortex loops. Clear structure of the Prandtl's horseshoe vortex is seen. Note, that this result is obtained without any additional hypotheses about flow separation line position or any equal suggestions.

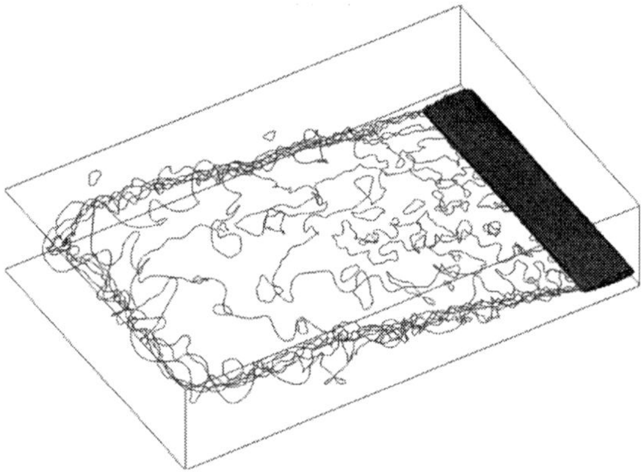

Figure 1. Vortex wake behind the wing.

The comparison of calculated aerodynamic loads coefficients with the experimental data is shown on the Figure 2. Points denotes the calculated results and lines denotes the results of experiments [13].

The error in determining coefficients of Cy and Mz does not exceed 8%, which is permissible for engineering calculations. The relative error of Cx is great due to the small value of the drag of the airfoil, but the absolute value of the error is also acceptable for engineering calculations. The reason for the such errors is probably the roughness of the profile model used for the calculations. Here, additional methodological studies are required.

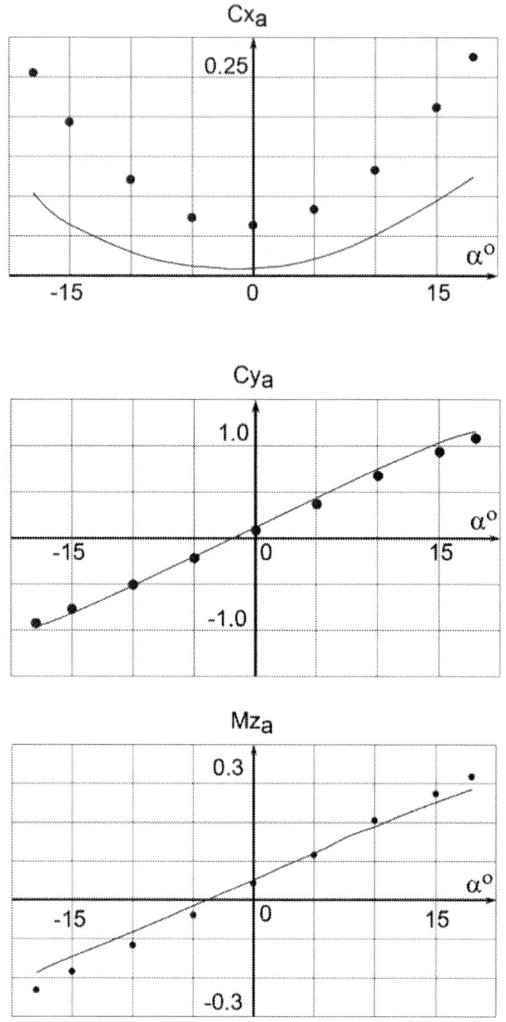

Figure 2. Aerodynamics coefficient. Points – calculated value, lines – experimental data [13].

For simulation of vortex wake behind wing-fuselage layout the wing model from previous case was supplemented by a blunt cylinder with diameter D=0,2 and elongation L/D=12,75. The numbers of panels on the streamlined surface are M~5500.

The change in the structure of the vortex wake in the flow behind the layout at different angles of attack is shown in Figures 3-5. Each figure consist isometric and top views of wake obtained after 500 time steps.

The figures show the interaction of vortex wakes behind the fuselage and the wing with increasing angle of attack, when, the effects of vortex flux from a smooth streamlined surface begin to appear more and more.

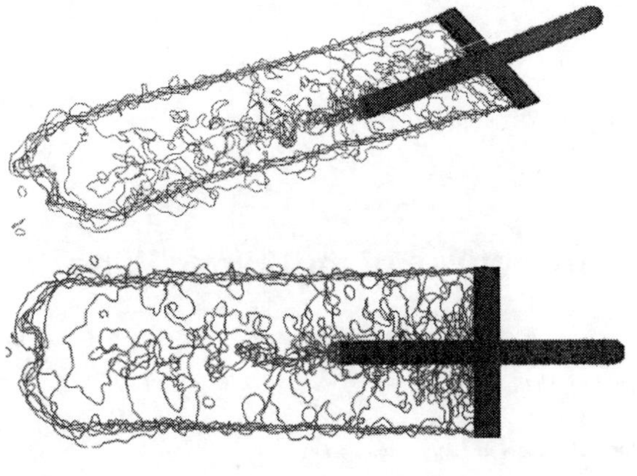

Figure 3. Vortex wake behind the wing-fuselage layout for angle of attack α=10°.

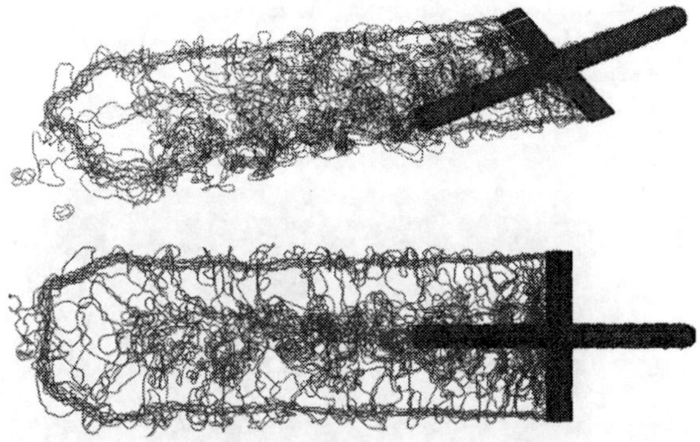

Figure 4. Vortex wake behind the wing-fuselage layout for angle of attack α=20°.

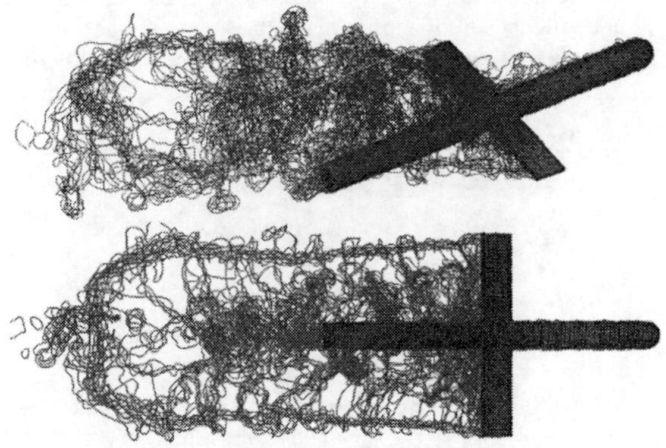

Figure 5. Vortex wake behind the wing-fuselage layout for angle of attack α=30°.

4 Conclusion

The application of meshfree pure lagrangian vortex element method modification based on closed vortex loops to calculation of aerodynamic loads, acting on wing of the finite span allow to obtain results acceptable for engineering calculations with minimal calculation costs.

The simulation of vortex wake behind wing-fuselage layout is shown that developed algorithm can simulate vortex wake evolution when flowing around a complex body geometry without any additional hypotheses about flow separation line position or any equal suggestions.

The developed software package can be used in the calculation of aerodynamic loads on unmanned aerial vehicles, as well as in solving the aeroelasticity problems of structural elements of these aircraft.

References

1. G.-H. Cottet, P. Koumoutsakos Vortex Methods: Theory and Practice. CUP, (2000).
2. R.I. Lewis Vortex Element Methods for Fluid Dynamic Analysis of Engineering Systems. CUP, (2005).
3. I.K. Lifanov, L.N. Poltavskii, G.M. Vainikko Hypersingular Integral Equations and Their Applications. CRC Press, Boca Raton, (2003).
4. K. Kamemoto J. of the Brazilian Soc. of Mech. Sc. and Eng., 26(4) (2004)
5. A.J.Q. Alkemade On Vortex Atoms and Vortons: Ph.D. Thes. (1994).
6. I.K.Marchevsky, G.A. Shcheglov Poc. Of ECCOMAS - 2012, (2012).
7. S. Weissmann, U. Pinkall ACM Trans. Graph, 29(4), (2010).
8. G.A.Shcheglov, S.A. Dergachev Proc. Of PARTICLES - 2017, (2017).
9. S.N. Kempka et al, SANDIA Rep. SAND96-0583 UC-700, (1996).
10. E.W. Dijkstra Numerische Mathematik, 1 (1959).
11. S.A. Dergachev, G.A. Shcheglov Heralds of MGTU GA, 1 (2016) (in Russian).
12. G.Ya. Dynnikova, P.R. Andronov Eur. J. of Mech. - B/Fluids V. 72, (2018).
13. B.A. Ushakov Atlas of airfoils. Edition of the BNT NCAP at TsAGI, (1940) (in Russian).

The Development of the Ideas and Perspectives of the Joined Wing Aircraft

Semenov Vladimir[1,2], Phone Myint Tun[1], Voityshen Vladimir[1,2] and Hardeep Zinta[1]

[1] *Moscow Institute of Physics and Technology, Department of Aeromechanics and Flight Engineering, 140180, Gagarina Street, 16, Zhukovsky, Moscow Region, Russia*
[2] *Central Aerohydrodynamic Institute (TsAGI). 140180, Zhukovsky Street,1, Zhukovsky, Moscow Region, Russia*

Abstract. The history of the development constructive aircraft scheme with a joined wing, the main sphere areas of their use and advantages in characteristics. Prospects for the application of a joined-wing scheme in adaptive constructions. The development of actuators based on shape memory alloys.

1 Introduction

In 1910 years, the designer L. Bleriot used the scheme of an aircraft with a joined wing for his aircraft Bleriot III (Fig.1).

Figure 1. Bleriot III, 1906

In the 1920 years received priority development biplanes. In the 1930 years, for speed aircraft the main place in aviation took monoplane aircraft scheme.

The biplane and the joined wing, which are sometimes mistakenly identified, are fundamentally different schemes. So, in biplane big building height constructions created on account of numerous racks and bracelets, working as rod elements, and in a joined scheme of the wings the elements work with the transfer of moment forces, and this creates favorable prerequisites for upgrade the characteristics aircraft, in the first queue, in the field of strength and weight reduction of the wing design.

In 1968 years, J. Wolkovitch wrote fundamental article [1], which compared aircraft with a monoplane and joined wing. This article substantiated the potential advantages of an aircraft with a joined wing (Fig.2) in many disciplines, including aerodynamics, flight control, strength and rigidity, specific weight of the structure.

Figure 2. Model by J. Wolkovitch. US. 1985

It was shown that in the transition from a monoplane to a joined wing design, in conditions of conservation the equivalent aerodynamic characteristics, weight of the power part wing design can be reduced by 20÷30%, maximum deflections of the wing reduced by 2÷3 times.

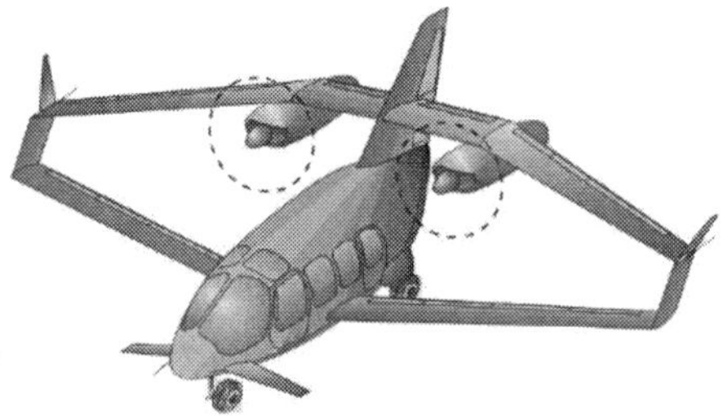

Figure 3. Aircraft "Stalker-232". 1996

Research, conducted in the 1990 years on the design aircraft "Stalker-232" (A.P. Rudometkin, Kramatorsk, Ukraine; V.N. Semenov, Yu.S. Mikhailov, TsAGI, Russia) confirmed also the possibility direct control of the lifting and lateral force aircraft by combination of flap deflection on the elements of the wing system, having a significant transverse V-imagery (Fig.3).

 https://doi.org/10.1051/matecconf/201822105005

In articles [2, 3] received and recommended relationships of geometry parameters joined wing, ensuring reduction its weight on condition ensuring strength, and also compared the aerodynamic properties. The next fundamental research was completed in Russia 2001 years in research work (scientific research work), performed at the Moscow Aviation Institute (MAI) together with other enterprises [4]. There were compared two aircraft concepts: with a classic monoplane wing (Fig.4), and with a joined wing (Fig.5).

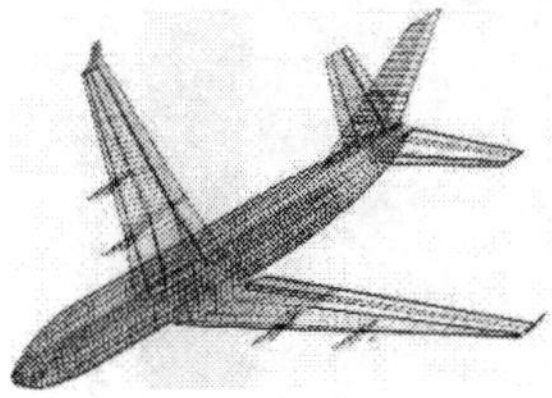

Figure 4. Monoplan;

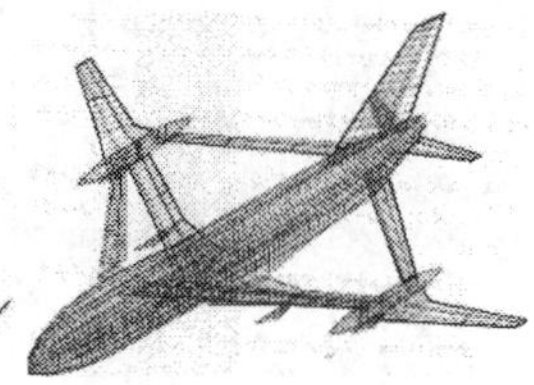

Figure 5. Joined wing by V.V. Malchevsky

The last two schemes, despite significant technical advantages before the first two, have problems with leaving the plane passengers in emergencies, and require a principle different design concept and long-term flight practice for development of new technical solutions, and also place increased demands on the runways.

Table 1. Comparative characteristic of schemes of aircrafts designed for the same requirement:
$Npas = 600$, $L = 13000\ km$, $M = 0{,}85$

Scheme Characteristics	Monoplan (fig. 4)	Joined wing (fig. 5)
Lift-to-drag ratio	Initial	- 4,0 %
Weight of empty	version for	-16,5
Fuel mass	comparison	-7,5
Take-off mass	(100%)	-10,0

The world's Aircraft building companies continue research and the design of an aircraft in a joined-wing scheme [5, 6]. Despite the advantages of a joined wing, their level is until insufficient for achievement large-scale economic advantages, such a level, so that air carriers began to order such planes, and accordingly, were made investments in the restructuring of factories and their equipment. Require additional arguments, demonstrating significant advantages on special requirements for the project.

2 Theoretical basis of weight gain in the joined wing design

In the figure 6 show that in a joined wing created by external forces bending moments M in each section are balance by the sum of the moments, acting in the upper (M_1) and the lower (M_2) wings, and moments, determined by the compression forces of the upper wing N_1, and stretch of the lower wing N_2 taking into account the shoulders of their applications [h_1 and h_2].

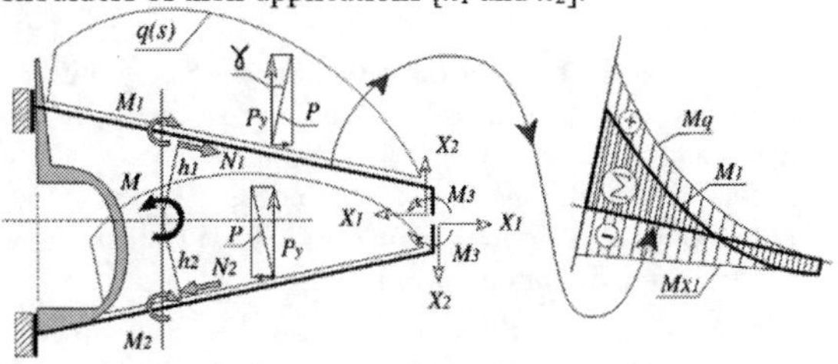

Figure 6. Superposition of bending moments

This equation can be written in the form of an equation of equilibrium:

$$- M = M_1 + M_2 + N_2 h_1 + N_2 h_2 \qquad (1)$$

In the figure 6 it is also shown (on the right) that in the element of a statically indeterminate frame due to the superposition of bending moments from external load M_q and the internal force of interaction of wing elements on the end washer M_{x1}, their total moment M_1 is significantly reduced.

Specific material costs, going on the formation of elements, working on bending, as a rule, much higher, then for elements, which reaction structure are represented by a pair of forces. In real constructions are present both specified forms perception of bend. In view of this, if by constructive transformations or changing the spatial shape of constructions manage increase the fraction of moments, perceived in the form of effort tension-compression, and respectively, reduce moments in elements, working on bending, possible to achieve a reduction mass in constructions and reduce the level of deformation. That is, at unchanged left side of equation (1) there is possible to change ratio of the first and the second pair of terms in right-hand side, and then higher the proportion of the second pair, more rational constructions. In the joined wings system, consisting from rectilinear sections, manage to 45÷55% bending moment from an external load to equilibrate efforts tensile-compression in the wings, spaced on high altitude, and thereby reduce the weight of the corresponding force elements.

Weight of the materials construction, going perception M_{bend}, can be estimated as an integral from the area of the corresponding diagram, taken module:

$$G = \frac{k\gamma}{[\sigma]} \int_L \frac{M_x(z)dz}{h_z} \qquad (2)$$

Where $[\sigma]$ – level permissible stresses,

γ - Specific weight of the structural material,

h_z – averaged caisson height in section z,

L – half-size of the wing line,

k – Statistical coefficient, accounting mass non structure components of the wing. Approximately $k = 2$.

This follows, what is the changing shape of the elastic axis of the joined wing, we change the value of the total moment in its section. Cross-section, in which acts the longitudinal force *N*, offset by the value Δy from the original rectilinear elastic axis, causes an additional compensation moment $N^*\Delta y$, which promotes to reduce the weight of the wing. Finally, belt of spars are in different conditions of superposition load, and redistribution of power material between them gives another direction of optimization.

2.1 Constructions with a curved elastic axis wing

In the publication [5, 6] it was proposed to give the elastic axis of the joined wing wave shape, minimize the required structural weight of the wing. Methods and program are developed, finding the optimal deviation of the elastic axis of the wing from its original linear shape.

In figure 7 presents shape guise of the aircraft with optimized by criteria Strength and weight of the shape of the elastic axis of the joined wing.

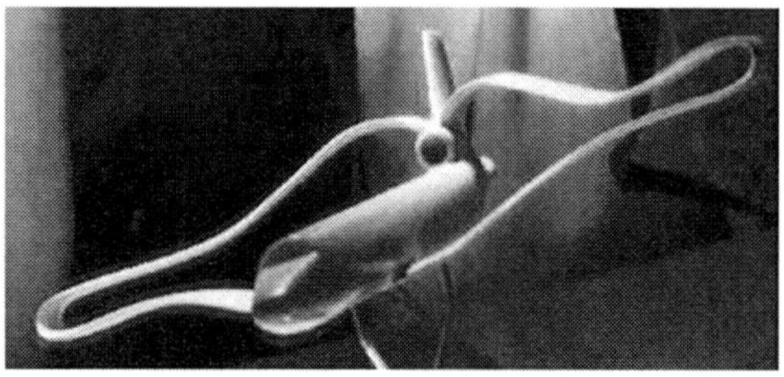

Figure 7. Aircraft by patent Ru N 2067948.
V.N. Semenov, V.V. Saurin. TsAGI, 1996

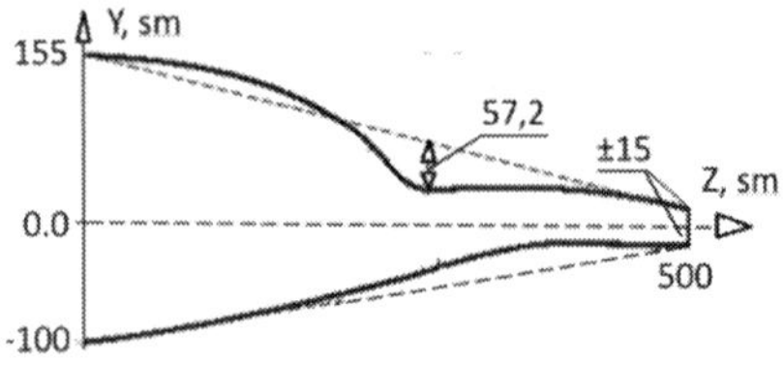

Figure 8. Optimal form of the elastic axes of a joined wing

In figure 8 presented rational curvilinear form of the elastic axes of a joined wing. It is not symmetrical relative to the horizontal plane in connected with geometric constraints, superimposed on the lower wing. Wing half-size 500 cm, height in the plane of symmetry 255 cm, the maximum deviation of the axis from the original position is equal to -57,2 cm.

Table 2. The vertical deviation of the optimal elastic axes from the original linear shape.

Z, cm	0 - uniform distribution - 500
ΔY_{up}, cm	0 2,9 4,1 4,3 4,0 2,8 -0,7 -9,8 -57,2 -47 -34 -23 -14 -6,9 -1,7 0
ΔY_{low}, cm	0 0,1 0,6 1,3 2,2 3,2 4,8 6,9 10,2 15 20 20 16 10 4,4 0

2.2 Structure influence shape of the axis joined wing aircraft on its weight

Mass of a classic monoplane wing approximately consider as consisting their two parts. Origins optimized force mass (weight) is taken as 100%. During the course of the synthesis shape of the elastic axis due to the joined contour managed to reduce wing weight by 25% and another 15% for calculate and finding the optimal curvilinear axis. This reducing weight is calculated by changing in the internal load structure.

At the same time in weight gain increases by changing the geometry of the wing. 4% adds an increase in the "linear" length of the wing axis due to the growth of transverse V-shape, and 8% - due to its curvilinear.

From statistical data it is known, that in the wing design there is present also a mass, on which optimization has no effect, and this fragment of mass is practically equal to the original wing effort. Therefore, in the total balance of the wing mass, the gain is approximately 14%.

2.3 Search for the optimal shape of the joined wing axis on a flat model

In the figure 9 shown 3D-model of the project of a convertiplane with a joined wing in flight. Turn engine is carried out around axes – actuators, manufactured from shape memory alloy, which are continuous twisted element. Show the scale is the attachment height of the root section of the rear wing and the point for variation of the height of the elastic axis, position of which are parametrically change vertically to obtain a piecewise linear solution.

Investigation search optimal shape of the elastic wing axis, carried out on a flat model aircraft in the form of a flight. At this accounted for only the main factor load, influencing on weight – M_{bend}, and also, that the stiffness of the wing box in the plane bending for two order less, than in the direction of flight.

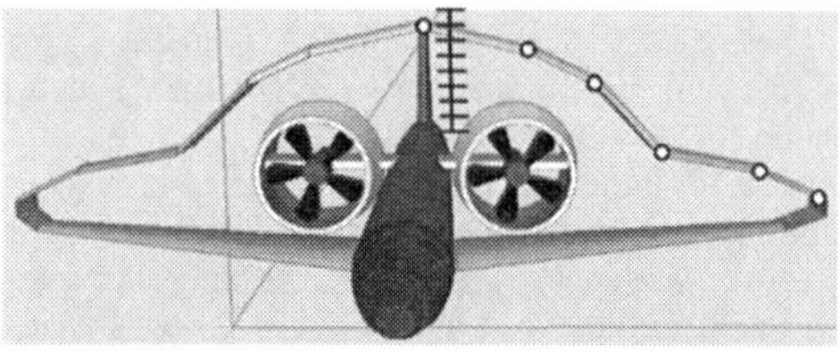

Figure 9. The 3D-model of convertiplane

At figure 10 compare the variants of linear and piecewise linear elastic wing axes, allow to reduce weight of a power part of a design of a wing on 23 and 39% respectively. When more complete accounting all force factors and cases of load, these profits are usually reduced.

https://doi.org/10.1051/matecconf/201822105005

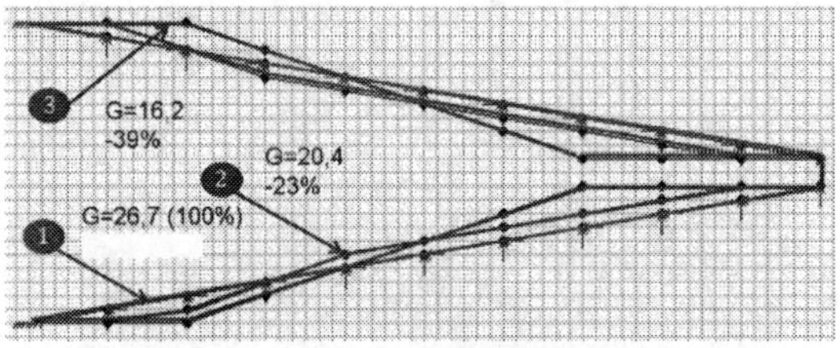

Figure 10. Variants of the forms of the elastic axes of a joined wing

At figure 11 comparison of the upper wing bending moments M_{bend} for variants 1 and 3 for the joined wing of the model (Fig.9). On the lower wing, not shown parts of beams, diagrams M_{bend} skew-symmetric the upper wing.

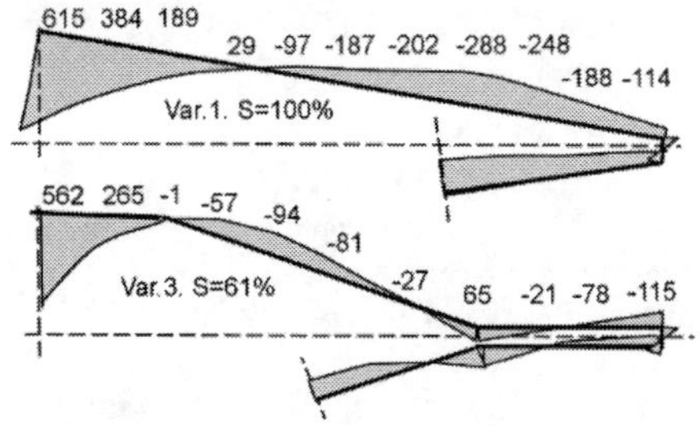

Figure 11. Comparison of the upper wing bending moments

2.4 Eigen forms and Eigen frequencies of an aircraft with a joined wing

There was performed modal analysis of a prospective aircraft with takeoff weight 45 tons and span of the joined wing - 28 m. The elastic axis of the joined wing consisted of straight sections. For analysis was created a FEM model of aircraft which was consist of the beam elements. This model had the stiffness and masses distributions along the wing and fuselage, which were obtained on the basis of airworthiness for Static Strength.

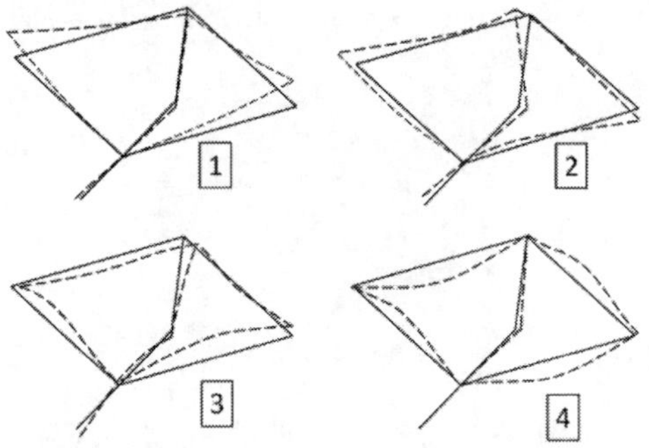

Figure 12. The first four Eigen forms of aircraft

At figure 12, are shown the first four Eigen forms of aircraft in order of increasing Eigen frequencies: 2.14, 3.32, 4.92, 6.85 Hz respectively. Of these, two forms are symmetrical (first and third) and two – anti symmetrical (second and fourth). The solid line represents the elastic axes of the undeformed aircraft.

Study has shown, that with increasing rigidity of the joined wing the Eigen frequency of the aircraft are increased, which positively affects the reserves at the critical speed of the flutter of the aircraft.

Changing the Eigen frequencies after optimization of the design by the criteria of static strength and the minimum weight has a different orientation, and can as create problems, so and help in solving problems of dynamic strength.

3 Shape memory alloys

The aircraft have many flight regimes and accordingly to calculated loading cases, each of has its own rational configuration. The transformation of the aircraft structure using mechanical drives makes the fragments of aircraft split and adds weight to it. Ideal option is, when the design itself change its configuration and perform self-adjustment, remaining in the state of a single continuous body. Shape memory alloy effect can achieve this possibility [7-9]. The shpe memory alloy consists of two or more phases (austenitic, martensitic, rhombohedral, and others) with different mechanical properties. Unlike conventional composites, shape memory alloys components can transform into each other with varying temperature and (or) stresses (thermoelastic martensitic phase transformations) or change the structure of the crystal lattices of some of the components under the action of stresses (structural transition). As a result, the shape memory alloys can accumulate or return deformations of large magnitude, that's why, it can be controlled to change its shape. Thanks to the phase and structural transitions occurring in the shape memory alloys, these materials have unique thermomechanical properties. They are characterized by such phenomena as accumulation of strains of direct transformation, monotonous and reversible shape memory, martensitic inelasticity and superelasticity, oriented transformation. The variety of unique properties of the shape memory alloys predetermines the broad prospects of their use for the construction of unique design with adaptive, intelligent and other new properties.

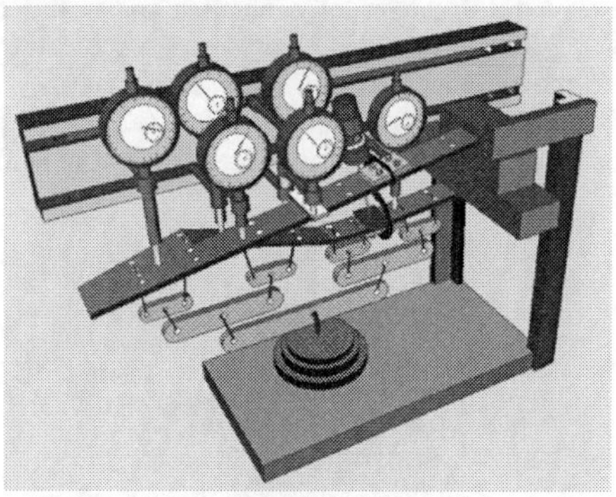

Figure 13. The 3D model of the stand for study of the joined wing

At the Faculty of Aeromechanics and Flight Engineering, MIPT, work is under way to demonstrate the capabilities of the actuators made from shape memory alloys for the transformation and control of structural deformations. The possibility of changing the angle of attack of a joined wing using an actuator is being studied. At the stand shown in figure 13, the model of the joined wing of the aircraft is affected by an *S*-shaped element made of a *NiTi* wire of 3 mm. This element is installed at the half-span of the joined-wing element system, and simulates a washer-actuator. Heating the puck leads to its extension, with the appearance of large forces deforming the joined wing. The local stress-strain state is investigated by a SPECKLE-holographic installation.

As a result of the action of the actuator, the angle of attack of the wing system increases by several degrees, which depends on the span from the conditions of closing the root sections of the wing, and the proportion of the cantilever part. Thus, it is possible to change the integral angle of attack of the aircraft wing relative to the axis of its fuselage. When the heater is removed from the actuator, the stiffness of the swirling joined-wing system serves as a return spring and brings it to its original position. A wire made of *TN-1* alloy (titanium nickelide) was used. The temperatures of the process of the martensitic transitions of the alloy are taken as follows: $Mn = 57\ ^{o}C$, $Mk = 17\ ^{o}C$, $An = 67\ ^{o}C$, $Ak = 107\ ^{o}C$.

Shape memory alloys properties allow creating devices that realize complex kinematics of deformation displacement of structural elements with high weight return of devices, their constructive simplicity and location in the minimum volume.

The problem in the technical implementation of massive devices from shape memory alloys is the difficulty in ensuring their rapid and even uniform heating (cooling), since the non-simultaneous start of the martensitic transition of the crystal structure in different layers of the device leads to the occurrence of large internal stresses, and this can lead to the appearance of cracks.

3.1 Design ratios for the *NiTi* (nickel – titanium shape memory alloys) tubular actuator

To solve the problems of adaptation of aircraft to the flight regime, tubular torsion actuators are considered to be the most acceptable, and at the same time they act as load-bearing elements of the structure, for example, sections of the tubular wing spar, or the axis of the whole-feather tail [6-9].

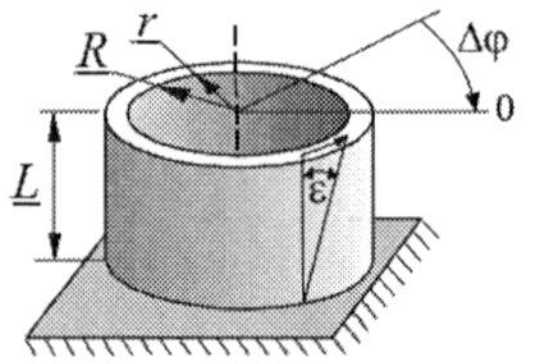

Figure 14. The cross-section parameters of the active element

The radius of the cross-section of the active element R (Fig. 14) is selected from the condition for providing the required operating torque M of the drive [7]:

$$R \geq \sqrt[3]{\frac{2\sqrt{3}M}{\pi\sigma\left(1-\frac{r^4}{R^4}\right)}} \quad (3)$$

Where L - is the length of the working cylinder,

D - is the diameter,

ε – is the relative pseudo plastic deformation.

For the *NiTi* alloy, $\varepsilon < 5 \div 10\%$

φ - torsion angle $360°$ rotation of reversible torsional strain is achieved with a *L/D* ratio of about 10.

(for $90°$ $L = 2.5\ D$)

The following relations and restrictions are also used:

$\varepsilon = D\varphi/4L$

$L/R \geq \varphi/\sqrt{(3\varepsilon)}$

$\varepsilon^* < (4 \div 8)\ \%$,

$\sigma^* < (400 \div 800)$ MPa.

The problem in the technical implementation of massive devices from shape memory alloys is the difficulty in ensuring their rapid cooling.

3.2. To the synthesis of biologically similar structures

Direct axes and regularity of structural elements underlie many aircraft projects. Often one can see an analogy in the decisions that nature and science have found. One of the promising areas of the development of aircraft designs is related to the science of Bionics, which studies the possibility of using the principles of organization, properties, functions and structures of living nature in technical devices and systems. When using bionic solutions, it is first important to identify the principal possibility of implementing a particular idea, and then, on the basis of certified methods, to prove its rationality.

The monoplane trapezoidal wing shown in figure 15 is calculated and the weight of its structure is taken as 1.

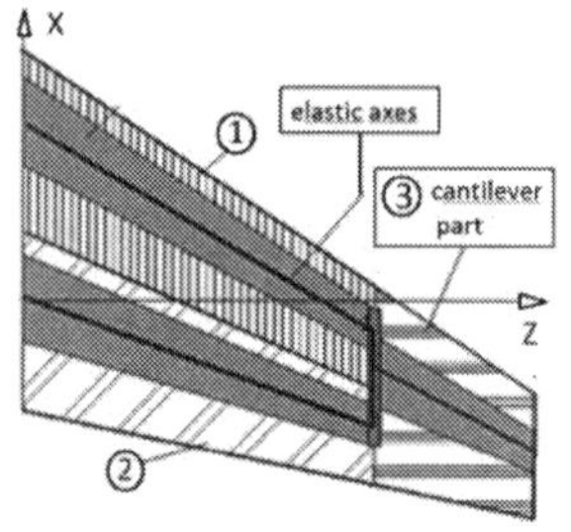

Figure 15. Equivalence of the areas of the reference monoplane wing and joined wing

Further, the monoplane wing is transformed into a joined one, consisting of a cantilever tip part and two planes, the root parts of which are spaced in height in the projection in the direction of flight, as shown in figure 16.

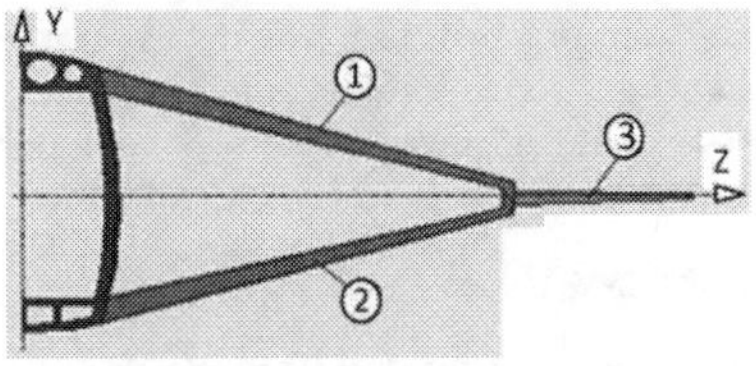

Figure 16. The monoplane wing is transformed into a joined wing, consisting of a two planes root parts 1, 2, and cantilever tip part 3

The figure 17 shows how the weight of the wing depends on the share of its cantilever part.

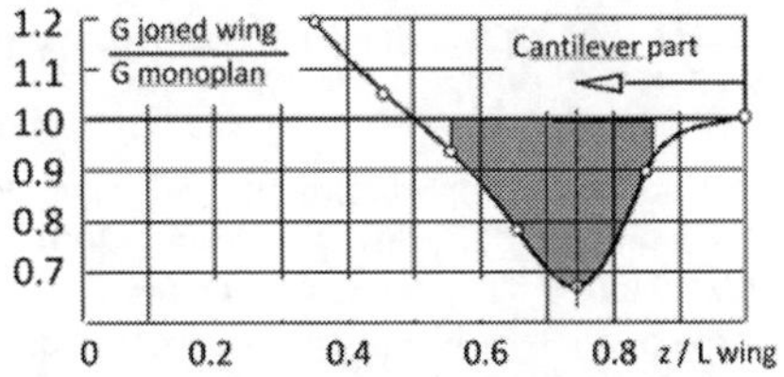

Figure 17. Dependence of wing weight from its cantilever portion size

Obtained during the synthesis of the structure by the criterion of its minimum weight, the shape of the aircraft with a joined system of wings with a cantilever part was naturally similar to the element of the spine of the Katran (the Black Sea shark) shown in figure 18.

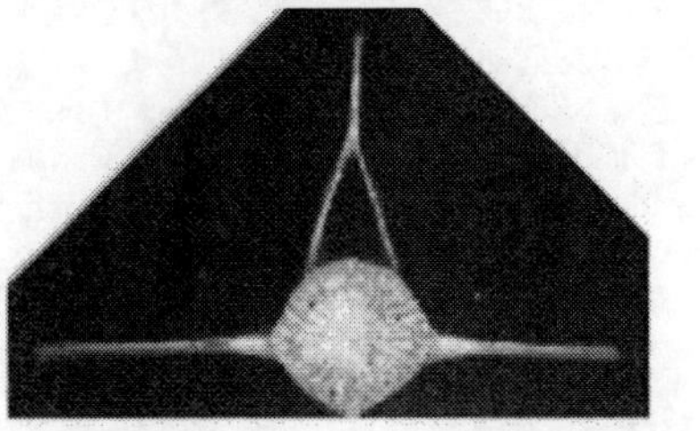

Figure 18. Shark's vertebra - a biological analogue of a joined wing with a cantilever part

4. Adaptive layout aircrafts

Adaptive designs that adjust their configuration to flight mode and aircraft loading have great prospects for improving technical and economic characteristics. In the past decades, the main progress was achieved through multi-criteria optimization of structures made from materials with permanent properties. As our knowledge and technological capabilities expand, the concept of design includes an ever wider range of tasks to be solved.

Figure 19. Airplane by V.S. Jaeger "E-1". MAKS- 2000

A whole-composite aircraft was constructed (Fig. 19) for which the technology of continuous winding of high-strength filaments in a joined contour of wings was developed.

The project is considered, in which the wing of the convertiplane is twisted in the takeoff and landing modes like the wing of a bird (Fig. 20, 21). The main problem is to find an economical way to convert the geometry of the wing. One of the solutions is the use of shape memory alloys compartments integrated monolithically in a continuous cylindrical spar. To solve the problems of adaptation of the aircraft to the flight mode, tubular torsion-active actuators are considered to be the most acceptable, and at the same time they serve as load-bearing elements of the structure, for example, sections of the tubular wing spar, or the axis of the whole-feather tail.

Conversion of the aircraft to a new optimal working configuration, in the transition from takeoff and landing regimes to flight configurations, can be achieved by turning the propellers relative to the fuselage of aircraft, or by changing the mutual arrangement of aggregates, until they are completely joined, changing the sweep angles and twisting the wing, changing the length and the twist of the propeller blades and other transformations [6-9].

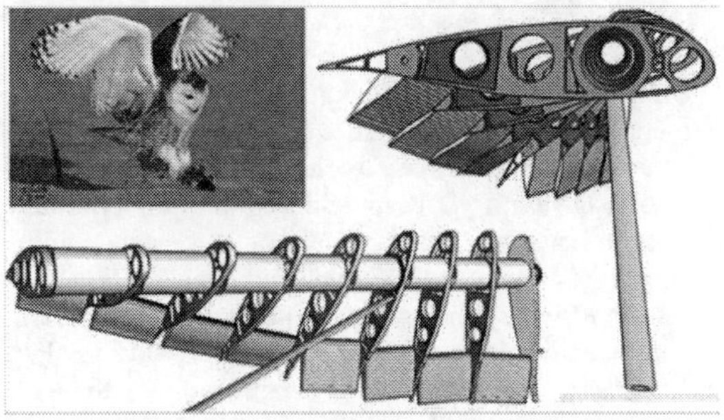

Figure 20. Deformation mechanism, making movements similar to the wing of a bird

Figure 20 shows the structure of the adaptive wing using the tubular compartments of the spar from shape memory alloys. The spar member between adjacent ribs has a section of shape memory alloys capable of making an autonomous rotation at a predetermined angle and a passive region of traditional metal.

Heating the corresponding rings from the shape memory alloys makes it possible to form any form of twist of the compartments, as shown in figure 20, 21. It is important to show specialists in aerodynamics and flight control of aircrafts on the promising realizability of such transformations, which can initiate their search for the rational use of adaptation of the shape of the structure.

The mechanical and functional properties of the shape memory alloys make it possible to form on their basis intelligent devices that react by their behavior to changes in the external environment.

The insertion of elements from the shape memory alloys into joined strong structure makes it possible to redistribute the shear flows into the structure, to act on deformations, to create, when necessary, large internal forces, and also to be an energy accumulator.

The use of shape memory alloys makes possible to construct adaptive constructions with gapless kinematics which is actual for a convertiplanes. The combination of these methods allows you to make advancement in the idea of creating aircraft with solar traction, and drones with a record low relative weight of the structure.

In accordance with the equation of the existence of an aircraft, the gain in the specific weight of the structure makes it possible to use this gain in any other components of the total weight of the aircraft: improving aerodynamics by increasing the wing extension, payload weight and taking additional fuel.

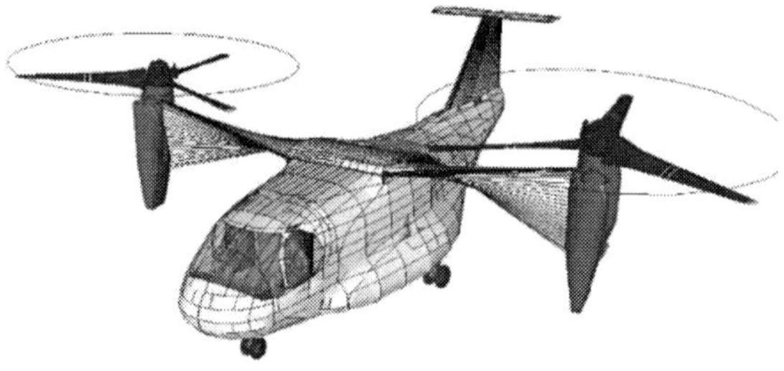

Figure 21. The convertiplane with sectional folded spars

One of the promising methods for creating complex structures with internal cavities is 3D printing. However, at the present time, the required dimensions of the products have not yet been achieved, and the continuity of the material being stacked and its required characteristics are not always ensured.

Aircraft with joined wings can be especially effective when used as high-altitude, working on accumulated solar energy. Especially large lengthening of the wing leads to their great flexibility and deformations. In this case, a joined system of aircraft substructures with a large building height is the most important factor preventing the destruction of the structure.

For an aircraft with a joined wing, rational proportions of the design parameters have already been formed. Thus, the transverse V of the wing is optimal at $\pm 15 \div 17°$, and the share of the console of the joined wing system, if present, is about 25% of the wing's half-span.

The idea of using a joined wing in aircraft designs from the field of science has already passed to the stage of design and practical use. The deterrent is still the high intensity of science projects and the lack of proven prototypes and well-proven technical solutions.

References

1 J. Wolkovitch. The Joined Wing: An overview.// Journal of Aircraft, v. 23, No. 3, 1986. P. 161-178.

2 V.N. Semenov. A comparison of the weight ration of the airframe designs of aircraft with a cantilever wing and with a joined wing system. // NASA TM-77842. Apr.1985.

3 S.V. Novoseltsev, V.N. Semenov. Investigation of the aerodynamic and strength characteristics of an aircraft with a joined biplane system of wings // M., TVF, No. 7, 1984.

4 V.V. Malchevsky. The design of very large commercial transport aircraft with nontraditional configuration, ISTS 501-98. Moscow 2001.

5 V.V. Saurin, V.N. Semenov. Optimization of the shape of a joined beam structure under a distributed load. TsAGI, No. 3, 1992. P. 85-93.

6 V.N. Semenov. Constructions of airplanes of joined and variable circuits. TsAGI, 2006. 228 p.

7 A.A. Movchan, Co Nyunt, V.N. Semenov Designing a torque exciter for an alloy with shape memory. // Proceedings of TsAGI. 2004. Issue. 2664. P. 220-230.

8 I.E. Vyakhi, V.N. Semenov. The use of alloys with shape memory as drives and active elements of adaptive designs of aircraft. // Proceedings of TsAGI. 2005. Issues. 1996. P. 208-214.

9 V.S. Voityshen, M.Yu. Kollerov, V.N. Semenov, V.N. Shcherbakov, Phone Myint Tun. Tendencies of using alloys with shape memory in engineering and medicine. // Mechanics of composite materials and structures. Moscow: IPRIM RAS. T.23. No. 2. 2017. P. 157-167.

A Mission Performance Evaluation Approach for Civil UAS Applications

Ekaterina Fokina[1, a], Jens Feger[2] and Mirko Hornung[3]

[1] *Research Associate and Ph.D., Institute for Aircraft Design, TUM, 85748 Garching near Munich, Germany*
[2] *Ph.D., Institute for Aircraft Design, TUM, 85748 Garching near Munich, Germany*
[3] *Professor, Head of Institute of Aircraft Design, TUM, 85748 Garching near Munich, Germany*

Abstract. The Unmanned Aircraft Systems (UAS) mission fulfilment grade is determined by performance capabilities of the system elements, such as UAV flight performance, sensor parameters, energy consumption and communication abilities. The mission simulation and evaluation tool chain developed at the Institute of Aircraft Design allows to assess the system effectiveness in terms of civil and commercial UAS applications and by this to evaluate trade off studies regarding the compatibility between the air vehicle, the sensor payload and the mission. The presented approach for mission performance evaluation is based on the calculation of an overall mission performance index implemented in the UAS design and optimization processes.

1 Introduction

In this work an approach for UAS mission assessment and evaluation is presented. The methodology is based on deriving the corresponding mission metrics and the mission performance index (MPI). At the beginning of the presented approach the key mission performance parameters together with the mission requirements are defined. Depending on the mission goal the key performance parameters are weighted, normalized und implemented into the metric. In the second stage the UAS mission is simulated in the Mission Simulation and Evaluation Tool. During the mission simulation in the virtual operational environment the key evaluation criteria such as sensor coverage area, probabilities of objects detection, number of detected objects, slant range between the unmanned aerial vehicle (UAV) and search objects, communication range, obstacles detection in the line-of-sight, time of communication losses and others are obtained. Based on the derived criteria the MPI is calculated and the designed system in terms of overall mission performance is assessed. In order to compare different system configurations the evaluation methodology is implemented into the UAS design and optimization processes. As a result the UAS tailored to mission requirements is obtained.

The state of the art, the simulation and evaluation environment, as well as the mission evaluation metrics for civil UAS applications are presented and demonstrated on a simplified case study in this paper.

2 State of the art

Approaches for military UAS selection and evaluation according to mission requirements are already presented in several research groups. For example, a technique for military aircraft design evaluation is presented at the Georgia Technical University in the group of Aerospace Systems Design [1]. Five most important attributes of a military aircraft, such as affordability, capability, operational safety, survivability and readiness are considered through an overall evaluation criteria. The approach results in a series of outcomes depending on the needs and preferences of the customer. The alternative concepts are compared through an evaluation matrix of overall evaluation criteria.

Owing to the system attributes which are taken into account, these approaches are not suitable for a mission performance evaluation of a civil UAS. The application variety of the latter is rising and therefore a method for a system evaluation and selection according to the operational requirements is needed.

In this paper the presented UAS evaluation approach is based on the system simulation in the geospatial environment and on the calculation of the overall mission performance index.

3 Mission performance evaluation metric

UAS mission performance success depends on the performance capabilities of the system elements such as sensor camera parameters, flight altitude and speed, energy consumption and communication capabilities. Since the mission performance can be expressed in a number of criteria, the design process of an UAS consists of conflicting goals and objectives. For example, required ground resolution and desired image quality set the

[a] Corresponding Author: ekaterina.fokina@tum.de

maximum altitude and speed for the UAV, sensor camera characteristics put limits on area coverage rate and give additional weight to the platform. At the same time bandwidth limits of the communications systems restrict the payload data transmission rate.

In order to assess and compare different UAS configurations an evaluation metric is introduced and implemented into the mission simulation and evaluation tool chain.

3.1 Key performance parameters

With technologies improvements UAS get a wider usage in civilian applications. Systems becoming more affordable and available and new classes of UAS consumers appear. According to key evaluation parameters the area of civil UAS applications can be divided into 3 groups [2]:

1. Mapping tasks, such as: crop performance, surveying of a construction site, environment monitoring, aerial photography and etc. For these type of missions important parameters are coverage area, quality of gathered data, cost, and noise.
2. Search and Track, such as: pipelines monitoring, traffic control, border patrol, wildlife monitoring, search and rescue missions, fishery protection, forestry monitoring and etc. In addition to coverage area and data quality, parameters like objects of interest detection probability, flight time to the search area, searching time and communication capabilities have to be taken into account during the mission evaluation process.
3. Delivery and Transport, such as: delivery of packages to customers, relay radio and wireless internet hot spots to restore service after a natural disaster. For these type of missions important parameters are operation time, UAV flight performance, payload capacity and costs.

According to Gundlach [3] a system requirements document can contain hundreds or thousands performance parameters that have to be met for compliance. The most important of them are called key performance parameters or top-level requirements and can include the following:

- Total endurance
- Mission time
- Time to target
- Mission availability
- Payload capacity
- Sensor performance
- Area coverage rates
- Communications range
- Turnaround time
- Number of UAVs required for the mission (Fleet sizing for target coverage).

Operational analysis represents the systems simulation in a geospatial simulation environments. It includes the following:

- Target coverage
- Payload effectiveness in operational environment
- Communication analysis
- Fleet sizing
- Systems comparisons
- Logistic analysis

In order to quantify UAS effectiveness the key performance parameters have to be defined. In the presented approach these parameters are obtained during the mission simulation and system overall assessment in the operational environment and are regarded as optimization goals in the UAS design process. The details of the design, simulation, optimization and evaluation tool chain are presented in the Section 4.

3.2 Mission performance index

Traditional system evaluation and optimization approaches focus on one or two system attributes independently [3]. The introduction of the overall mission performance index (MPI) allows to take into account the correlation between the performance capabilities of the system elements, air vehicle design and system effectiveness.

In order to incorporate a mission evaluation method into a multidisciplinary design process the following requirements have to be fulfilled [4]:

- MPI has to be quantified with a scalar value in order to be used as the result of an objective function for numerical optimization
- Each contributing parameter should be normalized by a referenced value in order to facilitate comparability between mission evaluation results
- Performance parameters should be normalized with reference values of the same physical unit
- Each performance parameter is weighted by importance coefficients, which have to sum together to unity.

The goal of the UAS optimization process is to maximize the scalar MPI by varying aircraft geometry and payload sensor parameters using a genetic optimization algorithm. By that a UAS configuration is tailored to mission requirements.

The most commonly used method to handle more than one decision criteria and which allows to obtain an overall performance score is the Weighted Sum Model (WSM). In WSM method each attribute is multiplied with a weighting factor and added together. This approach is used as a base in MPI calculation.

According to the UAS performance attributes derived from the mission simulation and evaluation tool chain, the MPI is defined as [5]:

$$
\begin{aligned}
MPI = \alpha \cdot \frac{ACR}{ACR_{ref}} \cdot f_{deg}(GSD) + \beta \cdot \frac{E_{ref}}{E} + \\
+\gamma \cdot \frac{P_{det}}{P_{detref}} + \delta \cdot \frac{T_m}{T_{m\,ref}} + \\
+ \varepsilon \cdot \frac{CA}{CA_{ref}} + \zeta \cdot \frac{UOI}{UOI_{ref}}
\end{aligned}
\tag{1}
$$

The area coverage rate (ACR) is defined as covered area per mission time T:

$$ACR = \frac{1}{T}\int_0^T w_{swath} \cdot V dt \quad (2)$$

where w_{swath} is the ground sensor swath width and V is the airspeed during the mission.

The degradation factor $f_{deg}(GSD)$ is introduced in form of an exponential function and it takes into account limits of the Ground Sample Distance (GSD) [6]. The latter is used as a measure of a distance between the pixels projected on the ground at a slant range [3] and is calculated as:

$$GSD_H = 2 \cdot tan\left(\frac{FOV_H}{2 \cdot H_{Pix}}\right) \cdot R \quad (3)$$

where FOV_H is the horizontal field of view, H_{pix} is the number of horizontal pixels of the camera and R the current altitude over ground of the aircraft.

The parameter E quantifies the amount of energy or fuel needed to fulfil the mission depending of the UAV engine type.

The object detection probability P_{det} is determined according to the semi-empirical method called Johnson Criteria and time of object detection [5]. Johnson Criteria calculates the probability Jp(N) of performing the respective task by comparing the number of sensor cycles N covered by a target to a reference value N_{50}, which stands for 50% probability.

$$Jp(N) = \frac{(N/N_{50})^{2,7+0,7\cdot(N/N_{50})}}{1+(N/N_{50})^{2,7+0,7\cdot(N/N_{50})}} \quad (4)$$

According to Johnson the reference values N_{50} is used with 0.75 for detection, 3.0 for recognition and 6.0 for identification. In an approach introduced by Schmieder and Weathersby the Johnson's criteria is adapted to different target scenarios by setting the respective N values [7].

Figure 1 depicts the dependence of detection probability and detection time. Based on the fact that in average a human needs 0,25s to notice visual changes [8], it is assumed that an operator needs minimum 2 second to detect an object of interest. At that moment the detection probability is considered as 70% and is rising further linearly. The 100% probability is achieved at a time of 10s [5].

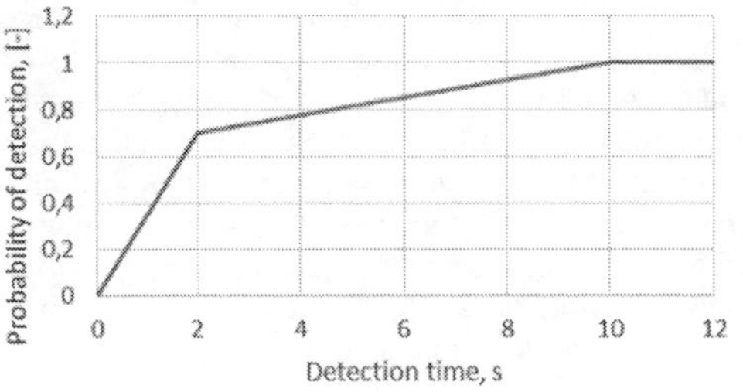

Figure 1. Dependence of detection probability and detection time [5]

The Communication Abilities (CA) parameter contains information about time of communication losses Tcl and the possibility to store data on board during the losses. It can be represented as ratio between the mission time and time of communication losses:

$$CA = \frac{T_{mission} - T_{cl}}{T_{mission}} \quad (5)$$

User Operating Issues (UOI) attribute depending on the mission requirements can contain information about information about launch/landing options according to the system weight, how easy it is to use the system, if it needs any special handling or storage facilities and how environmental sensitive the systems is. For instance, for an agriculture application the most attractive UAS would be robust, hand-launched and belly-landed, not sensitive to environment conditions and of course easy to operate. The UOI attribute is introduced in form of a scalar value within the range from 0 to 1 [5].

3.2 Assignment of weights

In general weighting approaches can be divided into sum-normalized (coefficients sum to unity) and scoring models. In both cases the more important the criterion the higher the weighting coefficient.

Weights coefficients of criteria can be determined in several ways. One of the methods is so-called weight decision matrix, obtained from Analytic Hierarchy Process (AHP) method. The AHP method was developed mainly by Thomas L. Saaty [9]. According to this approached each criteria is weighted in terms of all other criteria with relative importance indices from Tab. 1. From these values a matrix of weights is derived and the total criteria weights are calculated based on the use of eigenvectors.

Table 1. Relative importance indices for weight determination in AHP.

Importance Index	Definition
1	**Equal importance** of both criteria
3	Criterion A is **somewhat** more important than criterion B
5	Criterion A is **notably** more important than criterion B
7	Criterion A is **significantly** more important than criterion B
9	Criterion A is **absolutely** more important than criterion B
2,4,6,8	Intermediate values where compromise is needed

In order to reduce the large number of ranking decisions made by the decision maker a Multi-criterion Analysis of Preferences by means of Pairwise Actions and Criterion comparisons (MAPPAC) method, which is sometimes also referred as PAPRIKA (Potentially All Pairwise Rankings of All Possible Alternatives) can be used. In this method each criterion has a limited number of possible expressions, such as "poor", "medium", and "good". The criteria are pairwise compared and ranked [10].

Another technique called a Rank-Order Centroid [11] does not require the weights assignments made by a decision maker. The criteria are ranked from most important to least important. Weights can be assigned using one of the following formulas, where ri is the rank of the i-th objective and K is the total number of objectives:

- Rank Sum

$$w_i = \frac{K - r_i + 1}{\sum_{j=1}^{K} K - r_j + 1} \tag{6}$$

- Rank Exponent

$$w_i = \frac{(K - r_i + 1)^z}{\sum_{j=1}^{K} (K - r_j + 1)^z} \tag{7}$$

where z is an undefined measure of the dispersion in the weights

- Rank Reciprocal

$$w_i = \frac{1/r_i}{\sum_{j=1}^{K} 1/r_i} \tag{8}$$

- Rank Order Centroid

$$w_i = \left(\frac{1}{K}\right)\sum_{j=1}^{K}\left(\frac{1}{r_i}\right) \tag{9}$$

$$w_1 = \left(1 + \frac{1}{2} + \frac{1}{3} + \cdots + \frac{1}{K}\right)/K \tag{10}$$

$$w_K = \left(0 + 0 + 0 + \cdots + \frac{1}{K}\right)/K \tag{11}$$

Other possible ranking methods are presented in Buede [12].

The assignment of the weighting coefficients has to be adjusted to every mission case by the design engineer based on engineering insight and experience. Despite the fact that there are many methods for weight assignment, this process still stays subjective [5].

4 Mission simulation and evaluation

A tool chain is developed at the Institute of Aircraft Design, TUM, in order to take into account all elements of the system, e.g. to bring together aircraft design, payload, communication and other elements into one multidisciplinary design process and to get an effective UAS solution.

4.1 Overview of the environment for UAS design, mission simulation and evaluation

The overall structure of the tool chain is presented in Figure 2. It consists of the UAS design, mission simulation, visualized operational environment, mission evaluation and optimization parts.

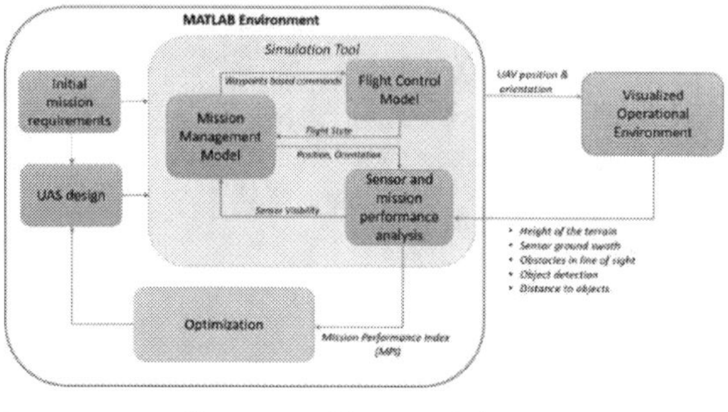

Figure 2. Overall structure of the UAS design, mission simulation and evaluation tool chain [5].

The UAS design environment creates a consistent design of an UAV taking into account general geometric data, masses, aerodynamic polars and propulsion system data. Simplified physical methods, especially for aerodynamics and structural weight estimation, are used where applicable. An overview of the UAS design process is presented in Figure 3 [4].

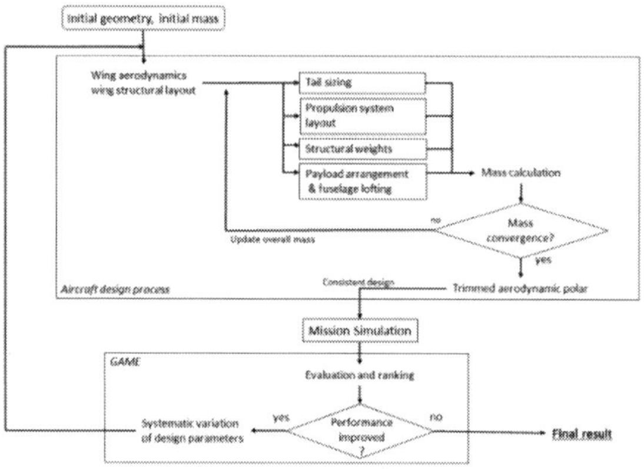

Figure 3. Design process for calculation of a consistent aircraft dataset [4].

The simulation part is represented by the Mission Management Model, the Flight Control Model and the Sensor and Mission Performance Analysis block. The Mission Management Model generates waypoint navigation commands, distributes input data for other model elements, delivers flight state information to the sensor model and after performing the simulation stores the results into the data base for further evaluation. The aircraft's current position, velocity and attitude angles are calculated using linearized equations of motions in the Flight Control Model block. The following data are provided for each sample time point [4]:

- Position in latitude, longitude and altitude in WGS84 coordinates
- Kinematic velocity with speed, course angle and climb angle in North-East-Down coordinates

- Attitude angles with heading, pitch and roll attitude
- Current control variable settings (angle of attack, bank angle, thrust lever setting)
- For combustion engine: mass, used fuel mass and fuel flow
- For electric propulsion system: Used energy
- Sensor data: Ground Sample Distance (GSD), detection probability.

The visualization part of the tool chain simulates the operational environment where the UAS performs a mission.

It receives information about the UAV position and orientation, as well as the orientation of the sensor installed on the platform from the simulation model. By means of high resolution texture data, elevation model based landscape, simulation of interaction of the UAS elements with each other and the environment the following information is obtained from the visualization part [12]:

- Sensor ground swath
- Detection probability of search objects
- Slant range between UAV and search objects
- Communication range and obstacles detection in the line of sight
- Time of communication losses
- Height of the terrain

These data are used in the sensor performance and mission effectiveness evaluation process which is based on the MPI.

In order to find the best suited UAS design by maximizing the MPI a genetic optimization algorithm is applied [4]. By varying design variables, such as aircraft wing geometry and sensor performance parameters, a finite number of UAS configurations is generated, evaluated and ranked by the MPI. By using methods of randomized recombination, mutation and saving best individuals a new generation of UAS configurations is created, evaluated and ranked. This process is repeated until no further improvements of the MPI is achieved.

4.2 Mission design study

A design study for search and rescue mission is presented in order to verify the capabilities of the presented tool chain and performance evaluation metric.

The objective of the mission is to detect within the shortest time a lost boat in a certain area of the lake Chiemsee, close to Rosenheim, Germany. The area of interest is 16 km^2. In order to take into account the uncertainty of the boat location three possible spots are defined at the beginning of the simulation. Therefore during the simulation it should be determined whether an object with an average size of 2.7 m length can be detected or recognized in one of these locations.

For the optimization process 4 design variables and their ranges are defined in Table 2. The possible camera types are presented in Table 3.

Table 2. Design variables.

Design variables	Min	Max	Unit
Wing area	0,2	3	[m^2]
Aspect ratio	5	15	[-]
Design speed	15	40	[m/s]
Camera index	1	3	[-]

Table 3. Payload camera types.

	Camera 1	Camera 2	Camera 3	Camera 4
Sensor resolution	1920 x 1080	1280 x 720	1280 x 720	1920 x 1080
Max Horizontal Field of view	59°	31.5°	30°	31.2°
Weight	1,2 kg	4 kg	6.8 kg	16.8 kg
Index in the tool	1	2	3	4

The required GSD is defined as 40 cm in order to detect and recognise the objects of interest. Sensors with bigger resolution allow to fly with higher altitudes and therefore with bigger sensor field of view, while sensors with smaller resolution achieve the desired GSD with lower altitudes. At the same time, cameras with better resolution are more expensive and can be heavier. This trade off should be taken into account when choosing a sensor payload for an UAS.

For the presented mission design study there are two important evaluation parameters: object detection and mission time. Therefore the MPI for the optimization process is represented in the following way:

$$MPI = \gamma \cdot \frac{\frac{\sum_i^n P_{det}}{n}}{\left(\frac{\sum_i^n P_{det}}{n}\right)_{ref}} + \delta \cdot \frac{T_m}{T_{m\,ref}} \tag{12}$$

where P_{det} is probability of the object detection in the i-th location and T_m is the mission time.

The object detection is assumed as more important attribute of the mission success and weighting coefficients are taken as γ=0.7 and δ=0.3.

Four generations, with 5 individuals each, of UAS configurations were created during the simulation. Each individual, e.g. UAS configuration, was evaluated and ranked by the MPI. As a result of the optimization process the unmanned aerial system with camera 1 was designed according to the initial mission requirements. Figure 4 and Figure 5 presents the results of the optimization process.

The mission is fulfilled in 24 min at an altitude of 600 m over the water level and with a flight speed of 27 m/s. The boat is detected and recognized in all 3 possible locations. Results of objects detection, recognition and identification according to Johnson Criteria are presented in Figure 6.

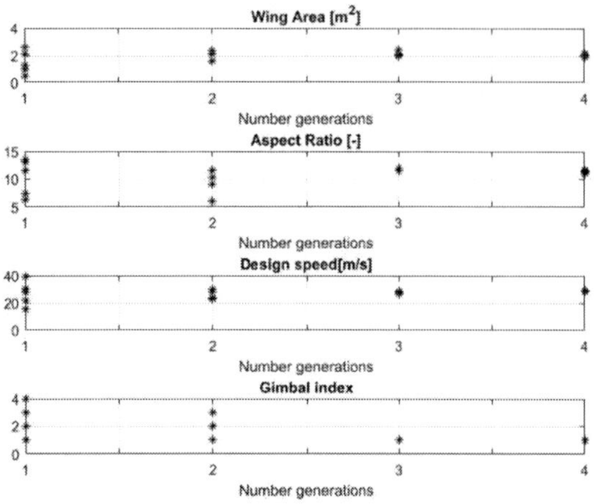

Figure 4. Variation of design variable values during. optimization for γ=0.7, δ=0.3 and search area of 16 km^2.

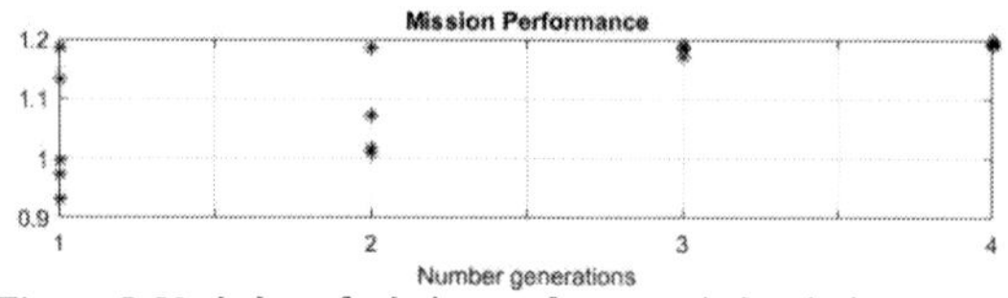

Figure 5. Variation of mission performance index during optimization for γ=0.7, δ=0.3 and search area of 16 km^2.

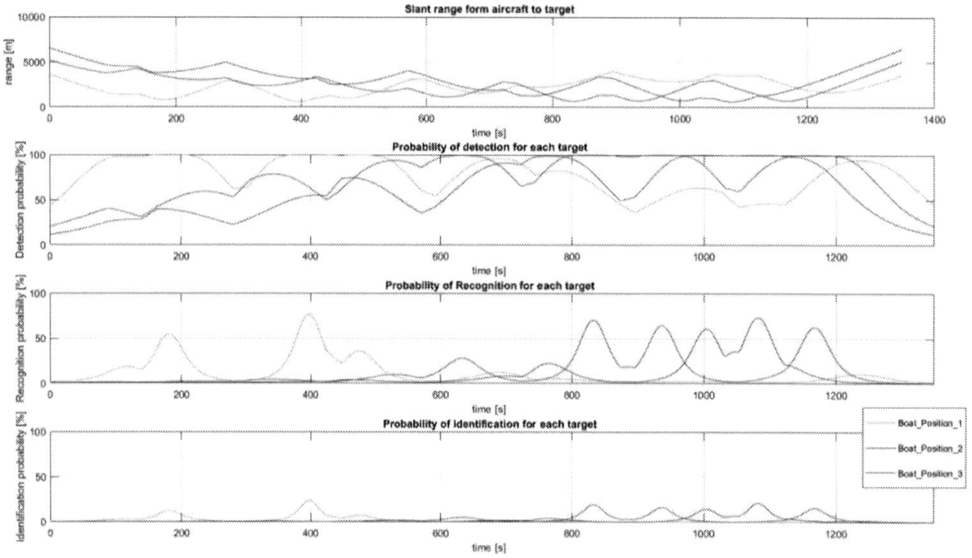

Figure 6. Results of objects detection, recognition and identification according to Johnson Criteria for γ=0.7, δ=0.3 and search area of 16 km^2.

In order to show the flexibility of the presented approach for UAS design and mission evaluation, a bigger search area on the lake Chiemsee and different weighting coefficients assignment in Equation 12 are chosen. The new search area is 41 km2 and γ=0.5, δ=0.5

The variation of the mission performance index during the optimization is presented in Figure 7. In order to obtain a converging process a bigger number of generations is necessary.

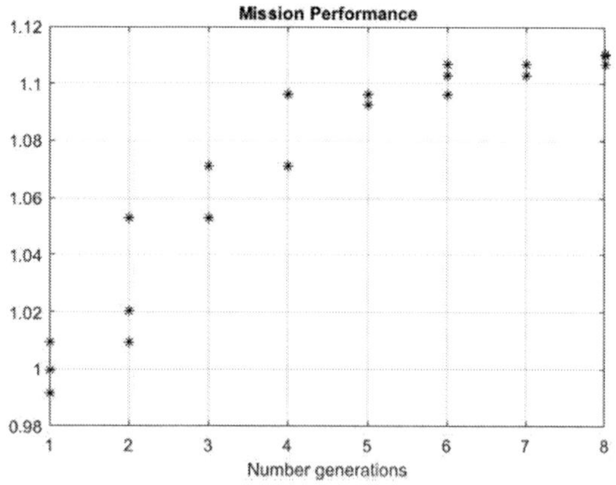

Figure 7. Variation of mission performance index during optimization for γ=0.5, δ=0.5 and search area of 41 km2.

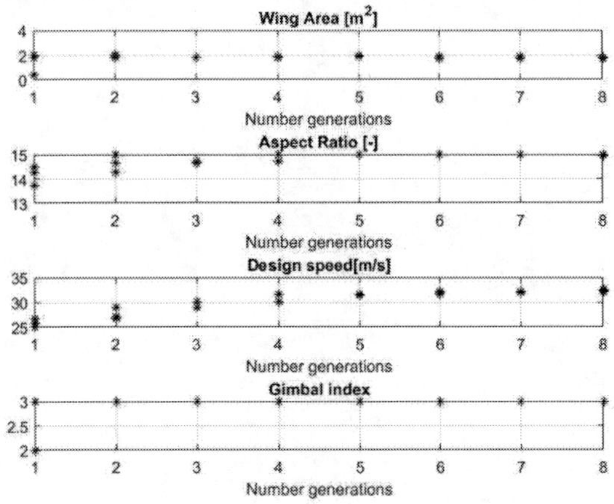

Figure 8. Variation of design variable values during optimization for γ=0.5, δ=0.5 and search area of 41 km^2.

The variation of the design variables for the new weighting scheme and search area is shown in Figure 8. The UAS configuration is designed with camera 3 and the platform is significantly bigger than in the first case. The mission is fulfilled in 58 minutes at an altitude of 960 m and with a flight speed of 32 m/s.

5 Conclusion

The presented tool chain allows to design, simulate, evaluate and optimize an UAS according to initial mission requirements. It is possible to perform within the simulation environment different UAS mission scenarios. By simulating the visualized operational environment it is possible to take sensor and communication requirements into account in the UAS design process. A genetic optimization algorithm is used to find an unmanned aircraft design that optimally fulfills a certain mission task.

The mission design study presents the flexibility of the tool chain and importance of chosen mission evaluation metric. Mission performance attributes selection and it's weighting assignment play a role in the UAS optimization process.

References

1. D. N. Mavris, D. A. DeLaurentis. An integrated approach to military aircraft selection and concept evaluation. 1st AIAA Aircraft Engineering, Technology, and Operations Congress, Los Angeles, CA, USA (1995)
2. J. Gundlach, *Civil and Commercial Unmanned Aircraft Systems*. AIAA Inc., Reston, Virginia. ISBN 978-1-62410-354-4.
3. J. Gundlach, *Designing Unmanned Aircraft Systems: A Comprehensive Approach*. AIAA Inc. Reston, Virginia. ISBN 978-1-60086-843-6.
4. J. Feger, E. Fokina, M. Hornung. An integrated design process for mission optimized design of unmanned aerial vehicles. SciTech Conference, Florida, USA (2018)
5. E. Fokina, J. Feger., M. Hornung. UAV, sensor and mission matching approach using the visualization environment. AIAA Aviation Conference. Atlanta, GA, USA (2018)
6. J. Feger, F. Pozatzidis. Simulationsmodell zur Bewertung der Missionserfuellung Unbemannter Fluggeraete im Vorentwurfsstadium. Deutscher Luft- und Raumfahrtkongress (2015)
7. J. Johnson, Analysis of image forming systems. Proceedings of Image Intensifier Symposium. Ft. Belvoir, Virginia. 249–273 (1958)
8. A. Jain, R. Bansal, A. Kumar, K. Singh. A comparative study of visual and auditory reaction times on the basis of gender and physical activity levels of medical first year students. International Journal of Applied and Basic Medical Research. 5(2):124-127 (2015)
9. T. L. Saaty. How to make a decision: The Analytic Hierarchy Process. European Journal of Operational Research. North-Holland. S. 18 (1990)
10. B. Matarazzo. Multicriterion analysis of preferences by means of pairwise actions and criterion comparisons. Applied Mathematics and Computation, 18. 119-141 (1986)
11. D. M. Buede, W. D. Miller, *The Engineering Design of Systems: Models and Methods* (Wiley Series in Systems Engineering and Management, Band 1). ISBN-13: 978-1119027904.
12. E. Fokina, J. Feger, M. Hornung. Application of a visualization environment for the mission performance evaluation of civilian UAS. Aerospace Europe Council of European Aerospace Societies Conference (2017)

AUTHOR INDEX

AUTHOR INDEX